ENVIRONMENTAL ECOLOGY

ENVIRONMENTAL ECOLOGY

Edited by

Prof. P.R. Trivedi

Chairman

Indian Institute of Ecology and Environment

&

Prof. Gurdeep Raj

Hony. Professor in Environmental Science

I.I.E.E., New Delhi

Reader in Chemistry

D.S. College, University of Delhi

AKASHDEEP PUBLISHING HOUSE

NEW DELHI - 110 002 (INDIA)

Akashdeep Publishing House
4374/4B, Ansari Road, Daryaganj
New Delhi - 110 002

First Edition, 1992

Reprint, 1995

Reprint, 2002

ISBN 81-7158-243-5

PRINTED IN INDIA

Published by J.L. Kumar for Akashdeep Publishing House, New Delhi - 110 002 and Printed at Mehra Offset Press, Delhi.

Preface

The title of this book is a precise indication of its contents and focus of attention. The main aim of this book is to aid the students in acquiring an understanding of various basic concepts of ecology.

The concepts used in this book has created the basic parameters or culture of environmental science, ecology, etc. Every attempt has also been made to incorporate the new developments on this area. The editing of this manual is more or less based on the research and teaching experience of editors who have been actively engaged in the different aspects of environmental science and conducting practical classes at all levels of environmental studies. Their valuable experience has been carefully condensed in compiling this book.

We do not claim any originality because in the preparation of this book large number of texts and research journals have been freely consulted. Efforts have been made in making this book readable.

As far as possible, the old, out dated and rejected views have been avoided and modern findings have been given.

The Editors

Contents

Contents

1

Concepts of Ecology

Scope and History of Ecology

The word ecology, derived from the Greek word : oikos meaning habitation, and logos meaning discourse or study, implies a study of the habitations of organisms.

Ecology was first described as a separate field of knowledge in 1866 by the German zoologist Ernst Haeckel, who invented the word oeckologie for "the relation of the animal to its organic as well as its inorganic environment, particularly its friendly or hostile relations to those animals or plants with which it comes in contact".

Ecology has been variously by defined by other investigators, as "scientific natural history", "the study of biotic communities", or "the science of community population"; probably the most comprehensive definition is the simple one most often given; a study of animal and plants in their relations to each other and to their environment.

Objectives

Ecology is a distinct science because it is a body of knowledge not similarly organized in any other division of biology; because it uses a special set of techniques and procedures; and because it has a unique point of view. The essence of this science is a comprehensive understanding of the import of these phenomena :

1. The local and geographic distribution and abundance of organisms (habitat, niche, community, biogeography).
2. Temporal changes in the occurrence, abundance, and activities of organisms (seasonal, annual, successional, geological).
3. The interrelations between, organisms in populations and communities (population ecology).
4. The structural adaptations and functional adjustments of organisms to their physical environment (physiological ecology).
5. The behaviour of organisms under natural conditions (ethology).
6. The evolutionary development of all these interrelations (evolutionary ecology).
7. The biological productivity of natural and how this may best serve mankind (ecosystem ecology).
8. The development of mathematical models to relate interaction of parameters and predict effects (systems analysis).

Relation to Other Sciences

Ecology is one of their main divisions of biology, the other two being morphology and physiology. The emphasis in morphology is on how organisms are made, in physiology on how they function, and in ecology, on how they live. These divisions overlap broadly. To appreciate fully the structure of an organ, one needs to know how it functions, and the way it functions is clearly related to environmetal conditions. The morphologist is concerned with problems of anatomy, histology, cytology, embryology, evolution, and genetics; the physiologist, with interpreting functions in terms of chemistry, physics and mathematics; and the ecologist, with distribution, behaviour, populations and communities in relation to the environment (ecosystems). The evolution of adaptation and of species is mutual interest to the ecologist and to the geneticist; biometeorology is a connecting like between ecology and physiology; and system analysis interrelates ecology and mathematics. All areas, in the final analysis, are simply different approaches to an understanding of the meaning of life.

Subdivisions of Ecology

Ecology may be studied with particular reference to animals or to plants, hence animal ecology and plant ecology. Animal ecology, however, cannot be adequately understood except against a considerable background of plant ecology. What animals and plants are given equal emphasis, the term bioecology is often used. Courses in plant ecology usually dismiss animals as but one of many factors in the environment. Synecology is the study of communities, and autecology the study of species. There is some confusion in these terms since Europeans commonly use "ecology" in narrower sense—meaning the environmental relations of organisms or of communities. The broader study of communities, including species interrelations and community structure and function as well as environmental relations (synecology), is generally termed "biocenology" or "biosociology" by Europeans.

We shall survey the fundamentals and basic facts of ecology as they relate to animals and have application to man. We will study community ecology, the local distribution of animals in various habitats, the recognition and composition of community units, and succession; ecosystem dynamics, the processes of soil formation, nutrient cycling, energy flow and productivity; population ecology, the manner of population growth structure, and regulation; evolutionary ecology, the problems of niche segregation and speciation; and geographic ecology, concerned with distribution, paleoecology, and biomes. We will be integrested throughout the text with how organisms respond and adjust physiologically to the physical factors of their environment, but a full study of physiological ecology must be left to another time and place. We will also be concerned throughout with systems ecology that is the possibility of translating ecological concepts into mathematical models, although we will not go deeply into the actual statistical manipulations.

History of Ecology

The modern concept that plants and animals occur in closely integrated communities began with the studies of Agust Grisebach, a German botanist, in 1938; K. Mobius, a German, investigator of oyster banks, in 1877; Stephen A. Forbes, an American, who described the take community as a microcosm in 1987; and J.E.B. Warming, a Danish botanist, who emphasized the unity of a plant communities in 1895 C.C, Adams recognized and described many animal communities in his ecological surveys of northern Michigan and of Isle Royale in Lake Superior, published in 1906 and 1909. V.E. Shelford presented a classic study of animal communities in temperate America in 1913, and Charles Elton puplished an outstanding analysis of community dynamics in 1927. Although an appreciation that the whole community is one biotic unit, rather than one unit plants and another of animals, may by described in the writings of some early investigators (for example, J.G. Cooper in 1859), the fact has been brought to modern emphasis in the work of F.E. Clements and V.E. Shelford, especially in their Bio-ecology, published in 1939. Much interest has been stimulated in recent years by D. Ramon Margalef, Robert Mac Arthur, and others for analyzing the structure of these biotic communities, particularly in respect to such phenomena as species diversity, niches, and how they come about through evolution.

Geographic ecology, in the modern sense, dates from the generalizations on the worldwide distribution of animals made by the French naturalist Georges L.L. Buffon (1707-1778), and the explorations of the German naturalist Alexander von Humboldt (1769-1859). There was lively interest and many important contributions in this general field during the nineteenth century; notably, the life-zone concept of C. Hart Merriam (1890-1910) needs special mention. During the present century the concept of biotic provinces is identified with L.R. Dice (1943) and the biome concept with F.E. Elements and V.E. Shelford (1939). The broad survey of ecological animal geography made by R. Hesse in 1924 exerted considerable, infleuence, and this treatise was later translated into English and revised by W.C. Allee and Karl P. Schmidt (1951).

The study of population dynamics, so important in modern ecology, dates back at least to Malthus, who pointed out in 1798 the limitation to population growth exerted by available food. Darwin, in 1859, recognized the importance of competition and predation in developing his theory of evolution. Pearl, 1925, analyzed mathematically the characteristics of population growth, and Lotka, 1925, and Volterra, 1931, developed theoretical mathematical equations to show the manner in which populations of different species interact. These studies led to the classic experiments of Gause, 1935, with interacting populations of predators and prey. Nicholson's publication in 1933 stimulated much thinking concerning the factors that stabilize populations at particular levels. Andrewartha and Birch, 1954, emphasized the importance of climate and other factors on determining the size of populations.

Physiological ecology had its historical beginning in the correlation of biological phenomena with various temperature stimulated by Galileo's invention of a herimetically sealed thermometer about 1612 A.D. The French naturalist Reaumur summed the mean daily temperatures for April, May and June in 1734 and again in 1735, and correlated the earlier maturing of fruit and grain during first year with the greater accumulation of heat : A discovery of parallel significance was of oxygen in 1977 that it was an essential part of air, Claude Bernard, another French physiologist, enunciated the principle of homeostasis in 1876. This concept originally referred to regulatory mechanisms which maintained the "internal environment" of the body constant in the face of changing external conditions. Later, the concept came to be applied also to maintenance of community interrelations.

The development of animal behaviour or ethology may be traced back through the n tural history of ancient times. More recently the 13 volumes of Thierleben, prepared by A.E. Brehm during the period 1911-18, are noteworthy. H.S. Jennings, 1906, and Jacques Loeb, 1918, made valuable contributions to the understanding to the behaviour of invertebrates. Precise modern techniques and concepts as applied to vertebrates began to take form about 1920 with the development of banding marking of individual animals by S. Prentiss Baldwin (1919) and the recognition of territories in the nesting of birds by H.E. Howard (1920). The formulation of the concept of releasers as controlling instinctive behaviour by Wallace Craig (1908), K. Lorenz (1935), and E. Thinbergen (1951) has produced a profound effect on present-day thinking.

Ecology, then, is of comparatively recent development as a distinct science, but its roots extend well back into the past. An early comprehensive treatment of the subject is Principles of Animal Ecology by Allee, Emerson, Park and Schmidt published in 1949. Since ecology is a young science, it should be emphasized that its concepts and techniques have not become standardized and that there is opportunity and stimulus here for many new investigators.

General Nature of Environmental Responses

Ecology, by definition, deals with the interrelations of ogranisms with each other and with their environment. These interrelations become established as organisms respond in various ways to contacts with one another and with the everchanging environment.

The term environment describes the sum total of physical and biotic conditions influencing the responses organisms. More specifically, the sum of those portions of the hydrosphere, lithosphere, and atmosphere into which life penetrates is the biosphere. There are no characteristic of permanent inhabitants of the atmosphere, although the air is traversed by many kind of animals and plant propagules. Of the hydrosphere, there are two major bicycles, marine and fresh water; of the lithosphere there is one, land.

A habitat is a specific set of physical and chemical conditions (for example, space, substratum, climate) that surrounds a single species, a group of species or a large community (Clements and Shelford 1939). The ultimate division of the biosphere is microhabitat, the most intimately local and immediately set of conditions surrounding an organism; the burrow of a rodent, for instance, or a decaying log. Other individuals or species are considered as part of the community to which the organism belongs and not part of its habitat. The term biotope defines a spatial or topographic unit with a characteristic set both of physical and chemical conditions and of plant and animal life.

In order for organisms to exist they must respond or adjust to the conditions of their environment. The first living organisms probably evolved in the sea and must have possessed very generalised adjustments to this relatively uniform and favourable habitat. However, these early organisms and inherent in them the potential for expansion, as they later spread into other and more rigorous habitats, particularly fresh water and land. As evolution proceeded, organisms became more and more limited in the range of their ability to respond as they became specialized in their adjustments to particular habitats. This led to the great diversification of species that we see at the present time, with each species restricted to its particular microhabitat and place in the community.

Organisms respond to difference or changes in their environment in four principal ways : morphological adaptations, physiologial adjustments, behaviour patterns, and community relations.

Morphological Adaptations

Morphological variations induced by preculiarities of habitat do occur in motile animals; thickening of the shells of clams subjected to strenuous wave action; variation in number of vertebrae, scales, and fin rays among fish subjected to different temperatures at critical periods in their growth (Barlow 1961); changes in the number of facets in the bar eye of the fly Drosophila as a correlative of temperature variations during a short critical period in larval growth (Krafka 1920); the many variations in form and size of internal parasites, depending on crowding and other environmental conditions (Baer 1951); pointed tails in certain flat worms crawling over a substratum during growth, contrasted with rounded tails occurings in the same species when individuals are experimentally prevented from crawling (Child 1903).

That individuals of the same species are so much alike attests the great extent to which the course and outcome of morphological development are genetically determined. But that there are variations between individuals, and between groups of individuals, of the same species shows that morphological development is also responsive to environmental influences. Modifications induced by the environment emerge as the individual develops and are not specifically inherited by the succeeding generation. These modifications are called growth-forms or ecophenes. If the generation following is similar in growth form to the parent generation, it is a similar morphological response to a similar environment (Schmalhausen 1949). If and when the growth-form becomes inherited as the result of evolutionary proceesses, if then becomes an ecotype. Life-form is a general term referring to the shape of appearance of an organism irrespective of how formed.

Specific life-forms are adaptations of plants and animals to live in particular habitats and to behave in particular ways (Klaauw 1948). The life-forms listed for mammals are largely adaptations to particular strata (water, subterranean, ground, tree, air) within a community rather than to the habitat as a whole; for instance, the subterranean adaptations of mammals living in the Arctic tundra are similar to the subterranean adaptations of mammals in the tropics in communities lacking one or more strata (for instance, the free stratum in grassland), animals specifically adapted to the missing strata are also absent. In communities in which all strata are present, a catholic variety of life-forms occurs.

In addition to adaptations to stratum and habitat there occur ecologically significant adaptations for food-getting and metabolism, protection and production. The variety of teeth found in mammals and lizards, the variation in shape and size of bills of birds, the different mouth parts of insects, the siphons of clams, the suckers of leeches, the water canal systems of sponges are but a few special anatomical features especially designed for food-getting. Associated with food-getting is a great diversity in structural adaptations for the digestion of the food, for respiration, for circulating food materials and gases through the body, for excreting wastes, for support and movement, and for nervous and hormonal regulation. All these internal organs and structures are necessary to the animal for utilizing the energy resources of the environment.

Physiological Adjustments

Nature of Adjustments. Probably the first response of any organism to a change in the environment is physiological. A physiological response must certainly precede any change in form or structure which requires growth. Even a change in behaviour must follow a change in some receptor or sense organ followed by nervous function; a fall in air temperature, for instance, brings a drop in the metabolic rate of cold-blooded organisms but a rise in the rate of warm-blooded organisms. Cold may stimulate nerve endings in the skin of birds or mammals and produce shivering and a search for protective cover. Transference from the dark to light may immediately initiate photosynthesis in resting chloropastids within a plant cell, or a change in turgescence on opposite sides of a sessile zooid may result in a turning movement, an orientation to or away from the light source. Physiological responses are thus internal responses to factors of the environment. Often they are difficult to defect.

Types of Response. Environmental factors influence organisms physiologically in various ways (Fry 1947). These effects may be classified as follows :

Lethal : Causing death; for instance, extreme heat or cold, lack of moisture and so forth.

Masking. Modifying the effect of some other factor. Low relative humidity increases the rate of evaporation of moisture from body surfaces so that warm blooded animals are able to survive at otherwise intolerably high air temperatures.

Directive : Producing an orienting response in relation to some environmental response so that the organism gets itself into favourable conditions.

Controlling : Influencing the rate at which some process functions, but not entering the reaction. Temperature, pressure and viscosity, for instance, affect metabolism, secetion, and looomotion.

Deficient : Curtailing an activity because some essential ingredient, such as a salt, oxygen or the like, is absent or at unfavourably low concentration.

The same environmental factor may produce different effects at different times and under different conditions. Temperature may be lethal, if extreme; masking as when cold reduces the demand of cold-blooded organisms for food; directive, by inducing a search for more favourable locations; or controlling, as a modifier of the rate of metabolism. Often the distinction between controlling and deficient factors is not made, or they are considered as together constituting limiting factors.

Law of Toleration. For each species there is a range in an environmental factor within which the species functions at or near an optimum. There are extremes, both maximum and minimum, toward which the functions of a species are curtailed, then inhibited. Upper and lower limits of tolerance are intensity levels of a factor at which only half of the organisms can survive. These limits are sometimes difficult to determine, as for instance, with low temperature, organisms may pass into an inactive, dorman or hibernating state from which they may again become functional when the temperature rises above a threshold. At high temperatures, there may be similar inactivation or aestivation before the lethal level is attained. Even without dormancy occurring, there are normally zones of physiological stress before the limits of tolerance are reached.

Behaviour Responses

Orientation. Behaviour responses to changes in environmental factors can usually be detected immediately as turning or locomotor activites on the part of the organisms (Fraenkel and Gunn 1940). These movements tend to take the organism away from points of danger and into more favourable locations, or to perform some task essential to existence, or to reproduction. If the movements involve curvature or a turning movement either toward or away from the source of stimulus, the movement is called a tropism. Motile organisms frequently respond by actual locomotion toward or away from the stimulus rather than mere turning and such guided or directed locomotor movements are called taxes. When the movements of the animal are random in direction, and there is no immediate orientation to the source of stimulus, but the frequency of turning or speed of the movements is dependent on the intensity of stimulation, such responses are termed kineses. As the result of kiness an animal may arrive by chance in a favourable environment, by which the intensity of the stimulus is responding, the following prefixes are employed thermo-temperature; photo-light geo-gravity; hydro-moisture; chemo-chemicals; thio-contact; baro-pressure; rheo-current; and galvano-electricity.

Innate Behaviour. Much of the behaviour or organisms is determined by heredity and is characteristics of the species in its proper environment.

The behaviour may be evident at birth or it may not develop until the nervous system, including both the receptor and effector mechanisms, is fully matured. Such innate behaviour is of various degrees of complexity. A reflex is a quick, automatic response of a single organ or organ system to a simple stimulus; for instance, the knee jerk in man. Tropisms taxes and kinesees may involve a series of reflexes and represent a higher level of integration. An instinct, or inherited behaviour pattern, is a complex fixed behaviour that is activiated, more or less automatically, when the animal is presented with the proper stimulus (Thorpe 1951).

Stimuli. Before an action will take place the nervous mechanism must be released by the reception of a stimulus. Stimuli may be either external or internal to the organism. Protoplasm is sensitive to any kind of stimulation, provided it is intense enough. In higher organisms, however, specialized tissues have become particularly sensitive to one kind of stimulus, and these tissues, or sense organs, are called receptors. There are several forms of receptors : photo-receptors, mechano-receptors, chemoreceptors, thermo-receptors and statto-receptors. Not all types of receptors are present in all organisms, and the structure and effectiveness of those present varies from one kind of animal to another. The efficiency of the receptor mechanisms is important, as they largely determine the environmental factors to which the animal will respond and the degree of sensitivity involved.

Ecological Life Histories. Developmental life histories trace the origin and growth of structures and functions of an animal from the egg stage until maturity is reached. Such studies are largely embryological in nature. Behaviour life histories attempt to analyze the activities of animals in terms of innate and learned behaviour, and the neutral mechanisms involved. In order to do this, it is often necessary to trace the origin of each activity to the manner in which it first makes its appearance in the young animal. Ecological life histories, on the other hand, are concerned with the activities of a species throughout its life cycle and in relation to its adjustments to natural conditions. Ecological life histories usually proceed with, first, analysis of the behaviour adjustment needed for the survival of the mature animal; then of its reproductive behaviour; and, lastly, of the development of behaviour and physiological adjustments of the young animal.

The following are important items that should be included in a complete ecological life history of a species :

1. Phylogenetic and geological history.
2. Geographic and habitat distribution with an analysis of adjustments to the physical environment and of biotic interrelations within the community.
3. Variations in population, through time and in space.
4. Changes in seasonal activities and physiological states : breeding, migration, hibernation.
5. Food, enemies.
6. Parasites, diseases.

7. Reproductive potential, mortality rate of population turnover.

8. Requirements for reproduction : home range, territory, nest-site, nesting materials, etc.

9. Breeding behaviour : mating nesting, etc.

10. Development of offspring ; rate, stages, generations per year, etc.

Ecological Niche. The ecological niche is the particular position in a community and habitat occupied by a species population as the result of its pecular structural adaptations, its physiological adjustments, and the special behaviour patterns that have evolved to make best use of these potentialities. Important factors in the niches occupied by white-footed mice and deer mice. Both mice are equipped with large eyes for nocturnal vision, large external ears for hearing, long vibrissae on the face for aid in running through dark underground burrows, and protective coloration.

Every species has its own peculiar niche. No two species can permanently occupy exactly the same niche in the same locality. The living together of many species in the same community is possible only because their various niche requirements are different.

Community nterrelations

The fact that species with similar tolerances and similar or complementary requirements aggregate into similar environments to form communities is a response of special interest. No organism occurs alone. Each must find its place in the community and establish relations with other members of it. The manner in which the response of species to each other is affected is shown in the structure and composition of the community and in its internal dynamics, succession and distribution.

Biotic Community

Structure and Dynamics. The concept of the biotic community is basic to an understanding of ecology. We will here be concerned only with laying a foundation of general principles. Details will come in later lesions, but for proper orientation we must know something about how the community is organized, how it functions, and how it may be recognized.

Internal Structure and Processes

Community and Ecosystem. A community, or biocenose, is an aggregate of organisms which form a distinct ecological unit. Such a unit may be defined in terms of flora or funa, or both. Community units may be very large, like the continent-wide coniferous forest, or very small like the community of invertebrates and fungi in a decaying log. The extent of a commmunity is limited only by the requirement of a more or lessuniform species composition.

A different community occurs in each different habitat and environmental unit of larger size, and in fact the composition and character of the community is an excellent indicator of the type of environment that is present. Since plants and animals, bacteria and fungi, all occur together in the same habitat and have many interrelation they can scarcely be considered independently of each other. Together they make up the biotic commmunity.

Communities may be distinguished as major or minor. Major communities are those which, together with their habitats, form more or less complete and self-susttaining units or ecosystems, except for the indispensable input of solar energy. Minor communities, often called societies, are secondary aggregations within a major community and are not, therefore, completely independent units as far as circulation of energy is concerned.

The biotic community along with its habitat is called an ecosystem. The term "ecosystem" has been loosely applied to units of various sizes and characteristics but had best be limited to distinctive ccmbinations of air, soil and water conditions with vegetative, animal, and microbic life.

Productivity. A characteristic of communities that has become of considerable importance in modern ecological search is productivity. The number of individuals of biomass in a community at any one time is the standing crop. At the beginning of the year or reproductive season the standing crop is usually small but as reproduction and growth take place there is an increase in the amount of organic matter making up the biomass of the community. The production of organic matter per unit of time and area is productivity. Productivity is thus a rate and is commonly indicated on a yearly basis, but it is also possible to measure monthly, weekly or daily production. Small standing crops may have a high productivity and large standing crops a low productivity, hence biomass alone does not give a measure of productivity. The largest standing crop which a habitat can support without deterioration, or the maximum number or biomass of animals that can survive the least favourable yet tolerable environmental conditions during a stated period of time, is the carrying capacity. Carrying capacity is determined not just by the amount of food available, but also by shelter, social tolerance, and other factors. A variety of methods are being used to measure productivity of different kinds of organisms and of different habitats. It is desirable to indicate productivity as accurately as possible in descriptions or analysis of community dynamics.

Species Diversity. Ecologists have long been aware that communities in different environments vary in the number of species that they contain, but only in recent years have quantitative indices been developed to show the relation between community structure not only (1) in number of species but also (2) in the relative number of individuals in each species.

Succession

Communities are in a more or less continual process of change (Clements 1916). These changess result in part form the reactions and coactions of the organisms themselves and in part from such external forces as changing physiography, changing climate, and organic evolution. The habitat is usually affected as well as the community, and as the habitat changes, new species invade it and become established, and old species disappear. These changes are especially noticeable in dominant species, since these species exert a controlling role over the composition and structure of the community as a whole. The replacement of one community or ecosystem by another is succession, and succession continues until a climax or final stage is reached.

Succession is a process. The series of steps or communities comporising a successional sequence leading to the climax is the sere. Seres are sometimes classified according to the predominant force that is bringing them about. These forces are biotic, climatic, physiography, and geologic and their resultant seres are commonly called bioseres, cliseres, eoseres and geoseres.

Biotic Succession. Biotic succession is brought about by forces inherent within the community and in the activities of the plants and animals themselves. The most important of these activities are the organismal reactions and coactions that produce modifications in the habitat and interrelations between species. Important reactions involve filling in of ponds with plant and animal remains, the addition of organic nutrients to sterile soil, and the reduction in light intensity by increasing density of plant growth (Horn, 1971). With progressive improvement of the soil and changing light and moisture conditions, a series of new dominants come into the area. When invasion of new species occurs, intense competition develops; if the invaders are successful, the old species disappear as a new community replaces the old one.

Succession is considerably influenced by the kinds of propagules available in the vicinity. Seeds, spores and the like are dispersed more or less readily, depending on the form. Some kinds of animals roam more widely or sprend more readily into new areas than do other kinds. The composition of the community that develops and the rapidity with which it becomes established depend, in large part, on the rates at which different species invade.

Bioseres may be broadly grouped as priseres and subseres, depending on whether they develop on primary or secondary bare areas. A primary bare area is a sterile habitat, such as rock, stand, clay or water. A secondary bare area is a denudation resulting from temporary flooding, fire, logging, cultivation, overgrazing, or other phenomenon that does not produce an extreme disturbance of the soil or substratum. Since in the latter the habitat has already supported community life, and since the soilsubstratum is already in an advanced stage of development, the resulting subsere progresses rapidly and the early pioneer stagess of the prisere are not usually required. A subsere will develop following the destruction of any stage in a prisere, and the species composition of the stages in the subsere will be influenced by particular priseral community that was destroyed.

The early stages or communities that make up the prisere drpend largely on the type of bare area on which the prisere originates. As succession proceeds, however, later stages in the various seres in any area having a relatively uniform and humid climate come to be more and more alike. The successional develcpment from widely diversified communities in initially different habitats to closely similar or identical climax communities in habitats that have also become much alike is called convergence.

Bioseres may occur on a small scale in microhabitats as well as in major ones. When hay infusions, prepared in the laboratory, are seeded with representative protozoans, the order of appearance of maximum or peak populations in the various species is bacteria and monads, colpoda hypotrichates, Paramecium, Vorticella and Amoeba. Disappearance for species is in the same order, except that Amoeba precede Paramecium and Vorticella. Algae may come in at the final stage, so that a more or less balanced community is established. The succession of species results from the higher reproductive rate of earlier species and from the fact that the excreta of at least some forms, especially the hypotrichates and paramecium, are toxic to themselves.

Recognition of Communities

The organismic concept considers the community to be a superorganism, a complex organism, or a social organism. As such, it is highest stage in the organization of living matter : namely, cell, tissue, organ, organ system, organism, species population, community. There is emergent evolution, so to speak, at each higher stage in this hierarchy; the whole is more than merely the sum of its parts. The species population has inherent characteristics of density, rate of natality rate of mortality, and age distribution, while the total community has such unique functions as dominance, cooperation, trophic balance competition, and succession which are beyond the characteristics of the individual organisms of which the community is composed. The community behaves as a unit in its competitional and successional relations with other communities, in its local and geographic distribution, in its seasonal activities and response to climate, and in its evolution. Although the community varies in its taxonomic composition and structure in different environmental situations, this variation is proportionally no greater than occurs in different cells of the same type or between individuals belonging to the same species.

The individualistic concept places emphasis on the species, rather than the community, as the essential unit for analysis of interrelations, activities, distribution, and evolution. Each species responds independently to the integrated influence of the various factors of the physical environment and biotic coactions. The environment may be conceived as a pattern of gradients with the intensity of the various factors changing gradually in space from one extreme to the other.

Physiognomy. The gross structure of a community or its physiognomy is an important basis for its recognition. In terrestrial communities, physiognomy is determined by the life-forms of the dominant plant species and their specing. The life-forms that prevail in a given area depend on the climate and tne substrate or other special features of the habitat and give character to the landscape. The distribution of animal communities is closely correlated with the structure of the vegetation-types need, to be recognized and defined :

Deserty. Hot, arid habitats with scattered scrubby or thorny vegetation or, in extreme cases, none.

Thron Scrub. Tropical dry thorny deciduous tall shrubs, succulents, ephemerals.

Steppe, Plains. Semi-arid grassland covered with short grasses.

Prairie. Semi-humid grassland covered with mid and tall grasses.

Temperate Shrubland. Semi-arid areas covered with bushes and shrubs, either deciduous or broad-leaved evergreen (chaparral).

Savanna. Grassland with scattered trees or groves of trees or shrubs covering 10 to 25 per cent of the ground (Dansereau 1960).

Woodland. Open stand of small deciduous or evergreen trees covering 25 to 60 per cent of the ground and with undergrowth of grassland or desert vegetation.

Forest-Edge. Mixture of trees, shrubs and open country, ordinarily occuring as a narrow belt on the margin of forests.

Forest. Closed stand of trees forming a continuous canopy over at least 60 per cent of the area.

Deciduous Forest. Broad leaves fall during cold or dry seasons.

Broad-Leaved Evergreen Forest. No regular season of leaf fall, leaves often sclerophyllous, warm climates.

Rain Forest. Tall luxuriant forests often with several strata of trees, foliage retained throughout the year, climate continuously warm and wet.

Needle-Leaved Evergreen or Coniferous Forest. Forests of pines, spruces, firs, larches, hemlocks and the like.

Forest-Tundra. Stunted open growth of coniferous forest in cold climates.

Tundra. Extensive flat or gently rolling treeless areas occurring in cold climate.

Alpine Tundra Treeless areas at higher elevations of mountains.

Bog. Wet areas in cold climates containing sphagnum, health, plants, coniferous trees.

Swamp. Wet areas in warm climates covered with deciduous trees or shrubs.

Marsh. Wet areas containing sedges, rushes, cat-tails and the like.

Preservation of Environmental Quality. Up to the advant of civilized man, the biosphere of the world consisted of a complex pattern of biotic communities resposive to a vari ty of environments. The harmony prevailing between the biotic and physical realms was the result of hundreds of millions of years of evolution and adjustment of one organism to another and of each to the habitat in which it lived. The relationship was a salient one. A sporadic outbreak of a species as the result of some temporary advantage, or a catastrophe to a population as the result of a change in the physical environment, soon became tempered and a balance restored. Primitive man occupied a niche and assumed a role comparable to other animals, and the ecosystem maintained itself in a vigorous healthy condition.

Modern man, however, has initiated a number of abrupt changes during the last few centuries beyond the capacity of the ecosystem to adjust. He has become the dominant organism in the biosphere, forcing changes to fit his needs and desires, and creating ecosystems different from any that have before existed. Emergence of new communities is a characteristic of the geosere, but the replacement of the biosphere by an "anthroposphere" or "noosphere" (from the Greek word, noos, mind) is unique in the speed and extreme changes produced. This acquisition of doominancee is the result of the extraordinary evolution in man of intelligence, ingenuity, and use of tools. He has greatly modified or replaced the natural processes the control stability and balance within the ecosystem with new processes and manufactured products which the environment cannot absorb. The result has been a general deterioration of the environment, a population, explosion of the human species, and a disharmony in the world ecosystem and will, unless resolved, lead quickly to catastrophe. Whether man will survive as the dominant organism be reduced again to the role he occupied as a primitive, or become extinct may well be decided within the next few decades. There is conflict here between the demands of man for expanding population, productivity, power and pleasures and the functioning of natural laws for maintaining stability and equilibrium between the living and physical environments.

Immediate problem of concern in the preservation of environmental quality are the elimination of air, water and soil pollution the disposal of solid wastes; and climate modification on a planetary scale. Perhaps noise pollution should also be added. Pollution is change in environment, caused by man's discharge of matter or energy into it, which alters the normal function of an ecosystem. Fundamental to the attainment of stabilized and balanced worldwide ecosystem are the control and regulation of the human population at a suitable level and preserving the habitat and the basic community of plant and animal life upon which human life depends. Considerable technical knowledge already exists for solving these problems. The immediate need is concerted attention for implementation of the necessary remedial measures. Satisfactory resolution of the problems will involve not only science and technology but also law, sociology, politics and economics. Emphasis on existing knowledge, however, should not obscure the fact that continuing basic research is required to elevate man's understanding of the environmental system. Scientists are just beginning to study ecosystems on the multidisciplinary basis that is clearly required.

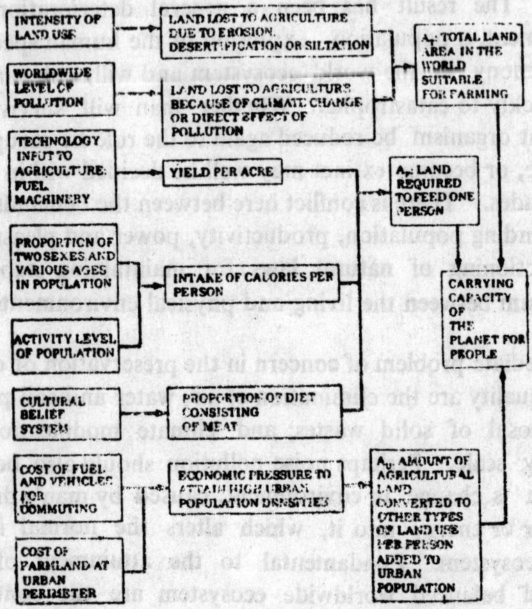
INTENSITY OF LAND USE
WORLDWIDE LEVEL OF POLLUTION
TECHNOLOGY INPUT TO AGRICULTURE: FUEL MACHINERY
PROPORTION OF TWO SEXES AND VARIOUS AGES IN POPULATION
ACTIVITY LEVEL OF POPULATION
CULTURAL BELIEF SYSTEM
COST OF FUEL AND VEHICLES FOR COMMUTING
COST OF FARMLAND AT URBAN PERIMETER
LAND LOST TO AGRICULTURE DUE TO EROSION, DESERTIFICATION OR SILTATION
LAND LOST TO AGRICULTURE BECAUSE OF CLIMATE CHANGE OR DIRECT EFFECT OF POLLUTION
YIELD PER ACRE
INTAKE OF CALORIES PER PERSON
PROPORTION OF DIET CONSISTING OF MEAT
ECONOMIC PRESSURE TO ATTAIN HIGH URBAN POPULATION DENSITIES
A: TOTAL LAND AREA IN THE WORLD SUITABLE FOR FARMING
A: LAND REQUIRED TO FEED ONE PERSON
CARRYING CAPACITY OF THE PLANET FOR PEOPLE
A: AMOUNT OF AGRICULTURAL LAND CONVERTED TO OTHER TYPES OF LAND USE PER PERSON ADDED TO URBAN POPULATION

2

Ecosystem Ecology

(A) Soil Formation

Communities cannot exist without a habitat, nor is a habitat likely to remain long without a community developing in it. The functional interrelations between community and habitat are many and complex, constituting an ecosystem. Most important are soil formation nutrient cycling, and energy flow. Human interference in these processes often causes pollution, and exploitation, often bring exhaustion of natural resources.

We have already considered many of the reactions of plants upon the habitat, such as reduction of light and wind intensities, mitigation of temperature extremes, interception of rainfall and increase in relative humidity. Plants also exert important effects on the formation, structure and characteristics of the soil or substratum produced by accumulation of dead plant remains; they further...the weathering of rock through acid excretion and the mechanical action of roots; they offer obstruction to wind and water borne materials; they help stabilize moving sand and talus slopes and help prevent erosion generally; they variously increase or decrease the water content of the soil; they foster decomposition of raw humus into usable untrients; and so forth. Water plants form marl. It is by these reactions that plants exert dominance interrestrial communities and establish the physical conditions of the habitat, which must be acceptable to all minor plants and animals that dwell there. Succession of plant stages eventually brings the interactions between habitat and community into equilibrium upon the establishment of the climax. In this chapter we will be primarily concerned with soils as dynamic components of ecosystems, including the cycling nutrients between the soil and the biota.

Soil Formation

Texture porosity, consistency arrangements of particles, chemical nature, and organic content of the soils are determined by three factors; parental rock material, the biota, and the climate. Differences in topography modify the relative effects of these three factors, and time is required before their full effects are realized.

Parental Rock

The basic rock from which the mineral portion of a soil is derived determines, to a large extent, not only its chemical composition but also its structure. For instance, soils derived from limestone are highly calcareous and more alkaline than soils derived from limestone are highly calcareous and more alkaline than soils derived from feldspar, sandy soils from quartite. Clay forms a finely textured compact, water-retaining soil. Sand is coarse-textured and porous. Loan is a mixture of sand and clay and makes the best soil. The presence of iron oxides and silicates produces the red and yellow colors of some soils. Humus produces black soils. Soils from swamps or bogs and very rich in organic material is called muck.

Residual soils are formed in situ from underlying bedrock. Soils may however, be formed in one locality and moved considerable distances. Soils transported and deposited by wind are called losses, by water alluvium; by glaciers, till.

Biota

Plants and animals have a highly important role in the formation of soil, both of them affect its structure and they aid in the production of humus. Plants contribute to the mechanical and chemical weathering of rock. Plant roots, especially those of trees, can split large rocks. Lichens, mosses, and even bacteria and fungi excrete acids in the course of metabolism, which dissolve the substances that cement rock granules together. When plant roots die, fungi convert them to dry, soft spongy materials (punk), used as food by saprophytic micro-arthropods. Usually the bark of the root remains in tact the longest. Hollow tubes are thus formed that permit water and air to penetrate considerable depths into the soil. These channels gradually become filled with silt and animal excreta.

The addition of plant and animal organic matter to heavy compact soils or clay tends to open the soil, making it more porous. Addition of organic matter to sandy soils binds the particles closer and together, making the soil less porous.

Earthworms may be divided into deep and shallow-working species. Deep working species dig narrow tube-like channels which may reach 2-3 metre down through overlying soil to parent rock. Earthworms ingest soil while borrowing, digest and absorb organic matter from it, and digest the residue in semi-liquid form which is used to cement the walls of the burrow or else is deposited at the surface as castings. Earthworms prefer easily digested succulent vegetation and dung for the purpose, but in the autumn may pull the freshly fallen leaves down into their burrows to use as food or nest linings. Ejected petioles may form midden piles around burrows. Earthworms are not, however, important soil builders in disturbed grassland; they may be absent altogether in and regions, the dry weight of casts brought to the surface annually by earthworms varied from 475 g/m^2 (2.1 tons/acre) in a moderately hot dry climate, to 24,000 g/m^2 (10^7 tons/acre). Earthworm casts compared with the surrounding soil show higher total nitrogen, oganic carbon, exchangeable calcium, exchangeable magnesium, available phosphorous, exchangeable potassium, organic matter, base, capability, pH and moisture equivalent. Only certain species make these surface castings; other species void the ingested soil into subterranean spaces.

The ant Lasius niger neoniger spends most of its time in its underground burrows and deposits excavated soil upon the ground surface around burrow entrances.

Humus in soil, organic matter that is partly or entirely decomposed is called humus. The amount of humus varies from less than 1 per cent to as much as 20 per cent of the soil; peat soil may be largely organic material, but much of it resists decomposition, and hence is not true humus. Decomposition breaks down compounds into similar ones that are washed back into the soil, thus becoming available again as nutrients.

The annual dry weight of leaves that fall to the ground in deciduous and coniferous forests varies from site to site and with the density of the trees, but is commonly in the range of 50 to 400g/ms². In mature climax forests the rate of decomposition of the liter and re-absorption by plants of the nutrients thus yielded keeps pace with the annual accumulation so that an equilibrium is established. In several stages, decomposition and utilization do not keep up with the annual accumulation, so that the ogranic content of the forest floor increases with time. Litter production is lower in artic than in tropical regions.

Metric Tons/Hectare-year

Region	*Leaves*	*Other*	*Total*
Arctic alpine	0.7	0 4	1.0
Cool temperature	2.5	0.9	3.5
Warm temperature	3.6	1.9	5 5
Equatorial	6.8	3.5	10.9

However, in the warm tropical regions the amount of humus that accumulates on the forest floor is low because of the high rate of decomposition, runoff and leaching.

The thick organic layer on the ground moderates extremes in the daily and seasonal rhythms of soil temperature, retards freezing of the ground in the autumn and thawing in the spring, and retains soil moisture. Because of humus formation (involving oxidation) and the respiration of plant parts and animals undergrounds, soil air contains little oxygen but much carbon dioxide, and it possesses a higher moisture content than does the general atmosphere above ground. This is especially marked in warm summer months when these processes go on more rapidly. The decay of organic matter usually makes the top soil somewhat acid (most commonly pH 5 to 7), but in the mineral subsoil, the acids are often neutralized by the basic salts commonly present.

The mineral content of leaf fall varies according to the species of tree.

Element	*Hardwood forests*	*Coniferous forests*
Calcium	7 3	3.0
Nitrogen	1.8	2.6
Potassium	1.5	0.7
Magnesium	1.0	0.5
Phosphorus	0.4	0.2

Silicon, cooper, manganese, carbon and zinc are also present in the leaves of hardwood trees. Carbon is relatively more abundant and nitrogen less abundant in coniferous than in deciduous leaves, but commonly the carbon nitrogen is 55 : 20.

Both plants and animals are important agents affecting the decomposition of organic matter and the formation of humus. An animal digests and mobilizes plants foods, much of which is returned to the soil, in parts as the excreta of the living animal, in part as the body of the dead animal, in part as gas. Fully formed human is in fact, derived mostly of fecal material. The larger herbivorous and carnivorous animals pass urine and feces containing simple nitrogenous compounds and compounds of phosphorous, potassium, and traces of calcium, magnesium, sulphur and other elements. Humus is but one point in a continuous cycles of decomposition of plant and animal organic matter, absorptian of decomposition products by plants ingestion and metabolization of plant matter by animals, decomposition of plant and animal organic matter.

The most important soil organisms concerned in the decomposition of the little are the bacteria, both aerobic and anaerobic forms. They are commonly divided into two types. Hetrotrophic bacteria obtain their energy from the oxidation of the carbohydrates and fatty substances as described above. They use this energy for the synthesis of cell substances and the production of enzymes that break down the complex compounds in the litter into simpler compounds, including proteins into ammonia compounds. They then use part of the ammonia compounds in synthesizing the amino acids they need in building their own proteins. Autotrophic bacteria, in turn, are of two types : chemosynthetic species, which obtain their energy from the oxidation of inorganic compounds (hydrogen, sulphur, hydrogen sulphide, iron and ammonia) and photosynthetic species which include purple and green sulphur bacteria, possess a form of chlorophyll, and utilize the energy of sunlight. Chemosynthetic bacteria convert ammonia compounds into nitrites and nitrates, part of which they use in their own anabolism, the rest becoming available for plant to absorb. Photosynthetic bacteria use the ammonia compunds in their own anabolism and do not render them directly available to plants. Chemosynthetic bacteria are more abundant than photosynthetic bacteria in soil; photosynthetic bacteria are more abuntant.

The decomposition products available to plants are reabsorbed by the roots and built up into plant tissue. Plants are eaten by animals and as plants and animals die the minerals are returned to the soil. All essentiel minerals cycle repeatedly through the ecosystem, the cycle of each mineral differing in various ways from the cycle of every other mineral.

Nitrogen cycle. In the nitrogen cycle proteins are broken down yielding ammonia (NH_3) compounds in the course of the metabolic processes of all animals and by the activities of heterotrophic bacteria, filamentous fungi, and actinomycetes. This process is called ammonification. Some of the ammonia is oxidized to form nitrites (NO_2) and nitrates (NO_3) through the action of autotrophic bacteria; the process is called nitrification. Other types of bacteria act on ammonia in the process of denitrification, by which nitrogen NO_2 is liberated into the atmosphere. Nitrogen is removed from the air by the nitrogen-fixing bacteria which live either freely in the soil or as symbionts in the root and leaf nodules of many legumes and some non-legumes. Ceanothus Elaeagnus, Alnus, and Myrica, among other Blue-green algae also fix nitrogen in moist soil and are more important than bacteria in this regard in aquatic habitats. Ammonia compounds, nitrates and other substances are added to the soil in small amounts with rainfall, sources of these nitrogen compounds are volcanic eruptions, terrestrial decomposition, and atmosphere nitrogen fixed by lightning. Man has developed an important chemical fertilizer by artificially fixing nitrogen from the air and much research is underway for the better understanding of the enzymatic mechanisms of biological fixation.

In a balanced or stabilized ecosystem, nitrogen molecules continue indefinitely to cycle among the soil, plants and animals with nitrogen fixation replacing that lost denitrification leaching and so on. In several communities, increase in soil fertility depends in large part on nitrogen-fixation exceeding denitrification so that there is a progressive increase of nitrogen molecules in the ecosystem as succession proceeds.

Carbon Cycle. Animals obtain much of their carbon, as well as nitrogen, from plants, although some forms are also able to fix carbon directly from salts dissolved in water. In photosynthesis, carbon dioxide obtained from the air and from dissolved bicarbonates in the substratum is combined with water to form carbohydrates, a portion of which may be converted to fats. Plants combine carbon with oxygen, nitrogen, hydrogen, and sulphur to form proteins. Carbon dioxide in the air comes chiefly from the respiration of animals, but small amount arise form the respiration of plants, the decay and fermentation of organic matter springs, volcanic action, and solution of sedimentary rock. Volcanoes were probably the original providers of carbon dioxide to the biosphere. Organisms tie up carbon dioxide as carbonates in skeletons and shells. Carbon is also tied up in the formation of peat, oil, shale and coal. When limestone and other carbonaceous sediments are exposed to water erosions, the carbonates may be hydrolysed to bicarbonates and thus become a source of CO_2. The concentration of carbon dioxide in the air is stabilized at 0.03 per cent by buffering action of bicarbonates and carbonates in the oceans and fresh water bodies (Hutchinson 1948) :

$$CO_2+H_2O \quad H_2CO_3 \quad HCO_3 \quad CO_3^{-2}$$

On the other hand, oxygen in the air (21 per cent) is derived almost entirely from the photosynthesis of plants. Studies are underway to quantify the various steps in the carbon cycle (Oslon 1970) as well as other mineral cycles (Duvigneaud and Denaeyer-DeSmet 1970).

Other Elements. In addition to oxygen, carbon, hydrogen, and nitrogen, animals require at least 13 other elements that are all derived from the soil : calcium, phosphorous, potassium, sodium, chlorine, sulphur, magnesium, iron, copper, manganese, iodine, cobalt and zinc. Only traces of some of these elements are required, but calcium is used in large amounts for skeletons, shells, antlers, and other organs, and in metabolism generally. Phosphorus is a constituent of nucleoproteins, phospholipids and skeleton. Goiter occurs in mankind and some animals in regions deficient in iodine. These elements are obtained from food, drinking water, salt ficks, and grit taken into the stomach. A salt lick is a local, usually clayey, area characterized by a high concentration of salts where deer and other animals foregather to lick the soil for the salt. Soils deficient in or lacking these various elements support sparse animal populations; individuals are in more or less poor health; reproduction rates are low.

Major Soil Groups

The interrelations between the basic mineral content of the parent substrate, biotic reactions, and climate can be seen through an analysis of the development of the great soil groups of the world. We will give only a simple classification of them. A detailed classification would include many subdivisions and intermediate categories (Lutz and Chandler 1946, Simnoson 1957).

Podzolic soils are formed in humid temperate climate, under forest vegetation. The A_2 horizon is moderately well developed, for there is sustained leaching. Soils are more or less acid and only moderately fertile. Podzols develop under coniferous forest and have a mor type of humus. Graybrown and brown podzolic soils are found under hardwood forests and have a mull humus.

Latosolic soils develop in humid tropical or semitropical forested regions. Humus is quickly oxidized by action of micro organisms and hence does not accumulate. The soil fauna is depauperate. Chemical weathering of the parental materials is intense. Water drainage through the porous soil is rapid, so eaching is extensive. In early stage of its formation, the soil is neutral or slightly alkaline, but as leaching continues, it becomes acidic. The soil has a thin organic layer (A_0 and A_1 horizons) on a reddish, leached soil (A_2 horizon) that extends to great depths below the surface.

Chernozemic soils occurs in humid to semiarid temperate climates under grass vegetation. The grasses on dying return considerable organic matter to the soil. The A_1 horizon is consequently dark in color and of great thickness. The soil contains more bases and hence is less acid than in the two types above. The B harizon in humid regions is indistinct, but where there is less rainfall, calcium salts may accumulate to form a hardpan. Prairie soils in temperate climates are among the most fertile soils of the world, but fertility decreases in the tropical and desert climates.

Desertic soil are characteristic of arid climates and contain very little organic matter. A profile is poorly developed. The surface soil is brownish gray, and grades quikly into the calcium carbonate horizon which usually forms a hardpan just below the surface. Wind erosion removes the finer particles leaving the coarser materials to form a har.. pavement. The soils are but slightly weathered and leached; lacking nitrogen, they are infertile.

Mountains and *Mountain Valley solis* vary shallow layer or eroding rocks to deep organic soils of valleys and swampy areas.

Tundra soils occur in cold northern areas where the substratum remains continuously frozen and the vegetation of lichens, mosses, herbs and shrubs makes a peaty surface layer. The region is poorly drained and characterized by many scattered shallow ponds.

Alluvial soils may be important locally. These soils are mostly without a developed profile and are the result of deposition by streams. The are usually very fertile and support luxuriant vegetation.

Saline soils are found in dry climates where rapid evaporation of water results in surface deposition and accumulation of salts leached from surrounding upland areas.

This classification shows clearly the correlation that exist among soil groups, climates, and vegetation. The compositiont and particularly the abundance of animals is also affected. The average biomass of soil animals under different categories of vegetation-soil-climate system is as follows (Ghilarov 1967):

Moss tudra : 3 g/m²

Coniferous forest : 20 g/m²

Broad leaved deciduous forest : 100g/m²

Short grass plains : 25 g/m²

Semi-desert : 1 g/m²

Soil Erosion

As rain water drains over the surface of the ground, it picks up and transports particles of soil more or less proportional to the rate of flow. This is a continuous process that has gone on throughout geological time, resulting in base-leveling of mountains, the filling in of valleys, and the formation of river deltas. Normally this is a slow process, with changes becoming apparent only in thousand years. With the ground covered with vegetation, the flow of water is retarded and more of it percolates into the ground or is evaporated. In fertile areas, new soil is formed faster than it is eroded or at least there is stabilization.

With removal of the native vegetation and cultivation of the soil in agiculture, surface particles are more easily moved and surface water drainage is more rapid. As a result, topsoil is being lost at an alarming rate. Much land has been so badly damaged that here is little likelihood that it will ever again be productive crops. Conservation of the fertile top soil is one of the major problems facing modern agriculture. Putting in underground drainage tiles in level farmlands helps to draw rain water into the soil and to decrease surface run off and erosion. Soil conversation procedures should be strictly followed,

Soil and Air Pollution

Soil, water and air pollution results from contaminants added to ecosystems beyond what can be readily absorbed through normal nutrient cycling. In undisturbed ecosystems, dead organisms and excreta quickly decompose and their nutrient elements recycle for reuse. The existence of plants and animals depends on the proper utilization of these "natural wastes' and there is no harmful accumulation of surpluses. Man, however, introduces into ecosystems excessive amounts of organic and other wastes and has created through chemistry an expanding quantity and variety of complex substances which are highly res stant to existing enzyme action and normal cycling processes. Consequently, there is general alteration and deterioation of the environment. Included are the chemicals than man uses for maintaining his dominance over "anthroposphere". The application of herbicides, fungicides, insecticides, fumigants; and rodenticides often produces far reaching and unanticipated disturbances between predators and prey producers and consumers, and in the various nutrient cycles. We need a world-wide system of careful monitoring of environmental quality as a basis for evaluating and controlling all kinds of disturbances to environmental quality (Wilson and Matthews 1970). Man's existence may depend on it.

Noise pollution results when undesirable sounds become irritating to the general public. Excessive noise may not only be annoying but may so cause permanent hearing loss, loss of efficiency in performing tasks, and other less-well defined organic, sensory or physiological disturbances. The loudness or acoustical effect of sound is nearly proportional to the logarithm of its intensity. An acoustical scale in decibels (dBH :.1 Bel) is commonly used, which is 10 times the measured intensity of a sound relative to the intensity of sound just perceptible to the average human ear, or approximately 10^{-16} watt/cm².

Soild Wastes

The United States generates 4.5 kg (10 lb) per person per day of municipal, agricultural, commercial and industrial wastes in the form of paper, plastics, garbage, textiles, wood, rubber, glass, ceramics, metallics, and so on. This figure does not include junked automobiles. Annual per capita increase is 2 per cent per year, and with population increasing 2 per cent per year, there is a compounding increase of 4 per cent per year. Many of these wastes are odorous and unsightly and are in mere mass, excessive. Furthermore, natural materials, foremerly used and easily decomposable, are being extensively replaced by synthetic ones that are resistant (nylon hose for cotton hose, for instances). Several methods are in use of under study for disposing of solid wastes.

Sanitary Landfill. The material is spread in depressions in thin compressed layers, alternating with dirt, to bring the ground level up to the general topography. Aerobic decomposition takes place at first and then is replaced by anaerobic decomposition. There is some danger from gases that are formed and of contamination of ground waters. Suitable landfill sites may not be available locally.

Incineration. Much of the solid, wastes may be reduced by incineration at temperature of 760-980 C. The residue may contain iron, aluminium, zinc, copper, lead and tin, which are recoverable, but has only limited marketing feasibility at the present time. Combustion gases consist of carbon dioxide, water, oxides of nitrogen and sulphur, and other gases is lesser amounts. Air pollution by these gases is usually not serious, but emission of particulate fly-ash may be important. In addition, considerable heat escapes into the atmosphere. New techniques are being developed to increase the percentage of combustion, prevent the scope of particulate matter and recover the heat for useful purpose.

Composting. The nitrogen phosphorous and potassium in the wastes if recovered could be used as fertilizer on agricultural land. Present procedures involve first removing noncombustible substances and those items that have salvage value. Grinding the remainder for faster decomposition and then subjecting the matter to bacterial action until it becomes humus. The material spread over fields in windrows commonly requires 6 to 7 weeks to decompose, but some mechanical systems require only 5 to 6 days. The humus is then bagged and offered for sale. There are some plants in operation, but the procedure requires more development to become generally economic.

Recycling. Junked automobiles present a special problem since there are 9 to 16 million cars abandoned yearly. Useful parts are commonly salvaged. Of the remainder, combustible parts may be stripped off and burned and the useful metals recovered. A difficulty lies in obtaining these metals in pure enough form for their reuse to be practical. The same problem exists in recycling industrial and mining wasted. Doubtless this can be solved if sufficient research and effort are made to do so.

Reuse. A different attack on the problem is the elimination of the waste, as far as possible, at its source. For instance, containers for holding liquids and foods could be standardized in a few different shapes and sizes and made in form that could be returnable and used repeatedly, rather than being discarded after the first use. Containers and items that cannot be reused could in many instances be made of degradable materials. Recycling glass and metal containers bears an energy cost that often ofsets the value of the material that are saved.

Certainly we must face up to the fact that space for storing solid wastes in the environment is limited. There is continuing pressure to use the deep sea as a natural sink for solid wastes, but this could soon mean the loss of the marine ecosystem for other purposes, such as a source of food. The more shallow waters estuaries and the continental shelf would be especially vulnerable. There is no choice but to find alternatives to prevailing practices at the present time (Cooke 1969, Breidenbanch and Eldredge 1969).

(B) Food and Feeding Relationship

Food-getting necessarily involves interrelations between organisms and between species; these interrelations are among the most important coactions in any community. Animals are adopted variously to capture and utilize certain types of food and to avoid being captured by other animals. One must understand these adaptations and interrelations to appreciate properly the role that food-getting plays in the dynamics of the community.

Feeding Behaviour

Free-Living animals are commonly classified on the basis of normal feeding behaviour, thus :

Herbivores : Feed principally on living plants.

Carnivores : Feed principally on animals that they kill.

Omnivores : Feed nearly equally on plants and animals.

Saprovores : Feed on dead plants and animals, and excreta.

The various categories are capable of further subdivision. Thus herbivores includes (a) those that feed chiefly on the foliage as the large cursorial grazers, such as bioson, antelope, muskox, caribou, sheep; small surface, living grazers such as rabbits, mice, grasshoppers; subterranean-living grazers, such as wood-chucks, prairie dogs, kangaroo rats, ground squirrels; browsers, which feed on foliage buds and twigs of trees and shrubs rather than strictly on grass or ground herbs, such as wapiti, deer, moose, grouse, and defoliating types of insects such as the hemlock looper, spruce budworm and larch sawfly; (b) those that feed chiefly on the products or non photosynthetic parts of the plants as seed nut and fruit-eaters, such as squirrels chipmunks gallinaceous birds, sparrows, plant-juice suckers, such as aphids leaf-hoppers, mosquitoes, chinch bugs, and cambium feeders, such as bark bettles gall flies, cympids : and (c) those that feed primarily on fungi and bacteria fungivores. Some forms that appear to be saprovores are actually fungivores, feeding on the fungi or bacteria that they obtain from decaying organic material.

Carnivores are also predators. Carnivores restricting their food chiefly to insects are called insectivores : those limiting themselves largely to fish are called piscivores; and so on. Parasitoids eventually consume their hosts, and hence are a special type of carnivore. Some plants are carnivourous. The picher plant. Venus's-fly-trap, and sundew, which grow in bogs or wet places, and bladderwort, which occurs in ponds, depend for their nitrogen supply largely on animals that they capture and consume. Some soil fungi trap and consume nematodes. Perhaps the bacteria fungi, and viruses that cause disease in animals also belong to this classification.

Many species eat both plant and animal matter, on occasion, or at particular season, but animals are considered to be truly omnivorous only if they feed on plants and animals in nearly equal amounts or indiscriminately. Omnivores occasionally also consume dead organic matter. Some aquatic organisms are filter-feeders and may consume everything within a particular size and range that passes through their feeding apparatus. However, filter-feeders may demonstrate selectivity by feeding in neighbourhoods where certain species predominate. Some copepods select particles of a particular size, rejecting larger ones, regulating the distance between the maxillae and the filter mechanism.

Feeding Adoption

Among kinds of mammals, teeth show considerable adaptive radiation, correlated with the type of food consumed. The molar and premolar teeth of insectivorous species, such as shrews and bats, are low and have sharp pointed cusps for crushing weakbodied prey. The piscivorous toothed whales have largely lost all differentiation in their teeth which are simple conical, grasping structures. The teeth of the carnivorous dogs and cats are high-crowned and tubercular, well fitted for shearing flesh. Herbivorous ungulates and rodents have teeth that are flat-crowned, suited to grinding harsh grasses and other vegetation. Their jaws are capable of considerable lateral motion. Omnivores may have both grinding and pointed teeth. Saprovores have rather blunt teeth. The ant-eating sloths and their relatives have no teeth, and the mouth is almost tubular in shape. The tongue has become long and prehensile for lapping up tiny ants.

The bills of birds display great variety in shape and size, adaptations to feeding in numerous quite specialized niches in the environment. The tongues of birds are variously modified to serve as long probes or spears (woodpeckers and nuthathches), as a strainer (ducks) as long capillary tube for obtaining nectar from flowers (humming-birds,) as a rasp (hawks and owls), as a finger for manipulating the food in the mouth (parrots and sparrows), and as a tactile organ (sandpipers and herons).

The mouth parts of insects are adapted primarily either for biting and chewing or piercing and sucking. Among marine invertebrates, adaptations for feeding on detritus include pseudopodia (Foraminifera) ciliated epithelium that maintains a flow of water through the animal (sponges, calms); prehensile, often ciliated arms (various polychaetes; holothuorideans) and soft eversible gullets (various polychaetes, sipunculids). Those that are herbivorous or carnivorous, as well as detritus feeders have prehensile tentacles armed with nematocysts (mollusks), eversible stomachs (starfish) and masticatory structures in the mouth or stomach, (crustacea, diptera larvae)

In addition to mouth-par adaptations, there are many modifications in other parts of the digestive tract for handling particular types of food. These adaptations occur throughout the animal kingdom, but are especially evident in birds. A crop is present in some species but not in others. The walls of the stomach are more muscular in seed-eating birds than in flesh-eaters. Owls and some other species form and regurgitate from the stomach pellets of indigestible matter. Gallinaceous birds are able to retain or shift the supply of gravel or grit in the stomach as an aid to grinding seeds. The length of the intestine varies with the type of food consumed; the caeca are longer in browsing and seed eating gallinaceous birds for digesting cellulose.

Associated with these anatomical and histological adaptations are adjustments in function and behaviour. Obviously, if an animal has morphological adaptations for ingesting and digesting flesh, it must behave as a carnivore and not as an herbivore. The possession of these adaptations and adjustments means that animals are generally restricted to the particular type of food that they can use most efficiently. The kinds of food eaten by animals is of fundamental ecological and economic importance.

Choice of Food

The kinds of food eaten by animals depends on their genetic heritage and parental training or conditioning while young. Involved in the evolution of the food habits of a species and the "profitability" of consuming one prey species rather than another are the animal's physical adaptations for ingesting and digesting particular types of food, the nutritional values of the food. Its palatability the size of it, its availability, or abundance, a d its ease of procurement, which depends in large on the various protective devices that it possesses.

Nutritional Values

Animals generally require proteins, carbohydrates, fats, vitamins, minerals and water. Proteins are used as the basic substance in the composition of protoplasm; carbohydrates and fats are oxidized to furnish energy for the body; vitamins serve as catalyst for specific metabolic processes, minerals are needed to regulate osmotic pressure and as constituent elements of various body organs; and water is used as a general solvent, lubricant, and circulatory medium. Species differ, however in their needs for particular substances. The beetles Tribolium, Lasioderma, and Ptinus, for instance, grow slowly but nonetheless satisfactorily on diets lacking carbohydrates. Hence, they may be disturbed more widely than are species which require carbohydrate in their diet.

Foods differ in composition. Foods staple to an organism's diet are those easily digested and of high caloric and protein content. They are adequate for sustenance of the weight and vigor of the animal, but usually need to be supplemented with vitamins and minerals. The bob-white and ringnecked peasant, for example eat certain cultivated grains and weed seeds, such as corn, sorghum, barley, wheat, rye, soya beans pigeon grass, and lesser ragweed, as staple foods, at least on a mixed diet.

Non-staple or emergency foods are not in themselves sustentative, and animals limited to them gradually lose weight and die. Such foods are, howerver, often abundant and easily procured in emergencies. When staple foods are covered with snow or ice and furnish sufficient energy to tide the animal over the critical period. In emergencies, the bobwhite and ring necked peasant eat black locust beans, fruits of bittersweet and sumac, rose hips, dried wild grapes, and sweet clover seeds.

Vitamins are necessary for the maintenance of good health in world animals, just as in domestic animals or man. The symptoms of vitamins deficiency, induced experimentally, are similar. Evidence has been difficult to secure, however that animals suffer from vitamin deficiencies in their natural environments.

Animals obtain most of their required minerals from their food and water. Additional salt must sometimes be given to caged animals to prevent cannibalism. The gnawing of cast off deer antlers by rodents is apparently for additional salts. The use of certain soil deposits and springs as natural "licking sites" by deer and other ruminants is apparently for sodium salts lacking in their general diet. Some birds, such as the evening grosbeak are also attracted to sources of salt supply. Grit in the form of gravel, sand, or small stones in the stomach of gallinaceous and other birds may not only aid in the grinding of seeds and hard vegetables matter but also be a source of minerals.

Digestibility

Different species of animals very considerably in efficiency of digestion and utilization of particular food substances. Thus, clothes-moth larvae can digest cloth and bird lice can digest feathers, because among other things they have an exceptionally high hydrogen ion concentration in their intestines.

Digestive enzymes occur generally throughout the animal kingdom although less is known about them in the protozoa. In the lower phyla the enzymes are generalized in respect to the kinds of foods on which they act; in the higher phyla, they become highly specialized. A specific enzymes, however, does not differ greatly from one animal group to another. Carnivores have strong, proteases and weak carbohydrates, correlated with their meat diet. Herbivores on the other hand, have weak protein, but active carbohydrate, enzymes. Herbivorous mammals and birds posses a bacterial flora in their digestive tract that makes possible digestion of cellulose. Omnivores have a full complement of enzyes and can utilize a wide variety of foods.

Practically all food contains some indigestible matter, which is ordinarily passed through the digestive tract and eliminated in the feces. If the indigestible material is excessive, or if it contains toxic substances, regurgitation or vomiting may result.

Size of Food Item

The size of the food in relation to the animal is not of major importance to many herbivores or saprovores as they normally feed in or on the organism or substance. With carnivores, the size of the prey must be within their power of conquest. Ordinarily the size of prey is less than that of the carnivore that feeds on it, but a high degree of ferocity and audacity, or pack hunting in the manner of wolves, often enables the carnivore to take prey larger than itself. On the other hand, a predator cannot profitably prey on species so small that the energy derived from its consumption does not equal the energy expended in its capture. Evolution of prey species subject to heavy predation tends to proceed either to very large size or very small size in relation to the size of their most important predators. Some very large aquatic predators, however, have become adopted to feed with a minimum effort on very small organisms occuring in dense concentration through the evolution of a filtering appartus in their mouth parts. A good example is the feeding of blue whales on plankton.

Availability

In order to determine if a species is fed upon in population to its abundance, it is necessary to find out what animals have been eating of that which is available in a habitat. The relation between the two may be shown graphically and expressed as forage ratio per cent of species in animals' food divided by per cent of species in habitat. A value of unity indicates that the food item is taken in proportion to its abundance, a value greater than unity indicates that it is taken more frequently, values of less than unity indicate that the item is either inaccessible, of the wrong size, too difficult to obtain, or is actually avoided. It is apparent that while there is a relationship between the relative abundance of various species and the degree to which they were taken as food, there are also several discrepancies.

The vulnerability to a predator of a prey species is directly proportional to its relative abundance among the other species available. Voles are preferred by the red fox over white footed mice, but both mice and voles are preferred over moles, shrew, and snakes. Predator preferences among them, however, occur only when all are abundant. When prey species are reduced in numbers and difficult to get, little or no predator preference occurs.

There is some evidence that predators-particularly insectivorous birds —when searching for prey concentrate on one or a few species at a time. By a kind of learning, they acquire what may be called specific searching images for these species and thereby mostly disregard other species. When a new species becomes numerous in an area they feed on it at first only as the result of chance encounters. To obtain a preference for the new species, they must become conditioned to it gradually and must learn the proper cues of where and now to search it out.

Predator species that remain in one habitat throughout the year must adjust their feeding to the kinds of food available in each season. Species that are unable to do this are compelled to migrate, hibernate, or make other adjustments to survive the unfavourable seasons. Thus birds that are strictly insectivorous may occur in a given area only during the warm part of the year, leaving it before insects disappear. Omnivorous species commonly changes from an insect diet in summer to diet of seeds and fruits in the autumn and are often non-migratory.

Protective Devices

Most, if not all, plants are equipped with defence mechanisms against the grazing and browing of animals. The most widespread defences are chemical – various kinds of better of toxic materials in plant tissues that make unpalatable to many animals. Animals evolve tolerances and neutralizing devices, however, so that most plants are fed on; their defence is relative, not complete. New leaves often contain less of the defensive chemicals, and are thus more vulnerable. Trees and shrubs may be deformed or killed be excessive browsing that destroys terminal twigs and buds. Grasses are not so affected, for they grow from the base of the leaves; they may even be benefited by moderate cropping Grasses also are protected by grains of silica embedded in their tissues; only by evolution of high-crowned teeth adopted to abrasion by the silica have some ungulates become able to graze on grass. Some species of trees, shrubs, and cacti are protected from browsing by prickles or thorns. The protection is important to the plants for survival in deserts and in grazing subseres of humid regions. Some plants are particularly noxious or toxic and animals quickly learn to avoid them. Seeds may develop alkaloids or other disagreeable substances that give them protection.

Various kinds of arthropods have delcusive secretions in the form of sprays that they emit when attacked. There is increasing evidence that attractment and repellant secretions are important in both predator prey and social relation of animals.

When coloration renders it inconspicuous in its normal environment, an animal is said to have concealing or cryptic coloration. When the coloration or markings reproduce the general tone or characteristics of background, it is called protective resemblance; disruptive, when the markings break up the outlines of an animal and replace it by some irregular configuration so that the animal is not recognized as prey.

These are many examples of mimicry among insects. Batesian mimicry is the resemblance of a palatable species in external features to an unpalatable one that in turn possesses warning coloration and is the more abundant of the two species. The palatable species derives benefit from the relation, since predators, specially birds avoid them as well as the unpalatable ones. The effectiveness of the mimicry decreases as the number of mimics equals or exceeds the number of models. There is considerable controversy about mimicry, however, and even the classical example of the butterflies is disputed. In Mullerina mimicry, both model and mimic are unpalatable. Pooling of numbers between the two species give more chances for inexperienced birds to lean to avoid them and reduces the losses per species during the learning process. The yellow acket and polisitine wasps are examples and many more can be found among butterflies.

Some animals posses bright spots of colors so placed on the body as to be deflective. The attention of pursuing predator is drawn to less vulnerable parts of the body; for instance, eyespots on the fins or tail of the fish. Eyespot on the wings of some butterflies and moths are concealed at rest, but when flashed out, spread wings may frighten away an attacking bird or other predator.

Range of Food Selection

There is considerable range in the variety of food eaten by most species. Herbivorous species are often more specific in their feeding habits than are carnivorous forms. However, even herbivorous forms have various degrees of restrictions as shown in the following analysis of 240 species of plant-juice sucking aphids.

	Percentage
Species restricted to single plant species	27
Species restricted to single genus but freeding on more than one species	40
Species feeding on several on several different genera	33

One may argue with considerable justification that animals such aphids, gall wasps, some bugs, and others that show host specificity are really parastic rather than herbivorous in their feeding behaviour, and hence are not good example of free-living animals on a restricted diet. Actually it is very difficult to find proven cases of animals that confine themselves to single species food. It is more common to have an animal feeding on a small group of related species, as do aphids.

Herbivorous species may be classified in respect to the diversity of their food into :

Monophagous : Restricted to single food plant,

Oligophagous : Restricted to a few very definite food plant,

Polyphagous : Feed on many species,

The restriction of animals to particular foods may be the result of chemical affecting odour or taste, or to structural adaptations, chemical stimuli are especially important with insects. The crossed bills of certain birds (Loxia sp.) are well fitted for prying seeds from between the practs of coniferous trees cones.

Energy Exchange, Productivity and Yield For Man

The food coactions of organisms are important since animals can obtain energy only from consumed food. All activities of organisms constitute work and require energy. The amount of works performed depends on the amount of energy the organism can mobilize. Hence the flow of energy through the ecosystem and the manner and efficiency with which it is used is an ecological process of utmost importance. All of biology is dependent on the earth taking in energy as sunlight, using, it, and then reradiating it back to outer space.

Energy Flow through the Ecosystem

Unlike nutrients, energy does not circulate indefinitely through the ecosystem. Energy is continuously and rapidly lost, although a certain amount of energy may pass through the ecosystem more than once before it is entirely dissipated. Hence energy must continuouly enter into the ecosystem form the outside.

Aquisition -

The basic source of energy for all tropic levels is solar radiation. A surface exposed at right angles to the sun's rays and outside the earth's atmosphere, would receive at the rate of energy 1.94 g-cal/cm^2/min. This is the solar constant. On the earth's surface, solar energy is effective only during the daylight hours and because of absorption in passing through the atmosphere scattering by smokes, dust particles and cloudiness, only about 40 per cent of daylight radiation reaches the earth's surface, although this varies with latitude, season and locality.

The basic means by which solar energy is trapped is the formation of sugar by plants that contain chlorophyll. In terms of moles

Photosynthesis : $6CO_2 + 6H_2O + 709 \text{ kcal} \rightarrow 6O_2 + C_6H_{12}O_6$

Respiration : $C_6H_{12}O_6 + 6O_2 \rightarrow 6CO_2 + 6H_2O + 674$ kcal (free energy) + 35 kcal (entropy).

Respiration is going on at all times to furnish energy for the plant's activities, and this energy is derived from oxidation of the sugars formed in photosynthesis. Under ordinary daytime light intensities, the amount of sugar formed by photosynthesis greatly exceeds the amount oxidized in respiration. Photosynthesis ceases during darkness, but loss of sugar at night, because of respiration, continues. Sugars may be converted to starch or fat, or the carbon, hydrogen, oxygen in combination with nitrogen, sulphur, and phosphorus may form into proteins. The amount of sugar or new organic matter resulting form photosynthesis represents the primary production of the ecosystem. Accurate measurements of the rate of primary production is one of the most important problems of tropic ecology, for the activities of all plant and animal organisms in the community depend on the energy thus supplied. It is necessary to distinguish between gross primary production, the total amount of energy captured, and net primary production, the amount that remains after that used respiration. Primary production is commonly expressed in terms of glucose or carbon, indirectly in the amount of oxygen released or carbon dioxide absorbed, or directly as calories of energy. Production during a period of time is productivity, and both production and productivity are commonly related to the standing crop or biomass (B) at a particular trophic level (n).

Use

Net energy a producer level becomes available for use of animals when it is transferred to the higher tropic levels through predation (1_{n+1}) here considered also to include consumption of plants (Lindeman 1942, Macfedyen 1957). When predators consume their prey completety as do fish feeding on plankot, there is no wastage. With many predators, however for example wolves feeding on deer, the prey is eaten piecemeal, and much is not used. Predatory kill must therefore be separted into the gross energy intake (i) and the energy wasted (W) Certain of the food taken in secondary productivity is indigestible or may simple be undigested, or if digested and absorbed is not completely metabolized in the tissues, so that it is eliminated in feaces, urine, and other excreta. This is designated as excretory (E). Assimilated energy, sometimes called metabolized energy, is the energy of the food actually absorbed and utilized (I-E)

A good portion of the energy ingested is used for existence, that is, for basal metabolism, temperature regulation, procurement and digestion of food, and other normal activities. There is almost continuous loss of heat energy from the body, and in homiotherms this must be compensated for by increased heat production. Energy is used for the production of eggs and sperms, reproductive behaviour, and other activities. Even the process of converting raw food into protoplasm is work and requires energy. In transfer of energy from one form to another or into work, there is always loss of free energy. No transfer is 100 per cent efficient except that it is degraded into heat. This is the second law of thermodynamics. The total energy that is utilized to perform work and to produce heat is called respiratory energy (R). Energy is continuously being lost from the ecosytem in this manner, hence needs to be repalced if the ecosystem is to continue to function.

Net production per unit time at the consumer levels is gross energy intake minus the losses of excreta and respiration (I-E-R). It is the actual production of animal tissue and involves growth.

Ecosystems are seldom entirely isolated one from another. Energy entering into an ecosystem by means other than photosynthesis is an import (im), energy leaving an ecosystem other than respiration is an export (ex).

When populations of different trophic levels are in balance, the total net preduction of each tropic level, after losses from non-prrdatroy deaths and wastage from kills have been subtracted, if consumed by predators of the higher tropic levels. In unbalanced populations, predatory consumption may not equal the available net production so that the population, of that trophic level increases. If the predatroy kill exceeds the available net production, the population decreases. A change in the biomass (b) of a population may therefore, be either plus or minus.

Measurement of Productivity

Productivity may be measured during any reasonable period of time. Because of essential metabolic differences between day and night, however, the 24-hour day is the smallest practicable unit. Similarily, because of seasonal changes in the environment and in community populations, the measurement of annual production is probably most useful. Since primary production is basic and concerns the capture of energy by plants, it will be considered first.

Primary Production

Various methods are employed for measuring primary productivity, each procedure having certain advantages and disadvantages (Ryther 1956, Mann 1969). Further work in evaluating and improving these methods or developing ones is desirable.

A common procedure for analyzing aquatic habitats, dating back to Gaarder and Gran (1927), is to suspend during daylight hours equal samples of green phyto-plankton, ordinarily inseparably mixed with bacteria and zooplankton, in both transparent and blackened bottles at the same depth at which obtained. Photosynthesis of course does not occur in the blackened bottle, and there is loss of oxygen, resulting from respiration, R, and decomposition, E + D + W. In the transparent bottle, photosynthsesis occurs in addition to respiration and decompostion, bringing a production of carbohydrates. There will either be an increase in oxygen concentration, or the loss of oxygen will not be so great as in the blackened bottle. The difference in the final oxygen content of the two bottles will be a measure of gross production : I_g.

If the oxygen content of the water is measured at the begining of the experiment, the loss of oxygen in the blackened bottle substracted from the difference in oxygen content of the two bottles at the end of the experiment will represent the net productivity. This net productivity may also be determined from the difference in the oxygen content of the transparent bottle between the begining and the end. The obtain net production for an entire daily cycle, the consumption of oxygen for respiration and decomposition over 24 hours must be subtracted from the gross photosynthetic output during daylight hours.

One may use the amount of carbon dixide absorbed during a period of times as a measure of photosynthesis if correction is made for the carbon dioxide given off in respiration and decomposition. Changes in the amount of CO_2 in the water may be calculated from the differences in pH of the hydrogen-ion concentration. As an illustration of the use of radioisotopes, net production during the daylight thours may be measured by introducing a known amount of $11CO_2$ into a volume of water where the amount of carbon dioxide already present is known. The amount of 11C absorbed by the phytoplankton can be accurately determined by use of counters applied to phytoplankton collected and dried at the end of the period. Then the proportion of the radioactive carbon absorbed to the amount introduced can be applied to the total CO_2 initially present to get the total amount absorbed.

Since nitrogen and phosphorus are metabolized more rapidly by plants in the manufacture of food during the growing season than they are regenrated from decomposing material, the rate and extent of the depletion of nitrates and phosphates is freely circulating bodies of water or in the epilimnion of stratified lakes serve as an index of an amount of organic matter produced. The rate of accumulation and regeneration of these substances is the hypolimnion from the dead organisms that sinks into it may also be used to get an approximation of primary production. These measurements are not exact since they do not account for the repeated regenration and neutrilization of the substances in the photic zone during the season or their transference to and storage in the bodies of animals.

The rate of photo ynthesis varies in relation to the amount of chlorophyll present and to light intensity. The amount of cholorophyll in the standing crop phytoplankton may be determined photometrically for all depths and calculated in terms of unit area of surface (Ryther and Yentsch 1957, Flemer 1959).

Determination of the rates of photosynthesis and oxygen use in streams and ponds may be made by direct measurement of changes in the concentration of oxygen and carbon dioxide in the water as between day and night. Because of photsynthesis, there is no increase in oxygen concentration during the day time. At night the oxygen loss gives a measure of the rate of respiration and decomposition and this p.esumably remains the same throughout the 24-hou daily cycle. Adding the average hourly night loss to the average daily gain during the day and multipling by the hours of daylight gives the total gross production for the 24-hours day. To obtain the net production for the entire day, the hourly loss a night must be multiplied by 24 and subtracted from the total gross production. Corrections need to be made, however, for possible diffusion of oxyg.n from air into the water, particularly at night when oxygen concentration in the water is lowered, and diffusion of the water during the day, if supersaturation occurs. Additional correction will also be necessary for import of oxygen from groundwater and surface drainage and export of oxygen and carbon dixoide downstream by swift currents.

Secondary Production

When an animals species is represented by a low overwintering population, or an immature stage, the maximum biomass obtained in each generation is the approximate not production is for that generation. However, is docs not account for continued repro duction and growth of individuals after the maximum biomass of the population is attained, nor does it account for excreta natural deaths, or the kill of predators. If the population of the species is maintained at a more or less uniform level throughout the year, the mean biomass times the number of generations gives the net production, again with the exception of the factors mentioned above. Linderman considered the phytoplankton turnover, or the production of a new generation, to occur every week form May to September and every 2 weeks through the rest of the year, the zooplankton to replace itself bi-weekly through the year, Chaoborus to have three generations per year; midge flies, two; and various acquat'c bettles and bugs, one generation per year. Juday (1940) estimated that the mean standing crop of both phytoplankton and zooplankton repalced itself every 2 weeks throughout the year. To obtain gross productivity, the respiratory rates of these animals must also be measured.

Although measurement of energy flow through the total phytoplankton, zooplankton, and smaller soil organisms is often practical and sufficient, measurements are required on individual species of larger size in the higher tropic levels before the total utilization of energy by the tropic level can be determined.

Energy for Man

Man's traditional sources of energy for food and industry are the current yield from natural and agricultural ecosystem founded on photsynthesis by green plants; on fossils fuel in the form of coal, petroleum and gas; and from harnessing physical forces in the environment, such as water power. There is considerable, concern that sources of energy will soon become inadequate in at least those areas where there will be continuous heavy demands.

One main purpose in the extensive damming of our river systems at present time is to derive power for industry. There is a limit, however, to this source of energy, and much of the potential new sources of hydro-electric power lies in relatively undeveloped countries or is not readily accessible. Some development is possible for deriving energy from the tides, wind and geothermal steam, but this also is limited.

Fossils fuels represents solar energy captured by photosynthesis and deposited during past geological periods when net production by green plants exceeded its utilization by animals and decomposers. This supply may be exhausted in the next 100 to 300 years. During the last 30 years, 90 per cent of the increase in power consumption has been due to per capita utilization, only 10 per cent to increase in population size.

Lack of energy may not, however be critical factor eventually limiting the population growth of man if two new sources of energy can be fully exploited. There is much research underway to unravel the chemical processes of photosynthesis. Once this is discovered, the artifical manufacture of food may supply more than enough for as large a population as the world can accomodate. In addition to energy, photosynthesis, of course, also requires the essential elements carbons, hydrogen, oxygen, nitrogen, and phosphorous, among others.

If atomic energy can be obtained on a fully efficient basis, this would be sufficient, it is estimated, for over 20 billion people. (Weinberg and Hammond 1970). This figure should not be seriously considered as potential world population, however, as other factors, such a population and pyschologica: effects, will limit human populations far below this level, Power from the fission reaction of Uranium-235 is already an accomplished fact, but if all the present day needs for power were derived solely from this source the world supply or uranium 235 would be exhausted within few years. If however, the breeder-reactor can be prefected where by non-fossile isotopes are converted into fissile ones, very much larger supplies of source material will be available.

Enormous amounts of energy are potentially available in the fusion of hydrogen isotopes into helium. This has already been achieved in an uncontroled explosive manner in the hydrogen or thermonuclear bomb. Research is underway to obtain a controlled fusion reaction, but it is not possible to estimate when, if ever, this can be accomplished. Inherent difficulties with both the fission and fusion processes for mobilizing atomic energy is dissipation of the heat that is liberated and the disposal of the radioactive wastes. Unless these problems can be solved, they may set limits on the amount of energy that can be used form the vast resource. The electric-generating capacity of modern industry is now increasing 7 per cent per year. It is estimated that at this rate the mean global temperature would rise 1 C in 90 years and 30 C in 10^8 years. Unless offset by compensating factors the rise would be sufficient to melt the ice caps and drastically alter the world's geography.

Until man mobilize other source of energy, feeding the increasing population becomes critical. The problem is partly one of transportantion and economics for supplying people in poor agricultural countries with surpluses from the high producity ones. At present there are 450 millions well fed people living in comparative luxury and 2,400 million who are under or malnourished. About one-fourth of the world's billion acres (128 billions hectres) are potentially suitable for cultivated crops, another one fourth is suitable for grazing, and the remaining half is in part for seed or of very limited use for food production. About one half, the better half of the potentially arable and grazing land is now being used. The role of grazing land is that harvesting is done by herbivores which are able to use plants that are otherwise wasted to man.

Increased food productivity depends on (a) expanding the amount of land under cultivation and (b) increasing the yield per unit area. Although there are extensive areas in Africa and South America not now under cultivation their utilization for crop production will depend on considerable improvement in our ability to manage tropical soils and to maintain their fertility.

Agriculture maintains predominatly monospecies seral communities of high energy and mineral productivity. They are however, ecosystems, although relatively simple ones. A comprehensive mathematical model worked out for, say a cornfield would be useful in simulating energy and material exchanges and for testing the influence of various factors. The same model would also be adaptable to the more complex ecosystem of nature. Such an expanded model would be powerful tool to help man order his priorities in regard to plant or plant-animal communities to determine how far they can serve his needs, be if food production, water conservation, climate modification or aesthetic enjoyment.

Supplying man with food is not only a problem in productivity but may also involve modifications in diet, man is an omnivore and enjoys a high proportion of meat, milk and eggs his diet. His per capita consumption of beef is said to have increased by 75 per cent in the last 20 years. These animal products come from the consumer levels in the food chain, man could harvest more energy from natural and agricultural ecosystems if he restricted himself to yield from the producer level. If he were to change to a die wholly of rice, wheat, corn, potatoes and other plant products, it is estimated that the amount of land now under cultivation could support a population six times the present size, assuming that there were no limiting factors on population growth. The fewer the links in the food chain from which man or any other predator takes his yield, the greater is the harvest that can be obtained. A farm pond, for instance, will yield a considerably larger crop of herbivorous crop than of fish eating bass, more muskrats than mink, and so on. Nutritional problems for man on a pure plant diet, however, involves the lack of certain essential amino acids, vitamins, as well as calcium that are deficient or absent in most plant food but common in animal food, Soyabean meal contains some of the missing amino acids or if the essential amino acids and vitamins were factory produced and supplied as supplements to the plants diet, health could be maintained (Brody 1952). Although providing all of mankind with adequat energy and nutrients in the future poses, serious proble ns, the solution of these problems is still within the realm of possibility if the human population explosion can otherwise be contained.

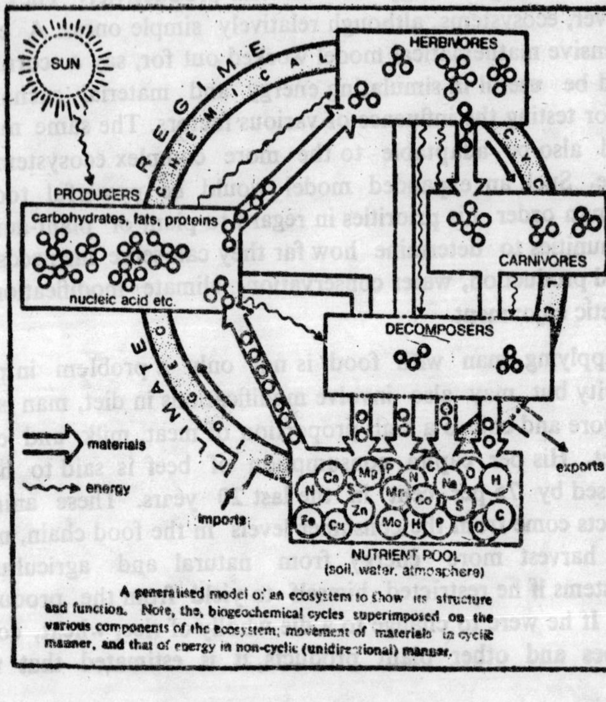

A generalised model of an ecosystem to show its structure and function. Note, the, biogeochemical cycles superimposed on the various components of the ecosystem; movement of materials in cyclic manner, and that of energy in non-cyclic (unidirectional) manner.

3

Community Ecology

Streams and Water Pollution

When rainwater falls on an uneven surface, it collects in depressions. As the water overflows them, the current erodes a narrow channel that deepens with each succeeding shower and may eventually drain the depression. There is usually also a lateral meandering of the stream, by which a valley is formed. The site of the head woters of such streams is impermanent, and continued erosion, forces the head waters and the channel farther and farther back into the upland. The stream is at first a temporary one, dependent for its waterflow. On rainfall runoff, but when its channel is cut below the level of the groundwater table the stream becomes permanent, fed by general, continuous seepage. The headwaters of such stream are therefore its youngest portion physiographically, and the stream is progressively more aged towards its mouth.

In hilly or mountainous terrain, water may accumulate in large basins until ponds or lakes are formed. Ordinarily, however, the water level in a lake will rise until it overflows of the lowest point on the perimeter. Then the waters continue to flow downwards until they reach the sea. Streams springing from fixed headwaters (melting snowfields and glaciers springs) carve valleys throughout that are essentially the same age.

A river system in youth is characterized by valleys that are narrow and steepsided; the flow of water is usually fast, there are few tributaries and there are many waterfalls, ponds and lakes along its course. As the river system matures, its valleys become wider, its slopes more gentle, and its tributaries more numerous and longer. Many ponds and lakes are drained, and waterfalls are worn down to rapids or riffles. The areas of upland are well dissected and the land is thoroughly drained. In old age, the river stream has reached base level. The upland has been worn down to low ridges between tributary river valleys and the region as a whole is called peneplain. There are no lakes, ponds or rapids, and the flow of water is sluggish. Strahler stream less than 3 m wide are called creeks of brooks. Rivers are stream 3 m or more wide.

Habitat

Exclusive of its lakes, the principal habitats in a stream are falls, rapids or riffles, sand bottom pools, and mud-bottom ponds. The character of the bottom depends primarily on the velocity of the water current, which along with the volume of the stream flow, can be easily measured. Warterflowing at the rate of about 50 cm-sec is considered swift-flowing; velocities greater than 300 cm sec rarely occur. Fast currents rolls or slide pebbles and rocks along the bottom, move sand partly buoyant transportation, and carry fine materials, such as silt and organic matter, in suspension.

In places where the topographic gradient is steep, the bottom will be composed largely of cobble and boulders too heavy to move, and smaller pebbles which are trapped by obstructions. This habitat is called a rapids, if extensive and turbulent; riffles, if of lesser order. Below the riffles there may be a raceway, which is intermediate between the riffle and pool in depth of water, rate of flow, and substrate and its depth is quite uniform.

When the gradient is less steep and the water current thus slower, gravel (particle size 2-64 mm) is deposited first, then sand (0.06-2mm), but the finer materials are carrred along. Only when the current becomes negligible downs the suspended materials settle so that silt (0.004-0.062 mm) or mud-bottom pools and ponds are formed. Clay has a particle size even smaller. These mud-bottom pools are fertile parts of a flowing, non-polluted stream. In a general way, riffles and sand and mud-bottom pools represent three stages in erosion cycle of a stream, and ecological study of them gives a good idea of what the eosere would be over a long period of time.

Stream Biocies

When quantitative sampling is made of the invertebrate populations of streams, one finds that there is a sharp distinction of species found in riffles and those found is mud-bottom pools. The raceway is here included with the riffles. The sand-bottom pool habitat has no characteristic indigenous invertebrate species, but is occupied by small numbers of individuals of species otherwise occurring abundantly in the other habitats. The unstable bottom apparently prevents the development of distinctive community. The high index of species diversity is caused by the uniformly low number of individuals in each species, which more than compensates for the smaller number of species present. There are, however, several fish species that find sandy pools a favourite habitat, although they depend in large part upon riffles organisms for their food. Stream fishes commonly segregate into more or less distinctive societies of species correlated with differencet in current type bottom, and depth of water. Many fish over-winter in deeper, more quiescent sand bottom pools, especially since low water temperature makes them sluggish to withstand rapid currents.

The stream biocies consist most typically, therefore, of the inhabitants of the riffles and sand-bottom pools found throughout the course of the river. The riffle and pool organisms make up two different facies in this community. These focies become more fully developed downstream as shown by an increase in the species diversity index, both for benthic organisms and for fish. Mud-bottom pools and sluggish streams are occupied by the pondmarsh biocies.

Plants are not abundant in the stream biocies, although the upper surfaces of rocks in a riffle may be completely covered with branched filamentous algae (particularly Cladophora) and a few species of water mosses (Fontinalaceae) may occur. Diatoms mostly sessile forms, may be numerous in early spring and again in the autumn. Dominance in the true sense, such as occurs in terrestrial communities, does not exist, although the algae and mosses passively provide food and shelter for active forms.

The most characteristic and abundant animal forms of the stream biocies are the caddisfly larvae, mayfly naiads, stonefly naiads, fly larvae, crayfish; snails and calms, sponges and bryozoans, and fish each occupying its own particular niche. Plankton is mostly sparse in swift-running water, but may be abundant in sluggish, pondlike stretches of large rivers.

Adjustments to Current

Probably the characteristic of stream most critical to the life therein is the current. All organisms that occurs in stream must adjust to it to maintain constant position. Torrential floods scour the stream bed, move rock and sand, cut new channels, and destroys entire populations. Recovery after such castarophes may take place within a few weeks, especially by those species possessing short life-cycles, but sometimes it may require years. (Moffett 1936, Surber 1936). Position is ordinarily maintained by clinging to the substratum, avoidance of the current, or vigorous swimming, and requires a good development of orientation behaviour.

Clinging Mechanism

The growth from fresh-water sponges is affected by a number of factors, but in riffles sponges are usually simple encrustations. In quieter water, long, slender, finger-like processes may form. The distribution of species depend both on current and organic content of water.

Plumatella is common bryozoan that forms an encrusting, plant-like, branching colony on the underside of rocks or fallen trees in swift water Pectinatella, on the other hand, forms a gelatinous spherical ball, and is more commonly found in ponds or slow flowing portion of streams.

Respiration and Oxygen Requirements

Oxygen is usually ample in streams, often saturating the water in turbulent riffles. The oxygen concentration is sometimes low, however, in sluggish streams and standing pools. The difference in oxygen tension of the two habitats is reflected in the respiratory adaptations of the organisms that inhabit them.

The lamelliform gills of the mayfly naiads inhabiting mud-bottom pools are larger in size than those of species, and are almost continuously flicked back and forth for better aeration. The gills of naiads living in riffles, or in waters in which the oxygen content is high, may have the surface area of the gills reduced by two third in proportion to body weight, compared with mud dwelling forms. They are never flicked, since the water movement continually brings oxygen to them.

The respiratory equipment of pond-in-habiting animals permits them not only to live in habitats with lower oxygen tensions but also to survive longer at high water temperatures, Often, these animals display relatively low rates of general body metabolism and oxygen requirement. Such relations between riffle and pond animals have been observed for mayfly naiads, caddisfly larvae, isopods, crayfish, and fishes.

Preservation of the Aquatic Environment

The chief problems in applied ecology related to streams and other aquatic bodies are those of maintenance of water supply for man, flood control recreation, fish management and control of pollution. Pollution takes four forms : silting, resulting from excessive erosion of the surrounding upland; industrial produced by inorganic chemicals from water lead and zinc works, tanneries, breweries, paper mills, gas plants, mines etc. (Parsons 1957); thermal as hot water effluents from atomic plants; and organic, principally municipal sewage from agricultural land.

Erosion and Silting

Stream erosion becomes considerable when upland vegetation is to be reduced that there is little or no retardation of runoff from heavy rains. Continuous erosion throws a heavy load of fine silt into the stream. This is detrimental. It makes the water opaque; reduces or prevents photosynthesis in algae, water most, and other plant life; handicaps those fish and other animals on sight finding and capturing food; and clogs the filtering mechanism of various invertebrates. Deposition of silt on rock or sand bottom may bring a considerable change in species composition of animals present. During the last several decades, greatly increased soil erosion in agricultural areas has reduced pan and game fishes in our streams.

Chronically muddy streams may often be cleared by reforesting the waters head, and by practicing modern erosion control in cultivated areas. With slower runoff, more rainwater soaks into the ground, and the water is raised. It is also desirable to maintain vegetation on the immediate stream banks to slow up undercutting. Streambank vegetation is also beneficial for shading the water and keeping it cool enough for such fish as trout. If artificial dams are necessary, they should be small and located where the drainage brings in the numerous head-waters of the streams.

Water Supply, Flood Control, and Recreation

As the human population has expanded during the last century, man has exerted more and more control over the stream flow to conserve it for drinking and home use, irrigation of arid lands, industrial and municipal use; navigation, swimming and boating, to produce electric power and to prevent flood damage. This extensive use commonly involves construction of dams and reservoirs, dredging of river channels, raising of levees along the shores to confine the course of water flow rates. As a result, the riffle habitats are eliminated, water levels are altered, and floodplains are permanently inundated. The stream biota disappears to be replaced by that of lake or pond, mud-flats are created with drawdowns or utilization of water in excess of the inflow, and consequently the environment is drastically changed. Unfortunately, sedimentation and filling of the basin with silt eroded from the upland often limit the life of such reservoirs to 50 to 100 years. There is a limit in height to which a dam can be repeatedly raised and to the number of reservoirs that can be built along the main channel of river. Alternative solutions are tapping of underground aquifers for water supply, construction of multiple dams on head-waters of tributary streams for flood control, and restricting water use to what can be rapid depletion of the ground water reserves.

Industrial Pollution

Industrial and mine wastes are often acid, and extreme acidity will kill fish and other organisms. Clams are greatly reduced to disappear altogether in acid waters. Industrial wastes contains a great variety of chemical compounds, including salts of the heavy metals, and many of them are toxic to fish. Young fish as species of small fish appear especially sensitive, and the pollution materials may cause physical or chemical injury to the gills without actually being absorbed into the body (Doudoroff and Karz, 1953).

Mercury pollution began to be a problem in Sweden and Japan in the mid-1950's and has since become of worldwide concern. Trace amounts of mercury naturally in environment, but the difference between tolerable natural levels and levels harmful to man and animals is very small. Mercury is used in agriculture and paper mills as fungicide. Seeds so treated have caused mortality in terrestrial birds. Mercury from these sources, from its use in industry and from combustion of some coal and petroleum becomes washed into fresh-water streams and lakes and eventually into coastal waters of ocean. Metallic or inorganic mercury does some damage to animals, but ordinarily it does not accumulate in the body to serious levels. When acted upon by micro-organisms, however, methylation occurs and methyl mercury may become so concentrated, via the food chain, in fresh and saltwater fishes as to be dangerous to man and fish eating birds

The controls of radioactive wastes from uranium mills and other atomic energy plants and from fallout has become an especially serious problem in modern times. No stream or other aquatic body can purify itself of these wastes through biological means. However, they become dilated downstream, undergo physical decay, settle out in the bottom, and are taken up by the organisms. Organisms assimilate elements at equal rates whether the elements are radioactive or not. Some radioactive elements have a life-life extending over several years and may thus accumulate and become concentrated in organisms to an extent many thousands times greater than their concentration in water. This is of potential harm to man when he consumes these organisms. Mammals are most sensitive to radioactive materials, than, in order, seed plants and lower vertebrates, insects, and micro-organisms.

Thermal Pollution

Water is required in atomic reactors for absorbing the waste heat that is generated, and heated water is then returned to the stream lake or ocean. Engineering devices are available to recool the effluent and either recirculate it through the reactor. Return it to the stream, or use it for other purposes, but to do this has been considered uneconomic (Smith 1972). A challenge to engineering is to use the heat generated for useful purposes, such as heating homes, in irrigation of crops, in industry, and so on.

Some bacteria, blue-green algae, rotifers nematodes, and protozoans can survive in water up 50° to 60° C, but very few higher organisms can tolerate aquatic temperatures above 35° to 40° C. Infact most animals come under stress at significantly lower water temperatures. With the rise in water temperature, the solubility of oxygen decreases whereas that of salt increases. Various physiochemical changes occur in the functioning of cellular membranes, hydration of enzymes, and so on. Chemical reactions such as those involved in decay of organic matter and other oxidation processes proceed more rapidly in warm water than in cool water. Likewise, metabolism of organisms and hence their demand for oxygen increases in warm weather. The augmentation of metabolism and oxygen requirements of organisms may well exceed the capacity of photosynthesis for producing oxygen. The resulting change in habitat brings a drastic change in the species composition of the biota or, when extreme, its elimination.

The heat produced by atomic reactors, as well as heat generated by man in industry and modern living generally, eventually reaches the atmosphere. The total heat now being produced by man is estimated at 1 per cent of the total radiated by the earth. This is too small to have a significant effect on the average temperature of the earth's surface. However man's output of heat is increasing about 7 per cent per year.

Organic Pollution

The introduction of small quantities of organic wastes may increase the size and productivity of plant and animal populations by adding to the basic nitrogen and phosphorus supply (Flemer 1970). The limit of the sewage load that a stream can carry without harm is, however, low and soon reached. As fresh organic material oxidizes, carbon dioxide and toxic gases are released into the stream, and there is a drastic reduction in the oxygen content. Fermentation is more rapid in summer in water, and may begin in wastes before they are discharged into the stream. The decomposing organic material continues to be oxidixed through bacterial and other action as it is carried down stream, and when this action is completed the stream is again pure.

There have been many attempts by means of chemical analysis of the water to determine the degree to which stream is polluted and the rapidity at which a stream is polluted and the rapidity at which it purifies itself. There is difficulty, however, in evaluating the extent to which each of the many chemical compounds to be found is harmful to the various organisms. There is considerable variation in this respect, even between different stages in life cycles of the same species Furthermore, the sewage load may vary from time to time, and infrequent heavy loads may wipe out the animal life in localities where chemical measurements made at other times do not indicate harmful pollution.

Various investigators (Hynes 1960, Tarzwell 1965) have attempted to use invertebrate animals as indicators of pollution. The presence of midge fly larvae Tendipes riparius, Glyptotendipes, mosquito larva Culex pipiens, rattail maggot, and sludge fly delimit zones of septic pollution. There are relatively few species that can tolerate septic conditions, but those that do may become very abundant.

Lakes and Eutrophication

Lakes are large bodies of fresh water usually deep enough to have a pronounced thermal stratification for part of the year.

Lakes are formed in youthful stages of river system development. Water from upland runoff, groundwater seepage, springs, and melting snowfields and glaciers collects in basins. As the basins overflow, erosion of outlets continues and the water level of the lake drops. Products of erosion from the surrounding upland are carried into the basin by wind and water, and along with the products of animals and plant decay fill in the basin in the course of time.

Morphometry aside, the essential distinction between lake and stream habits is the characteristic of water movement; continuous rapid flow is characteristic of the stream, the lotic habitat. The lake is a lentic habitat; the water is essentially a standing quiescent body, although at times wind action stirs surface layer and margins into great turbulence.

Habitat

The aquatic habitat offers a variety of unique conditions to which organisms must be a apted before they can occur. We must understand these conditions in order to inerpret how they are distributed.

Pressure, Density and Buoyancy

The pressure imposed on a lake dwelling organism is the weight of the column of water above it plus the weight of the atmosphere. Most lakes have maximum depth of less than 30m (Welch 1952). Maximum pressure are much less than in ocean, and organisms appear to adjust to them readily. The absence of animal life from deep water is ordinarily a consequence of low oxygen supply, or low temperature rather than pressure.

The density of water varies inversely with temperature and directly with the concentration of dissolved substances. Water is most dense at approximately 4°C and becomes progressively less dense as it is cooled below +4°C. Ice also expands markedly (that is, becomes less dense) the colder its gets. It is because the coldest water is at the surface in winter that ice forms there, rather than at the bottom. In summer, the coldest waters of deep lakes are at the bottom. Dissolved salts increase the density of water; the density of most inland water bodies is much less than that of the ocean.

By the law of Archimedes, the buoyancy of an object is equal to the weight of the water it displaces. Buoyancy varies with density of water and is influenced by the factors that affect density, viscosity, the measure of the internal friction of water, varies inversely with temperature and also influences buoyancy.

An organism will sink unless it keeps station by swimming movements, or unless it has special adaptation to decrease the specific gravity of the body and take advantage of turbulence in the water. Such adaptations take several forms : absorption of large amounts of water to form jelly-like tissues; storage of gas or air bubbles within the body; formation of lightweight fat deposits within the body, or oil droplets within the cell; increase of surface area in proportion to body mass, which increases frictional resistance. When an organism so equipped dies, the special mechanisms quickly cease to function, and it sinks to the bottom. If dead organisms did not sink to the bottom, living organisms, with the exception of some bacteria, could not exist in an aquatic habitat.

Light

The daily alternation of light and darkness establishes a rhythm in the activities of many aquatic organisms Light is essential to plant photosynthesis; some fish require light by which to feed. Many organisms orient to light and some are sensitive to light of particular wavelength, notably ultraviolet. Small, soft bodied, bottom-dwelling organisms are particularly sensitive to light, and it is thought that the evolution of pigmentation, chitinous exoskeltons, shells, cases and similar structures may have helped certain otherwise photosensitive species to survive in shallow, well lighted areas.

Suspended material in water produces turbidity, and reduces light penetration. In western Lake Erie, the depth to which 1.0 per cent of surface light penetrates varies from 9.7m, when turbidity is 5 ppm, to 0.8m, when turbidity is 115 ppm. Since photoplankton requires light for photosynthesis, abundance varies inversely with turbidity. Light penetration is also affected by the abundance of organisms themselves, both phyto and zooplankton.

An appreciable amount of light passes through ice in water. This enables phytoplankton photosynthesis to continue. In eutrophic lakes many fish suffocate when snow overlies surface ice, preventing phoiosynthesis.

Photosynthesis releases oxygen into the water; respiration and decomposition absorb it. The upper layer of a lake, where photsynthesis predominates, is called the trophogenic zone. Below this zone there may still be considered photosynthesis, but oxygen absorbtion is greater than oxygen release. The deeper portion of a lake is called tropholytic zone. The two zones are separated by a thin layer where the oxygen gains photosynthesis dur ng the daylight is balanced by the respiratory and decomposition losses during the day and night. This is the compensation depth, to which generally about 1 per cent of the full sunlight at the water's surface penetrates. The compensation level in a dark-coloured bog may lie less than 1 m below the surface; in a deep clear lake it may be 100m down.

Wind and Currents

Wind is an important environmental factor of lakes because of the water currents it generates. The effects of wind action depends largely on the extent of the exposed water surface, the presence or absence of protecing upland, and the configuration of the lake relative to the prevailling wind direction.

Waves may become sizeable in large lakes, but the forward motion of a wave does not involve any great mass water. The rate of movement of surface water is usually less than 5 per cent of the velocity of the wind.

The water washes back off the beach as an undertow, only to be carried forward again by the incoming waves. The problem of maintaining position here is similar to the problems of maintaining position in streams. The depth of wave action in the open lake and along the shore depends largely on the strength of the wind (Ruttner 1953).

In summer, surface water is warmed by solar radiation and its density, weight, and viscosity decrease. In deep lakes the warm water piles up on the exposed shore until, moving down along the bottom, it encounters colder and denser waters, which resist mixing. The warm water is then diverted horizontally to the oppsite shore. Thus the lake becomes stratified horizontally into an upper epillmnion, where the water circulates and is fairly turbulent, and a lower hypolimnion, which is relatively undisturbed. This difference in circulation in deep lakes is closely correlated with differences in temperature and oxygen characteristics; it is of considerable importance in the distribution of the biota.

Temperature

The thermal conductivity of water is very low; but because of the thorough mixing of the waters in the epilimnion during the summer by wind action, the temperature is nearly uniform down to the thermocline. The thermocline is zone of most rapid temperature decrease, generally involving a drop of at least 1°C per meter depth and accasionally as much as 7°C/m. The thermocline, as here defined, equals the metalimnion of some authorities, who have a different conception of the thermocline (Hutchinson 1957). When the thermocline forms, early in the season, it is close to the surface. As the season progresses, it sinks lower, increasing the volume of the epiliminion increases in thickness until it includes the entire lake. The waters are then uniform in temperature and density at all depths. This is the autumn overturn which may last for several weeks or until ice forms. As ice melts during spring and surface of water warms up, a spring overturns occurs when water at all depths is at the same temperature when a lake has to overturn during the year, it is called dimatic lake.

Oxygen

The distribution of oxygen at various depths depends upon the presence or absence of thermocline, the amount of vegetation, and the organic nature of the bottom. The amount of oxygen in water is only one forth to one-twentieth of that present in equal volume of air when the two are at equilibrium, although their partial pressures are the same. Diffusion of oxygen from the air into comparatively sedentary water occurs very slowly; agitation of the water increases the surface areas and promotes a faster rate of equilibration.

The oxygen supply of lakes is reduced in various ways, most notably through the respiration of animals and plants and the decomposition organic matter. As lake waters warm up during the summer, their capacity to hold oxygen may be released into the atmosphere. The saturation capacity of water at 0°C is 10.2 cc/litre, but at 25°C it is only 5.8 cc/litre. In some lakes, decomposition of organic material at the bottom may deplete the hypolimnion of its oxygen content for several weeks during the summer, lower than the level minimal to the support of many forms of life. This is called the summer stagnation period. During the winter, if the lake is covered with ice and snow, there may be a winter stagnation period. The oxygen supply of the waters is renewed with the autumn and spring overturns. Before decompositionc an proceed very far there must be calcium in water. Hence, decomposition is slow in soft or acid waters.

At temperature of 15-26°C oxygen concentration of less than 2.4 cc/litre (3.5 ppm) are fatal within 20 hours to several species of fish. From 3°C-4°C, oxygen concentration can decline through 48 hours to 1.4 cc/litre (2.0 ppm), or even to 0.7 cc/litre (1.0 ppm) before the same mortality results. This is related to the reduced metabolic requirements of cold-blooded organism at low temperatures. Some planktonio invertebrates can tolerate oxygen concentration as low as 0.2 cc/litre (0.3 ppm). Some bottom dwelling protozonas, annelids, mollusks, and insects larvae may survive actual anaerobic conditions for periods of days, even weeks. Organisms that tolerate a lack of oxygen do so by creating an oxygen debt; that is the lactic acid and other breakdown products produced in consequence of muscullar activity simply accumulate until conditions permit oxidation of them.

Carbon Dioxide and Other Gases

Carbon dioxide is required by plants for photosynthesis. Its presence in lake waters tends to vary inversely with oxygen. Carbon dioxide is derived form atmosphere, the respiration of both animals and plants, decaying organic matter, groundwater, and bicarbonates (dissolved CO_2), half bound state (HCO^{-3}) or fixed state (CO_3). These states are associated respectively with pH value 7, 7 to 10 and above 10. Algae and some rooted aquatic vegetation are able to obtain the half-bound CO_2 from the soluble bicarbonate salts, thereby converting them into less soluble carbonates.

$Ca(HCO_3)_3$ $CACO_3 + CO_2 + H_2O$. Molecules a few insects and some bacteria are also able to precipitate carbonates. Methane hydrogen sulphide, nitrogen and ammonia are other gases present in a lake which affects are biotic environment.

Lake Biocies

If we reserve ponds and peat bogs to separate consideration, there remain two major lake communities, abundance or organisms, distribution of niches, productivity, and physical characteristics. In as much as these two communities correspond fairly well to the oligotrophic and eutrophic types of lakes, we may name them simply oligotrophic and eutrophic lake biocies. Various facies of each community, or intermediate types (Deevev 1941) are affected by variations in the abundance of component species and correspond to differences in temperature, depth, fertility and other features of the habitat. The communities that occur in dystrophic lakes, for instance, are an impoverished facies of the eutrophic lakes biocies. In spite of taxanomic differences in constituent species, each lake, biocies, contains organisms belonging to the same life-forms and with similiar mores, so that they may be discussed together.

Depending largely on their morphological adaptation and behaviour, aquatic organisms are, for convenience, divided into plankton, neuston and benthos, although the differences between the groups are not precise. Seston is a collective term that includes all small particulate matter, both living and non-living, that floats or swims in the water. Plankton are free floating or barely motile organisms, either Plant (Phytoplankton) or animal (Zooplankton) that are readily transported by water currents. Most plankton are microscopically small, although some forms are visible to the unaided eye. Species that can be caught with a net are called net plankton to distinguish them from the minute varities that pass through No. 20 silk bolting cloth meshes. The later, less than 0.06 mm in length, include most protozoans, bacterial, and fungal forms, collectively called nannoplankton. Organisms that depend on the surface film for a substratum called neuston and are more important in the quiet waters of ponds than the lakes. Nekton are larger animals that are capable of locomotion independent of water currents. Aquatic birds swim and dive are included in this groupl Benthos organisms are attached to or dependent on the bottom for support; there are sessile, creeping and burrowing forms. Periphyton the assemblage of organisms attached to surfaces, either living or non-living, submerged in water above the bottom (Hanson 1962). Organic detritus, important in terrestrial as well as aquatic ecosystem, is "partiuclate material originating from disintegrating and decomposing biomass". All these major categorics may be subdivided into smaller units.

BENTHOS

Divisions

The lake bottom can be divided into littoral zone and profoundal zone.

The littoral zone extends from the water's edge to the limit or rooted aquatic vegetation. It may be subdivided into the eulittcalzone, between high and low water marks at the water's edge where the beating of waves is most effective, and the sublittora zone, extends from the lower limit of rooted vegetation. Where such vegetation is absent, the subtittoral zone may be considered the bottom of the epilimion down through the thermocline.

The profoundal zone is the entire bottom below the rooted vegetation or commonly the bottom of the hypoliminon. The boundary lines between the zones are variable and change with the depth of thermocline. The open water of the lake above the bottom is known as the limnetic zone.

Plankton

Fresh-water plankton (Welch 1952 Pennak 1946, Davis 1955) includes representatives from photosynthetic algae, Bacillariaceae (diatoms), Myxophyceae (blue-gree), Choropheceae (green) and occasional other form such as Wolffia among the higher plants, the non-photosynthetic bacteria and other fungi; and among zooplankton, all classes of protozoa except Sporozoa, Rotatoria, Entomostraca (especially Cladocer, Copepoda and Ostracod), some immature, Diptera, the statobiasts and gemmules of bryozoans and sponges, the rate fresh-water jellyfish Crospedacusta, and occasional aquatic mites, gastrotrichs and others. Fresh water plankton lack many forms common in the plankton of the ocean. On the other hand, the rotifers, aquatic insects and water-mites are mostly absent from the sea, and the Cladocera are only poorly represented. it is probable that the plankton evolved from benthonic forms occurring near the shore (Rutter 1953, and many species of groups listed above, notably Ostracoda and Rotatoria, are still largely benthonic in behaviour.

Nekton

The nekton of lakes consists principally of fish. There is an interestsng small shrimp, Mysis relica, found in the deeper waters of many northern lakes of North America that is often included with the nekton. This species is believed to be relic of a marine founa that happended to get cut off from the sea in some past geological period, yet was able to survive as waters become fresh.

Numbers and species of fish are more concentrated in the littoralzone of lakes than in the open, deeper waters of the limnetic zone. Limnetic species also invade shallow water for spawning. In deep waters, fish tend to remain close to the bottom, where their food supply is located, unless there is s ome deficiency of oxygen there. Caged fish, lowered to various depths in a eutrophic lake, did not survive long below the thrmocline (Smith (1925). Fish may however, make short excursions into the hypolimnion.

In the littoralzone, fish species are segregated according to the composition of the bottom, as are the invertebrates. The species living over rock and gravel bottoms in lackes are mostly different from those inhabiting similar bottoms in streams, but the mud-bottom forms are nearly the same as in ponds (Shelford and Boesel 1942, Nash 1950).

Amphibians and reptiles do not commonly occur in lakes except around margins supporting attached aquatic vegetation, and here pond species occur. Such pond mammals as the muskrat mink, and other are not the typical of lakes as such, although they are frequently found in shallow littoral waters. There are a number of bird species, however that occur most commonly in lackes.

Eutrophication

Eutrophication is a natural process, as oligotrophic lakes change to eutrophic ones in the course of centuries. Eutrophication becomes of concern when its rate is greatly accelerated by products of human activities that bring equivalent changes in decades or years. Modern agricultural practices cause increased erosion of uplands and correspondingly siltation of lakes. Excessive organic wastes from cities and farmlands becomes destructive pollutants even in large bodies of water.

Correlated with changes in the physical and chemical characteristics of water, there has been a change in the predominant species of phytoplankton, and the may fly naiad. Hexagenia libata, has been superseded in the benthic fauna by the midge fly larva. Tendipes plumosus. In the early nineteenth century, muskellunge, northern pike, and lake trout were commonly taken in the commercial fisheries but the harvest of these species began to decline rapidly towards the end of the century. Cisco, whitefish, sauger, blue pike, walleye, ling, fresh water cod, and channel catfish were among the principal commercial fish during the first half of the twentieth century but have decreased considerably since the 1940's and 1950's.

The remedy for situations leading towards accelerated eutrophication, particularly as it rises from the use of detergents, would seem to be either the reduction of the phosphates content in the product, replacement of phosphorus by some other chemical to produce the same cleaning effect, return to the use of soap. development of special sewage-treatment methods to remove the excess chemicals or a combination of alternatives. There are difficulties with each option (Rukeyser 1972). Water treatment technology is improving and phosphorus removals up to 90 per cent or more have been achieved, but conversion of existing sewage-treatment plants to these new techniques will take time and money. Some combination of these different procedures may be required for quick improvement of lake habitats.

Reduction of phosphorus from inflows into bodies of water may not always bring a corresponding decrease in excess algae present, since at the time the organic growth is controlled by the nitrogen supply rather than the phosphorus (Ryther and Dunstan 1971). It is even more difficult to remove nitrogen from water effluents than to remove phosphorus. There may be need, as with phosphorus, for regulation of the use of nitrogen at its source, which is often fertilizer. Certainly cultural eutrophication can be arrested, but this will require ingenuity, time and money.

Ponds, Marshes, Swamps and Bogs

Ponds is a popular terms for lakes of the third order that are small, shallow, and when mature, have rooted vegetation over the bottom. There is no clear distinction between the ponds and lakes of the first and second orders. The littoral zone of eutrophic lakes, for instance, is pond like in habitat and organisms. Floating and emergent vegetation commonly occurs around the margins of ponds to form extensive tracts of marsh, the pond habitat may originate as shallow basin, as a large pool in a stream as the result of the filling in of a lake, or from a stream dammed by beaveas, man, or landslide. Slow flowing rivers are essentially elongated ponds, and have a similar fauna. Because of the slight water movement in ponds, the surface film becomes an important microhabitate for some species, pond water temperature is often uniform at all detphs, but during warm sunny weather, ponds well protected from the wind may show considerable stratification, not only in temperature, but also oxygen content and other characteristics (Wallen 1955). Daily and seasonal variations in temperature may be great because of the small volume of water. Ice forms earlier and lasts longer in ponds than in lakes freezing shallow ponds to the bottom in severe winters. Light penetrates to all depths, encouraging growth of vegetation except in high turbidity.

Plant Sere

The plant hydrosere, or pond, sere typically contains the following stages and characteristic species.

Submerged vegetation. Water weed, pondweed, milfoil, hornwort, naiands, buttercup, bladderwort, eel-grass, and the hereb-like algae Chara.

Floating vegetation. Water lily, pond lily, pondweed, smartweed, duckweed, and water hyacinth. All except the last two are rooted in the mud, often at a depth of 2 to 3 meters and may have rhizomes from which long petioles extend to the leaves floating on the surface. Duckweeds and water hyacinths are unattached floates, and cover the surface extensively in some localities.

Emergent vegetation (marsh). The dominant species are cattail, reed wild rice and sawgrass. They invade water of over 1 meter depth, but in shallower waters or in secondary subession they are replaced by the sedge meadow composed or sedge rush and spike rush.

Swamp shrubs. Buttonbush, alder, dogwood, swamp rose and sometimes shrubby willow and cottonwood.

Swamp forest. Red and silver maples, elm ash, swamp white oak, and pin oak.

Succeeding stages depend on the climate region. In arid regions the swamp forest may be poorly developed, and grassland or desert vegetation may come quickly. In the mesic climate of the Estern stages, an oak-history associes follows the swamp forest, replaced in turn by climax sugar maple-beech or mixed mesophtic forest. The hydrosere in the board leaved evergreen climax area of south-eastern North American brings in cypress and a number of other unique species.

Animal Sere

Animals characteristic or marshes and ponds constitute a distinct pond-marsh biocies which extends into sluggish or base leveled streams and the littoral zone of eutropic lakes. Most animal species are restricted to single stage or community of the plant sere, but commonly occur in several stages in varying abundance and for activities. Fish, for instance, feed in open water but spawn in shallow water among the emergent vegetation. Submerged, floating and emergent vegetation represent different levels or strain in a single biotic community, and each stratum has about the same degree of distinctiveness as forest community strata.

With the invasion of swamp shrubs and the ground well above the water table most of the year, ponds and marsh species largely disappear, replaced by many characteristic new species. This animal community represents the swamp facies of the deciduous forest-edge biocies, which will be discussed later. The swamp forest is often quite open at its outer margin, and forest-edge species remains common. But as this forest develops a closed canopy and drier ground stratum, it is invaded by the swamp facies of the deciduous forest biociation. In the ponds sere there is a succession of animal adptations from aquatic to amphibious to terrestrial.

POND MARSH BIOCIES

Neuston

The supraneuston, organisms which move on top of the surface film in pursuit of most of their life activities, consists of the water strides Gerridge, Valiidae; Mesoveliidae, the water measures hydrometridae, the whiriligig beetles Gyrinidae; the sprintails Coloembola; some spiders; and occasional other forms. The gyrinids of several species commonly occur in social groups. They are remarkable for having each eye divided so that the upper portion looks into the air and the lower portion into the water. Several of the forms listed have long legs that distribute the weight of the body over a large area of surface of film. The portions of the legs or body that contact the surface film are water repellant.

The undersurface of the water film supports on infraneuston of Hydra, planarians, obstracods, cladocerans, snails and insects eggs, larvae, and pupae, (mosquitoes, certain kinds of midge flies, and so forth). For all except the insects, however, the use of surface film in this manner is usually transitory. Some cladocerans such as Bosming and Baphnia, occasionally break through the surface film from below, fall over into their sides and cannot return.

Plankton

The species of plankton found in ponds differ somewhat from those in lakes (Klugh 1927) but the transition from lake species to pond species is a gradual and progressive one Protozoa and Rotal oria are usually more abundant in ponds than in lakes.

Although plankton distribution does not vary with depth as much in ponds as in lakes, seasonal fluctuations are extensive and similar in nature.

Benthos

Subaquatic animals dwell not only on the bottom but also on the stems and leaves of submerged plants. Aquatic plant species have diversified animal periphyton of insects, amphipods, mites and snails, using them for food, shelter, or reproductive sites denied them in mud bottom below. Other kinds of insect larvae and oligochaetes are more abundant in the mud than on the plants. The undersurface of lily pads often contains many small organisms, including Protozoa, Hydra, Flatworms, Rotifers and Snails. The biomass of animals varies directly with the blomass of vegetations, and the quantity of invertibrates is especially great on those plants possessing finaly dissected leaves Very few species found in lake bottoms are found in ponds, but the pond-marsh biocies contains many species not found in lakes.

The number of species and individual found in the bottom fauna increases with the age of the pond from the time the pond is formed till the attached vegetation becomes excessive. Coincident with the development of the bottom fauna is an increase in variety and abundance of plankton and fish.

Terrestrial Invertibrate

The terrestrial insects found in marsh vegetation are in the main adults mosquitoes, midges, dragonflies, demselflies, mayficies and alderflies whose immature stages live submerged. On bare ground around ponds may be found taod bugs, shore bugs springtails, tiger beetles and sometimes ground beetles and pigmy, locusts. Spiders become numerous throughout the vegetation, and the snail Succimea appears in addition to these marsh and ponds species, invertibrates belonging to the forest-edge biocics may occasionally be found.

Fish

Fish are often abundant, included among the species that occur several bottomfeeders - suckers, bullheads, buffalo and carp. By Feeding on the submerged vegetation and stirring up the bottom, they may control the habitat and the composition of both animals and plant species present in the community. This condition, however does not last indefinitely. Vegetation encroaches on the margins of ponds, and the fish are gradually eliminated because of the disappearence of suitable breeding sites.. The mudminnow, bowfin and bullhead are usually the last to disappear before the pond becomes a dry marsh.

Amphibians and Reptiles

Salamanders and frogs are basically aquatic animals, although they show varying degrees of adaptation to terrestrial life. Siren and Necturus have permanent external gills and spend all their live in the water. Most other forms lay their eggs and pass through their early development in water, but the adults are air-breathing and wander over the land. Since their skins must be kept moist, they are confined to the victinity of water to humid climates, or to damp humus.

Reptiles are terrestrial. Desert reptiles never go to water, Painting geographic, and snapping turties bask in the sun on the shore or on protruding logs, but quickly plunge into the water to escape danger, to cool off, or to feed. The alligator, musk and soft-shelled turties spend nearly all their time in water. The softshelled trutle is able to utilize dissolevd oxygen in the water and hence has evolved epecial readaptation to water. Like other turtles, however, they lay their eggs on land, placing them in holes excavated in sand, loose soil, much, or decaying stamps or logs. Water snakes give birth to living young that enter the water immediately. Water snakes feed on insects, small fish and amphibians, crayfish, or whatever other animal food they can find.

Birds

Bird populations are high, and the pondmarsh, swamp shrub of forest-edge, and forest communities are especially clearly defined. Both the number of species and the equitability index increases with succession. There is an abundance of nest-sites and food, but the aquatic and terrestrial species exploit different niches to avoid competition as much as possible. The aquatic species feed in all stages of the plant sere, beginning with the open water, but nest for the most part in the emergent vegetation.

Food Chains in Ponds

Many animals in ponds depend for food on floating phytoplankton, bacteria and bottom detritus. Filamentous algae are more abundant in ponds than lakes and are a source of food to various immature insects are frog tadpoles. Ponds, unlike lakes, have additional producers of organic matter in the rooted pondweeds. Pondweeds are consumed by insects, ducks, and herbivorous mammals. The periphyton of bacteria diatoms, green and blue-green algae, and microscopic animals is important as food to many small crustaceans, immature insects, oligochaete worms, and snails Ffrohne (1956). Creeping predators among the pondweeds and alage are leeches, dragonfly and damselfly naiads, and water mites. Small swimming predators are the dytiscid beetles, most of the hemipterans, and the swimming leeches Erpobdella and Macrobdella. At the top of the food chains, feeding on all these small animals, and often on plants as well, are the fish and other vertebrate groups.

Mammals

One of the most characteristic mammals of the marsh is the muskrat. A well developed marsh may contain one of their haycock-shaped lodges on each acre (2.5 per hectare), with perhaps five animals per lodge during the atumn. The diet of the muskrat is largely the leaves and roots of marsh vegetation although they also feed to some extent on crayfish, clams, snails, and sluggish fish. Overpopulations of two or three lodges per acre (5.0 to 7.5 per hectare) may lead to "eat-outs" or local destruction of the marsh vegetation.

The mink is probably the most common mammalian predator of the marsh; it is an enemy of the muskrat. Foxes, reccons, and coyotes may invade the marsh when the water level is low. The other preys on fish and crayfish on the marsh; it has now been exterminated from most of its former range.

The breaver makes its own marsh habitat by damming small streams, flooding the surrounding lowland. Here it builds its large lodge and feeds on the bark and twigs of adjacent aspen, willow, and cottonwood, and on the roots of aquatic plants. When supply of aspen and other food is exhausted, the colony disappears, the dam decays, the water level subsides, and marsh vegetation invades. After some years the pond is converted into a beaver meadow.

BOGS

Characteristics

Bogs or moors typically develop in the hydrosers of cold nothern regions; while marshes and swamps, which are markedly different from bogs, are characteristically southern in their location. Bogs commonly develop into a coniferous forest climax; swamps succeed to deciduous forest or other southern climax types. Several thousand years ago, when glacial climates gripped the northern states, extensive bogs developed and have persisted as relic communities in spite of the warming of the climate. These bogs are slowly being replaced by pondmarsh species at equivalent seral stages by cliseral succession.

Bogs occurring in the Great Lakes region are ordinarily small in area and have little or no drainage. There may be oxygen present in the open water of the larger bogs, but it is characteristically in very low concentration, at the most, in small bogs or in the marginal zones. Bog water has a distinct brown colour a low nitrogen content; a low temperature beneath the surface; a low pH at least in the mariginal vegetated zones; and a low dissolved salt content.

A false bottom is characteristic of bogs. It consists of finely divided plant material of a light brown colour, held suspended in the water at varrying depths below the surface. This false bottom may be extended downward several meters before a true solid bottom is reached. The material disperses on slight disturbance and may render all the open water turbid. Ordinarily, the surface waters are quiet and clear. Dead vegetation does not completely decompose; as it accumulates, it becomes compressed to form peat.

Plant Bogs Sere

In the early stages of development of the bog, organic detritus may accumulate mostly in the deepest portions. In some bogs the first plant stage may be composed of floating vegetation (Nuphar, Nymphaea Potamogeton, Sparganium), but floating vegetation is often absent and the first stage is a sedge-mal composed of sedges cotton-grass, and buckbean. The rhizomes of the sedges grow out into the water and become so interlaced that they from a floating mat.

The next plant stage is dominated by low shrubs, which encroach on the floating mat. The leatherleaf, bog rosemary, laurels, labrador tea, sweet gale, and cranberries are important species.

A high shrub stage commonly follows the low shurbs at such times as the mat becomes thicker or grounded. Common shrub species are holly, willow, chokeberry, alders, and dwarf birch.

The first tree of the bog forest to invade the shrubs is commonly the tamarack, but this species is now less common than formerly because of fire, logging, and the deredations of the sawfly larvae Lygaeo nematus erichsonil. Black spruce may either invade the shrubs directly or follow the tamarack. Later, the northern white cedar may become dominant and persist for a very long time, but the ultimate fat of the bog, upon addition of upland soil or the water table, is to be covered with the climax forest of the region.

Animal Life

In bogs that have a large body of open water, or an inflow of water, entraining oxygen, and in which the pH is not extreme, invertebrate life comparable to that found in ponds and marshes occurs. True bogs, however, have little oxygen and a low pH, and many pond species do not appear. Mollusks are characteristically absent; sphaeriids may persist but their shells become very thin. Bottom organisms in general are poorly represented because of the tenuous physical nature of the substratum.

Desmids predominate among the phytoplankton, although dinoflagellates, Chlorophyceae, and Myxophyceae are common. Rotifers and a variety of Protozoa are the principal zooplankters.

Amphibians and reptiles are not characteristic of bogs, although the leopard frog is sometimes numerous on the sedge mat of bogs in Minnesota. Proceeding northward, boreal species of birds become progressively more prominent in the bog avifauna. The muskrat and beaver persist into northern Canada.

In general, the productivity and economic value of bogs is very low compared with ponds and marshes. Liming experiments, calculated to improve productivity, are being made. Calcium combines with the humid colloids, which then flocculate and fall to the bottom. This clears the water, light penetrates deeper, pH is raised and greater algal, zooplankton, and fish growth is induced. Peat is a special product of bogs. It is of importance in northern Europe. It is cut out in blocks, dried, and used as fuel and for mixing in the soil of gardens and lawns.

Rock, Sand and Clay

The origin of life was undoubtedly in the sea. Physiological adjustments were necessary before organisms could occupy fresh water. Although some organisms may have become air-breathers and invaded land habitats directly from the sea, most evolution of terrestrial forms has doubtless come from fresh water. Relatively few major groups of animals have been successful in this invasion of land, the most notable being oligochaete worms; gastropod mollusks many; arthropods, especially the insects and spiders; reptiles and their evolutionary offspring, the birds and mammals.

Adjustments to the Terrestrial Habitat

Living on land presents many problems. Our present concern in to analyze the ways in which animals have met these problems and to trace the succession of communities in the extreme terrestrial habitats of rock, sand, and clay. A variety of techniques have been developed for measuring environmental factors.

Gravity

In water, organisms counteract gravity by means of various floating and swimming mechanisms. Fluid buoyancy permits water-dwellers to attain huge size; consider the whale. A land animal, on the other hand, must support its entire weight. Some terrestrial animals gain a modicum of support by burrowing into the soil. Others drag their bodies over the ground surface. But the supportive advantage they thus gain is costly in other directions, for they are slow moving and relatively helpless before predators. The animals best adapted to terrestrial life have evolved appendages in the form of legs or wings that not only raise the body above the ground but are also the means of more or less rapid locomotion and adroit movements over the surface or through the air. Terrestrial adaptation has involved the development of a tough body covering to hold fluids and internal organs in place; a skeletal framework to give permanent shape to the body and, as a system of levers, to furnish means of locomotion; and powerful muscles to lift and move the heavy body.

Moisture

In sharp contrast to aquatic forms, terrestrial animals are not constantly enveloped with a continuous watery medium, with the limited exceptions of protozoans, nematodes, and other small organisms living in moist soil. Land animals, lacking constant contact with the water medium, are faced with the problems of obtaining water and preventing excessive water losses from the body.

Water becomes available to animals in varying amounts in the forms of rain, snow, hail, frost, and fog. Whatever the form it arrives in, the significant things are the amount of free, liquid water added to the substratum, accessible to plants and animals, and the humidity of the air. Considerable amounts of moisture are lost to organisms as runoff water flowing into streams, by evaporation back into the air, and as water bound in snow and ice.

Water is lost from the body through the skin and lungs as insensible moisture perspiration. Rapid and largely uncontrolled loss of moisture through the skin of amphibians, snails, annelids, and insect larvae is a limiting factor confining these animals to moist habitats or to activity only in times of high humidity. Adult insects and other arthropods, reptiles, birds, and mammals have evolved body surfaces of chitin and waxes, scales, or cornification of the surface layers of the skin that largely prevent uncontrolled loss of moisture. Moisture loss through the respiratory surfaces in these forms remains considerable, however. Water is also lost with the feces, although in some species much water is reabsorbed by the large intestine before the feces are ejected. The amount of water removed from the body by excretory organs, particularly the kidneys. The kidneys, are critical to maintenance of proper concentration of salts in the blood and body fluids. It is important that water intake balance water loss. Organisms are very sensitive to disturbances in body water balances and this factor is significant in determining the type of niche which a species comes to occupy.

In order that animals could exist in terrestrial habitats, they had to acquire the ability to carry on reproductive activities in the absence of water. The chief reproductive adaptations involve the following (Pearse 1950) : internal fertilization; a shell covering the egg to conserve moisture and salts; food provision to the embryo and young, by yolk in the egg cell, placenta in the uterus of the mother, or direct feeding by adults; reduction in number of young with more efficient parental care; reduction or ellimination of free-swimming larval stages; and greater segregation of species into different niches to avoid interspecific disturbances.

Temperature

Aquatic animals are not ordinarily subject to temperature below freezing and are in a relatively stable temperature environment, but terrestrial species are exposed to highly variable temperature that may reach extremes of about 68°C and +55°C. No single species is required to withstand such a wide range of temperatures, however. Optimum and tolerance limits vary from one species to another.

No aquatic species has evolved control over its body temperature. Aquatic warm-blooded mammals and birds are derived from terrestrial forms. An ability to maintain a constant body temperature has survival value in terrestrial habitats, however, and consequently physiological mechanisms for homoiothermism developed independently in birds and mammals. All other land organisms are poikilothermal; that is, they have no physiological mechanism for maintaining a constant body temperature. Some poikilotherms, such as bees, have developed special behaviour patterns that enable them to maintain fairly constant conditions in the hive by cooperative efforts; some moths and other insects with hair-like body coverings maintain a degree of constancy in body temperature by shivering their wings; some lizards, snakes, and turtles are able to exert some control over their body temperatures by moving into and out of sunlit areas.

Rates of activity, food consumption, metabolism, growth, and other physiological functions increase to a certain limit with rise of body temperature. Homoiotherms maintain a continuous high rates of functioning because of their constant high body temperature, but the rate at which poikilotherms function varies with the temperature of their habitats.

Oxygen

Dry air at 760 mm Hg pressure contains approximately 21 per cent oxygen, 0.03 per cent carbon dioxide 78 per cent nitrogen and traces of other gases. Oxygen is thus much more abundant, constant, and available at all times in air than it is in water. Oxygen availability seldom becomes a critical factor for lanc animals, with occasional exception of forms that live in the soil or invade high altitudes.

Although terrestrial organisms have evolved simple moist chambers, branched trachael systems, or complicated lungs to replace the gills found in many aquatic forms, the fundamental requirement of moist membranes for the exchange or oxygen and carbon dioxide between body fluids, tissues, and surrounding medium remains the same. The skin still serves this purpose in some terrestrial forms-annelids and some amphibians-but in most forms the moist membranes are within the body. Internal placement decreases the loss of water through evaporation. The evolution of an ability to take oxygen directly out of the air apparently preceded the actual invasion of land, and may have been induced invasion in the pond and marsh habitpt when oxygen dissolved in the water becomes reduced or absent during summer stagnant periods. The evolution of internal air-breathing organs was probably concurrent with the evolution of mechanisms to prevent excessive water loss from the exposed surfaces of the body.

ROCK SERE

Plant Communities

Stages in the plant sere on bare rock include lichens, mosses, herbs and grasses, shrubs, and forest. The species composition of each stage varies with the chemical nature of the rock, the prevailing climate, and the locality.

In the first stage, various kinds of lichens compete for a foothold, but crustose types usually precede foliaceous types. Mosses such as fruticose lichens as Cladonia follow foliaceous lichens; or may initiate the sere telescoping the earlier lichen stages.

Lichens and mosses soak up moisture in wet weather. They derive mineral nutrients from the underlying rock. Carbon dioxide secreted from the rhizoids forms a weak acid with water and dissolves the binding material of the small rock particles. Rhizoids may penetrate rock for several millimeters. These plants trap wind blown dust and obtain nitrozen from organic compounds in it. When the plants die, they become an addition to the accumulation of organic matter. Herbs, grasses, ferns and later stages invade to continue the crumbling of the rock and buildup of the soil.

Animal Life

Animal life in the pioneer plant stages on rock is scanty. Ants and spiders roam over the bare rock, and insects of various sorts may stop there, temporarily. Spiders may construct webs and inhsts in rock crannies or among the lichens. Some tardigrades find preferred niches in lichens. Mosses offer a somewhat more substantial micro-habitat, but only those animals that can tolerate great extremes of flooding, dessication, heat and cold can survive. Such forms are found in the rhizopod protozoans, nematodes, bdelloidid, rotifers, tardigrades, copepods, small insects and mites (Heinis 1910). They often have special means of attachment to keep them from being blown away by the wind, such as strong claws or cement glands on the feet, long bristle-like threads to entangle among the moss filaments; stickers or spines covering the eggs. Since wet periods are often too short to permit complete development, all stages must be tolerant of dessication, at which time activities and growth are largely suspended.

Animal life in general and land snails in particular are usually more abundant in vegetation (grassland and forests) established on calcareous solis derived from limestone than in the vegetation established on soils derived from sandstone, granite or volcanic rock.

CLAY SERE

Plant Communities

Erosion or calculated removal of overlying material in strip mining for coal may leave bare areas of clay exposed and impound small bodies of water, often with low pH values. These water bodies undergo a modified pond succession (Riley 1960). In clay above pH 4.5 annual plants, of which smartweed is particularly important, appear within a few weeks to 2 years; the higher the clay pH, the quicker the appearance of vegetation. Within 2 to 5 years thereafter sweet clover invades and develops nearly complete dominance over large areas. Sweet clover is a biennial, and an exotic species unimportant in the sere in some parts of the country. Prior to its introduction, this stage in the sere on bare clay may have consisted of the perennial grasses still found in small scattered patches. The first trees begin to invade early in the sere, out they are scattered and slow of growth, and do not attain dominance for 25 to 30 years. Willows occur in wet spots. Herb species of the first two plant stages disappear, for the most part, in the shrub stage. The herb stratum now consists largely of wood nettle. Advance forest stages of oaks, hickories, basswood, and sugar ample will likely invade in the future, as they occur now in adjacent areas.

Animal Life

The number of invertebrate species tends to increase as the sere advances, although not always regularly. In a study of a formerly stripmined area (Smith 1928), 18 species were found to be important in the annual stage, 41 species in sweet clover, 40 species in shrubs, 32 species in the early forest stage, and 67 species in the unplan climax. More species would be found in advanced stages because of the greater variety of niches available. Thus, in the initial bare area there is only the ground stratum, in the annuals and sweet tree strata. Since the early forest, is on a flood plain, the, ground is frequently swept by floods, and the shrub stratum is poorly represented. The climax forest has all strata, richly developed, and possesses the greatest number of animal species. There is an increase in the abundance of individuals per m^2 with the progression of the sere; annuals 268; clover, 531; shrubs, 532; early forest, 748; climax, 2445 (Davidson 1932).

Beetles, spiders, ants, and mites are the most abundant animals in the annuals stage, and along with aphids remain most abundant also in the sweet clover community. Grasshoppers are fewer in number but especially characteristics of the first three stages; they practically disappear in the forest. Earthworms are absent in the annuals and scarce in the sweet clover, as are the springtails; as the amount of soil humus increases with the development of the sere, both groups become more and more numerous. Snails first appear in the sweet clover stage and increase in importance in the forest stages.

Starting with bare ground and continuing through early and the shrub stages to bottom land forest, there is a progressive increase in bird species diversity. The first small mammal to invade the area is the praise deer mouse followed by cottontail rabbit and fox-squirrel.

Grassland, Forest and Forest Edges

We have seen that succession of animal communities in humid climates passes through three terrestrial stages before attaining climax : grass, shrubs and scattered trees (forest-edge), and forest. In arid climates the climax may be reached at first or second stage. It is important for us to examine each community in more detail, therefore, if we are to gain an understanding of the ecology of animals prevailing locally, in different parts of the world.

Vegetation

Grassland vegetation differs from forests in that the above-ground vegetation is completely renewed each year. Grasses may be divided into three categories on the basis of height; tall grasses (1.5-3m tall), such as big bluestem and slough grass; mid-grasses (0.5-1.5m tall), such as little bluestem and needle grass; and short grasses (less than 0.5m tall), such as buffalo grass and grama grass. The taller grasses grow in wet habitats, the short grasses in arid habits. Most native grasses are bunch grasses in that they grow in clumps with the areas between the clumps either bare ground or occupied by the species. Broad-leaved herbs occurring between the dominant grasses are called forbs. A few species are sod formers in that their growth in continuous over the ground surface. The leafy aerial parts of perennial grasses die in the winter or in dry season, leaving the under-ground stems or rhizomes to propagate the plant the following year.

Forests are composed of trees growing sufficiently close together to dominate the entire area of ground surface. In cold climates most forests are needle-leaved everygreen; in seasonally warm moist climates, they are mostly broad-leaved deciduous; and in continuously warm, moist climates, they are board-leaved evergreen. In spite of these secondary differences in life-form, the structure and internal dynamics of all forest communities are quite similar.

Between forests and open country, scattered trees are interspersed with shrubs and grasses. This transition area is usually narrow around the margins of a stabilized forests, but where succession is occurring, large areas of mixed grass, shrubs, and trees may provide in close proximity a variety of habitats which animals take advantage of in a variety of unique ways. Likewise, in agricultural areas, hedge and fence rows, or narrow strips of trees and shrubs along streams, are really edges without the adjacent forest and have the same characteristic animal life.

The grassland community typically contains subterranean, surface, and herb strata of vegetation. The forest in addition has shrub and tree strata. Correlation with the differences in micro-climate, both plant and animal species show segregation of their activities and presence to one or more of these strata.

The herb stratum in mature grassland is much better developed in biomass than in the forest, but the woody vegetation of the forest is so great that the total biomass of the community is considerably greater. In one study, it was 29 times greater. Good comparative data on the biomass of all animals are not available, but class studies in central Illinois during early autumn indicate no great differences, in the biomass of the macroscopic invertebrates. Seral grassland communities may, however, in contrast to climax ones, be relatively impoverished.

GRASSLAND COMMUNITY

Invertebrates

Snails, earthworms, and myriapods are not numerous in most grasslands because of the dry habitat. Insects, however, are abundant.

Spiders make up about 7 per cent of the total arthropod population in grassland.

In grazed pastures and in grassy meadows in New York State, invertebrates average 777 individuals per m^2. Of this population, ants make up 26 per cent, leafhoppers 15 per cent, other insects 34 per cent, spiders 9 per cent, millipedes 9 per cent, sowbugs 2 per cent, snails and slugs 2 per cent, earth-worms 2 per cent, and large nematodes 1 per cent.

Some insects show structural adaptations for living in grassland. May beetles in forested regions commonly feed at night on the foilage of trees and have well-developed wings, but closely related species in grassland area feed on lowgrowing plants during the dry and are the flightless. The development of pilosity and thick integuments in some insects appears to be an adaptation to prevent evaporation. Prairie May beetless pupate in the spring rather than autumn, probably in correlation with their change in food habits, and adults appear in mid-summer rather than late spring.

An insect microhabitat of special interest is the doing of the larger mammals. Bison formerly occurred in frequency one to 10 12 hectares. In as much as the output of each animal is about 25 droppings per day, the number of these microhabitats available was considerable. Some 83 species of arthropods have been collected from cow dung, mainly beetles and flies, but including annelids, nematodes, and protozoans. There is a regular succession of insect species breeding and maturing. The microsere is completed in about 8 days the length of time required for the droppings to dry. The first species that arrive are the obligatory breeders on dung. They have the shortest life histories, and remain for the shortest time. Species disappear as the dung disintegrates into the general surroundings. A comparable microhabitat and succession occurs in carcases of dead animals.

Vertebrates

Birds are not numerous in grassland. In north-western Towa, grasshopper sparrows western meadowlarks, bobolinks, ringnecked pheasants, marsh hawks, and shortearned owls averaged less than one pair per hectare (2.5 acres). Prairie chickens and sharp-tailed grouse formerly occurred where now is to be found only the introduced pheasant.

Several species of snakes are generally distributed over the prairies. To the east, the blue racer, messasauga, bullsnake, and garter snakes are frequently found.

The lizards like cremidophorus, sexleneatus its commonly occurs in grassy areas at forest edges. The most characteristic of grassy area is the toad.

Forest Edge Community

Grassland animals are usually restrained from penetrating forests in the same way that true forest animals are restrained from penetrating grassland, although the home ranges of these species may overlap at the forest margin and in shrubby areas. Since shrubs are especially numerous at the forest-edge and animals have an opportunity to make use of these as well as both grassland and forest, the forest-edge biocies is well developed for some groups of animals.

There are probably no soil or small ground animals characteristic of the forestedge. There are some foliage insects that find their preferred niches here. Many insects of grassland and agricultural crops that overwinter as adults migrate into the forest-edge to hibernate. Since many game species of interest to man reach their greater abundance on the forestedge, he has become impressed by this edge effect. When total populations of all species are measured, however, the density of birds or mammals is not always higher than in the forest. When two forest types come in contact, for instance different deciduous forest types or deciduous and coniferous forest there is no consistent change in the density of animal species. The forest-edge is the preferred nesting site of many birds.

Forest Community

Since the censusing of eachgro up of animals furnishes special problems, there have been no studies of total animal-populations in single forest communities. The total measofauna would number in the tens or millions and the microfauna in numbers so large as to be scarcely conceivable. The biomass of bacteria would probably be somewhat less than that of fungi but exceeding that of algae, protozoans, and nematodes combined (Burger and Rau 1967). In general, the number of individuals representing a species varies inversely with the body size characteristic of the species. There is, however, considerable variation in population levels both geographically and temporally.

Stratification

Animal life in the community is separated into different niches in relation to strata (Elliot 1930). Food, shelter, and microclimatic differences are the chief limiting factors. Because microclimate in each stratum varies from hour to hour, day to day and season to season, classification of a species by stratum must be in terms of the stratum it is observed to frequent for the major portion of a relatively long period of time. The inhabitants of the five strata divide into two major groupings, or societies. The soil invertebrates and some mammals move freely back and forth between subterranean and surface strata and may be considered a ground society, distinct from a foliage society which occupied the combined herb, shrub, and tree strata. The two societies are not mutually exclusive for ground animals such as millipedes and snails climb up onto herbs and tree trunks during humid weather and foliage animals often rest on the ground, and search for food, hibernate, and lay eggs, there.

The majority of arthropod species carry on their main activities within a single stratum for the major activities. The tree stratum spans a greater vertical distance than any other. Within this broad stratum, arthropods often show segregation to particular heights above the ground. There is a striking difference in species composition of spiders between the shrub, hurb, and surface strata. Likewise, several species of homopterans in grassland communities show segregation between the litter and different heights in the grass. Arts and beetles appear to move more freely between different strata than do other species.

Population density of invertebrates ranks highest in the surface stratum, followed by subterranean and herb strata, while shrub and tree strata rank lowest. The largest mammal populations occur in the subterranean and ground strata.

In a European oak-hornbeam forest (Turcek 1951), 15 per cent of the bird species neste on the ground, 25 per cent in the herb and shrub strata, 3 per cent in or on the trunks of the free, canopy. The largest number of individuals (32 per cent) occurred in the forest canopy, although the biomass of these birds constituted a smoller percentage of the total (16 per cent) than did the ground and herb population (67 per cent). In respect to feeding, however, the distribution was different : 52 per cent found their food on the ground. 9 per cent in the herbs and shrubs, 10 per cent on the tree trunks, 23 per cent in the tree foliage, and 6 per cent in the open spaces between the canopy, trees, and shrubs.

Seasonal Changes

Outsidé of the tropics, the forest community changes drastically with the seasons such that six aspects may be recognized. The total population of the soil macrofouna in the temperate deciduous forest is highest during the winter because of the migration of many foliage insects into this stratum to hibernate. However, many species overwinter only as eggs or other immature stage. Forest species hibernate in densities that vary randomly throughout the forest, except where there are differences in topography or substratum. Forest tracts adjacent to grass-or formland, however, receive an influx of non-forest species that hibernate principally on the forestedge generally, and long the south edge in particular, where exposure to solar radiation and protection from cold northerly winds produces warmer temperatures.

During the vernal aspect, insects and other invertebrates come out of hibernation, and the adults of firms variously frequenting the herb, shrubs and free foliage return to their characteristic stratum. The population of ground animals remains relatively high throughout the year, however, which can be attributed to the reproduction of the geobionts and to the fact that the immature stages of many foliage arthropods, particularly Diptera, Coleoptera, and Lepidoptera, occur in the ground. Various groups of these geodionts and geophils reach peak numbers of adults at different times during the year.

An insect species may show more than one population peak during the year, depending on the number of generations produced and the specific life span. A specier may appear attain to very large numbers, and disappear, all in a matter of a few days, or a few weeks at the most. Considerable variation also occurs from year to year in the population fluctuations of individual species and of whole groups, correlative with differences in weather, particularly temperature and moisture.

Bird populations in temperate zones reach peak populations with the passage of transients during the seasons of migration. These migration peaks are inconspicuous of absent in northern coniferous forests, since most birds that arrive stay to nest. In the tropics, birds are most abundent during the winter period of the north temperate zone, since the fauna then contains many migrant species from the north.

Mammal populations in temperate regions commonly reach their maximum numbers in the autumn, at the end of the breeding season. Populations decline progressively as winter wanes, the result of mortalities from severe weather, lack of food and predation.

In this lesson, we have very briefly tried to see the biotic environment (look plant and animal) in streams, lakes, ponds, marshes swamps, bogs, rocks, sands, clay, grassland, forest and forest edges. We have also tried to find out briefly the several factors which constitute and influence these environments.

4

Population Ecology

Introduction

Population ecology is concerned with inter-relations or coactions between individuals within and between species. Coactions may either be beneficial to the participants, cooperation (Allee et al. 1949, Allee 1949, 1951), or harmful, disoperation. Interspecific cooperation includes mutualism, commensalism and many of the interrelations that establish the community as a dynamic unit. Coactions that are harmful to at least one of the participants, include parasitism, predation, and competition. Many of these coactions act to regulate density or, along with climate, cause fluctuations or even more drastic changes in number of individuals from time to time. Established behavioural interrelations between individuals, characteristic of a species, are its ethics and as such are found in animals as well as in man.

INTRA AND INTER SPECIFIC COACTIONS BETWEEN INDIVIDUALS

Intraspecific Cooperation

An early manifestation of cooperation in the evolution of animals is the grouping of free-living protozoans to form colonies, and the further development of such colonies into multi-cellular metazoans that thereafter behave and respond as unit organisms. Whether the first gathering of protozoan cells to form colonies developed for better protection from some predator or environmental condition, improved utilizalion of food supplies, or more efficient reproduction, it is impossible to say. The colonial form, however, must have had survival value to persist.

Aggregations

All organisms react on their habitat in one way or another, and when they occur in numbers these reactions produce a conspicuous effect. Water conditioning occurs when physical or chemical changes occur as the result of organisms living in it. Compared with unconditioned water, these changes may have either a harmful or a beneficial effect on organisms introduced into the water after the original organisms have been removed. Water is said to be homotypically conditioned when the changes were previously produced by individual of the same species as being studied and heterotypically conditioned when the changes were produced by a different species.

Mutualism

Mutualism is an association between two or more species in which all derive benefit in feeding or in some other way. The term symbiosis has often been applied to this relationship, but symbiosis properly refers to the intimate association of two or more dissimilar organisms, regardless of benefits or the lack of them, and hence includes mutualism, commensalism, and parasitism.

Mutualism, as is true also with commensalism, and parasitism, may be facultative, when the species involved are capable of existence independent of one another, or obligate, when the relationship is imperative to the existence of one or both species. Considerable study and experimentation is sometimes required to decide whether a particular relationship is facultative or obligate, or even whether it is truly mutualistic.

Mutualism in plants is demonstrated in the associations of fungi and algae to form lichens, of nitrogen fixing bacteria with the roots of legumes, and of fungal mycorhizae with the roots of many flowering plants.

There are many intimate relations between plants and animals. Mutualism is suspected in the presence of photosynthetic algal cells in the protective ectoderm of green hydra, and those associated with turbellarians, mollusks, annelids, bryozoans, rotifers, protozoans, and the egg capsules of salamanders. The algae give off oxygen, benefiting the animals, which in turn supply carbon dioxide and nitrogen to the plants. The thick growth of algae often found on the carapace of the aquatic turtles is important mostly as camouflage for the turtles. Certain beetles, ants and termites cultivate fungi for food. Bacteria in the caeca and intestine of herbivorous birds and mammals aid in the digestion of cellulose. The cross-pollination of flowers by the agency of insects and birds seeking nectar and pollen is of such great importance that many structural adaptations in both plants and animals fit the one to the other to ensure the success of the function.

Commensalism

Commensalism defines the coaction in which two or more species are mutually associated in activities centering on food and one species, at least, derives benefit from the association while the other associates are neither benefited nor harmed. It is often difficult to establish definitely the nature of the relations between species; and phenomena considered at one time to be commensalism have been later found to be parasitism or mutualism. The concept of commensalism has been broadened, in recent years, to apply to coactions other than those centering on food; cover support, protection, and locomotion are now frequently included (Baer 1951).

The remora fish are remarkable for having the spinous dorsal fin modified to form a sucking disk on top of the head by means of which they become attached to the body of the shark, swordfish, funny, barracuda; or sea turtle. They are of small size and are not burdensome to the host. The host benefits the remora, however, for when the host feeds, the scraps of food floating back are swept up by the remora.

Parasitism

Parasitism is the relation between two individuals wherein the parasite receives benefit at the expense of the host; parasitism is therefore a form of disoperation. Parasitism is mainly a food coaction, but the parasite derives shelter and protection from the host, as well. A parasite does not ordinarily kill its host, at least not until the parasite has completed its reproductive cycle. Where the parasite kill its host immediately on infecting it, the parasite would be unable to reproduce and would quickly become extinct. The balance between parasite and host is upset if the host produces antibodies or other substances which hamper normal development of the parasite. In general, the parasite derives benefit from the relation while the host suffers harm, but tolerable harm.

Social parasitism describes the exploitation of one species by another, for various advantages. Old world cuckoos and the brown-headed cowbird of North America do not build nests of their own; rather, they deposit their eggs in the nests of other species, abandoning eggs and young to the care of foster parents.

Parasitoidism and Predation

Some Diptera and Hymenoptera deposit their eggs in the immature stages of other insects; the larvae on hatching feed on the host until they are fully grown. The relation of the larva to its host is frequently described as parasitism. It is fundamentally different from parasitism in that the host generally dies of the larva depredations before the larva emerges, but the larva lives inspite of the death of the host. The relationship resembles that of predator to prey, except that, unlike the true predator the larva lives within the body of its prey and kills it slowly as it feeds. Such larvae are called as parasitoids.

Predation is a form of disoperation, at least in point of immediate effects, since one animal kills another for food. Predation is important in community dynamics in so many ways.

Competition

Competition is the more or less active demand in excess of the immediate supply of material or condition exerted by two or more organisms (Clements and Shelford 1939; p. 159). The materials and conditions sought by animals include food, space, cover and mates. When these materials are in more than adequate supply for the demands of those organisms seeking them, competition does not occur; when they are inadequate to satisfy the needs of all the organisms seeking them, the weakest, least adopted, or least aggressive individuals are forced to do without, or go elsewhere. Competition may result in death for some competitors, but this is from fighting or being deprived of food or space rather than being killed for food as in predation, or by disease as in extreme parasitism.

Competition may be either direct or indirect. It is direct where there is active antagonism, struggle or combat between individuals; indirect, when one individual or species monopolizes a resource or renders a habitat unfavourable to the establishment or other organism having similar requirements. Direct competition, or interference, is evident in the fighting of bull seals for larger harems and of grouse for a better position in the social hierarchy; in chasing and color displays (a sort of saberattling) by fish and birds for defense of territories; in the singing and calling of birds, some mammals, and frogs as bids for mates; and in the excretion of chemicals that effect the behaviour or health of other organisms (allelo-chemistry).

Competition is usually keenest between individuals of the same species, intraspecific competition, because they have identical requirements for food, mates and so on. And because they are more nearly equal in their structural, functional and behavioural adaptations. Interspecific competition occurs where different species require in common at least some materials or conditions. The severity of competition occurs where different speeies require in common at least some materials or conditions. The severity of competition depends on the extent of similarity or overlap in the requirements of different individuals and the shortage of the supply in the habitat. It is generally the case that the more unlike the kinds of competing organisms, the less intense the competition.

Allelochemistry

Included here are the coactions whereby chemicals secreted by one organism affects the growth, health, or behaviour of other organisms. Allelopathy is produced in plants when toxins are liberated that inhibit seedling growth in the vicinity. This may affect succession of plant species, especially important in the early stages. Some pioneer species in the abandoned field sere produce substances inhibitory to nitrogen fixing and nitrifying bacteria. This retards invasion of other species that require higher nitrogen concentration in the soil. Volatile inhibitors are generally more prevalent than water-soluble ones and relatively more prevalent in arid than humid climates. Antibiotics produced by bacteria, fungi, actinomycetes, and lichens are widespread in nature and may be one of the reasons why bacteria pathogenic to man cannot multiply well in soils. A number of antibiotics, such as penicillin, have been used extensively in human medicine. The use of allelochemic coactions in agriculture has possibilities but has yet to be exploited.

Among animals, overcrowding of tadpoles in culture dishes is associated with the occurrence of peculiar round vacuolated cells in the intestinal tract and feces that appear responsible for curtailment of further growth. Under laboratory conditions, killer stocks of Paramecium aurelia produce a toxin, paramecin, at the rate of one unit particle per 5 hours. One unit particle is enough to kill one individual of sensitive stock of the same species as well as being lethal to other species of Paramecium. Conditioning that becomes unfavourable homotypically may sometimes be favourable, or at least tolerable, heterotypically. Thus in protozoan infusion there is a microsere of one species succeeding another.

Allelochemic effects are of great variety in both plants and animals; repellants, escape substances, suppressants, venoms, inductants, counteractants attractants, signals, stimulants, autotoxins, antoinhibitors, and so on. Pheromones are chemical messages between members of a species especially important in

nities. Benefits derived from cooperation are physiological and behaviorall and may affect survival, reproductive success, and more efficient use of natural resources. Cooperation between species that is intimate and beneficial to both participants is called mutualism; where only one participant benefits, commensalism. These relations may be either facultative or obligate. Where one or more of the participants is harmed there is disoperation, of which parasitism, parasitoidism, competition, and predation are the examples. Distinction is made between true parasites, social parasites, and parasitoids. True parasites and their hosts have evolved adaptive interrelations so that coexistence occurs for varying lengths of time. The host is generally weakened, however, and virulent strains of the parasite may cause high mortalities. Causes of mortality or disease among organisms are predators parasitoids, worm parasites, protozoan parasites, bacteria, viruses, fungi, external parasites, nutritional deficiencies, toxication, physiological stress, and accidents.

Competition may be exerted directly through interference in the activities of one organism by another, or indirectly in the form of excessive exploitation of natural resources. It may be either intraspecific or interspecific. Allelochemic secretions from one organism that affects the growth, behavior, or health of other organisms are a subtle means of competition, cooperation, or communication. Competition may result in establishment of social hierarchies, establishment of territories, regulation of population size, segregation of species into different niches, or speciation. The overall effect of competition is to relegate the individual and species to an orderly place in the structure and organization of the community with the result that there is decrease in tension and disturbance.

Reproductivity and Population Structure

The rate at which a species reproduces, or its reproductivity, and the frequency of its population turnover can affect the speed with which it occupies new areas, becomes adapted to new niches, or evolves into new races. In order to analyze the population dynamics of a species, it is necessary to know its 'ife history. This involves the stages in its life cycle, mortality rates of each stage, longevity, sex and age ratios, age at which individuals become sexually mature, fecundity, factors causing mortality, and so forth. The proportion of different ages and sexes gives the population a definite structure. All these essential data may be conveniently summarized in the form of life tables.

Fecundity

Species vary greatly in the characteristic number of generations, broods, or litters produced per year, and in the sizes of them. Protozoans often divide so rapidly that they produce a new generation every few hours. Plankton organisms, less fecund, may produce a new generation every few days. Many vertebrates breed butonce a year, some large animals only once every two or three years. Several species of small birds and mammals have two or more broods per year. The female woodland white-footed mouse in Michigan may produce three litters between early April and early June, and two more between middle August and early October. Under favourable environmental conditions, rodents may continue to breed throughout the winter, so that their reproductive potential is enormous.

Parental Care

The number of eggs or young produced per litter is correlated inversely with the amount of attention that they require. When parental care is altogether lacking invertebrates may lay 1,000 to 500,000,000 eggs at one maturation; where there is some protection afforded by brood puches 100 to 1,000 eggs may be laid; with a high degree of broad protection, 1 to 10 or more eggs may be laid. Mammals seldom have more than a dozen young in a single liter and, in larger species, usually only one. Characteristic clutch size among birds varies from 1 to 15; rarely, 20.

Death Rate

Death rates vary among species and are correlated with rates of reproduction. The death rate of a species is influenced by a number of factors, but of fundamental importance is the number of young that are born in relation to the carrying capacity of the habitat. When more young are born than the habitat can support, the surplus must either die or leave the area. When populations are stabilized at a constant level, the death rate must fluctuate with the birth rate. Evolutionary adaptation tends to lower the frequency at which the population replaces itself and to raise reproduction to the highest rate compatible with the energy resources both of the species and of the habitat.

Survival of Young

Success in raising young depends not only on the ability of the adults to care for the young, but also on the vitality of the embryo and on th· chance destruction of nests, eggs, or young by storms wind, floods, predators, accidents, and desertion of the parents. Considerable data are available in this connection with birds.

Nest failures in birds are most frequent early in the nesting proceeds : 2.4 per cent per day during nest-building, 2.2 per cent per day during incubation, and 0 5 per cent per day while the young are in the nest. Location of the nest is a factor in the successful raising of young.

Life Tables

Species differ widely in the number of young produced each year, in the average age to which they live, and in their average rate of mortality. When sufficient facts about a species are known, a life table that tabulates the vital statistics of mortality and life expectancy for each age group in the population may be formulated (Pearl 1923). Age is usually represented by the subscript index and is some convenient fraction of a species mean life span such as a year or stage of development. The life table is set up on the basis of an initial cohort of 100, 1000 or 100,000 individuals; and the number living to the beginning of each successive age interval is symbolized as I_x.

Life tables are also useful for computing the average longevity of a population, for showing the age composition of a population, for indicating critical stages in the life cycle at which mortality is high, for showing differences between species. for showing the success of the same species in different biotopes, for furnishing information of value in game and fish exploitation (yield), and in control of pests.

Information for constructing life tables may be obtained from a knowledge of age at death of a random but adequate sample of the population; information on the age ratios of the living population, provided it is stablized; or from data on a single cohort, adequately identified, followed throughout its life span.

Sex Ratio and Mating Behaviour

The primary sex ratio at the time the eggs are fertilized should be approximately 50 males : 50 females in most species, although it has seldom been measured. This ratio may be displaced in one direction or the other by differential mortality of the two sexes during the period of growth, become manifest in the secondary sex ratio at the time of hatching or birth, and even more pronounced in the tertiary sex ratio of the adults (Mayr 1939). Recognition of sex and age in living animals is often difficult, although criteria have been worked out for many species.

Breeding Age

The age at which young animals first attain the ability for reproduction affects the reproductive capacity and rate of growth of populations. Planktonic entomostracans are sexually mature in a few days; insects, often n a few weeks. Among birds, a tropical sparrow is khown to reach full reproductive level in 6 to 8 months. Small non-tropical song birds commonly nest during spring and summer of the year following that in which they hatched but banding of nestling house wrens indicated that 12 to 18 per cent failed to do so until the second or third year. Upland game birds probably nest as yearlings; geese and wild turkeys do not nest until they are 2 years old; common terns, commonly only after 3 years to reach sexual maturity; turtles much longer.

Non-Breeding Populations

Although the breeding population of a particular mammal, bird, or other animal is only fraction of the total population of a species concerned with its reproductivity, there is often present in an area a substantial, though inconspicuous, non-breeding population that must be considered in any understanding of community dynamics studies have shown that non-breeding population especially of birds, consists principally of young that have been slow to reach sexual maturity, of surplus individuals of either sex in monogamous species, and of adults which, for one reason or another, have lacked reproductive vigour or have been unsuccessful in establishing proper breeding conditions.

Longevity and Mortality Rate

When protected in capacity, animals are capable of living surprisingly long periods. Definite physiological limits of life are characteristic of each species and are occasionally realized under natural conditions, but invariably the potential longevity of a species in many times greater than the mean longevity actually attained by wild populations. Protozoa and some coelenterates and flatworms are potentially immortal when they reproduce by fission.

Age Ratios

The life table gives the number or percentage of individuals in a broad or litter surviving to the next age level From such data, as well as by occasional direct observation, it is possible to determine the age structure of a population at any one time, the percentage of each age class in populations of adults having three different mean survival rates in all age classes. It is at once apparent that the number of age classes in a population is greater when survival rates are high than when they are low. It is also evident that there is less difference between number of individuals in succeeding age classes when survival rate is high than when it is low.

The exact age of the sexually mature adult is usually a best difficult to determine unless one can band or mark, th young when they first appear, or unless there are growth rings such as in the scales and otoliths of fish and in the shells o clams, or other criteria of age that can be used. Immatur animals are often distinguishable from adults, so that adul young ratios are usually obtainable and often yield importan information. The proportion of immatures to adults is highes at the end of the breeding season, and then usually declines unti the beginning of the next period of reproduction, because of th higher mortality rate of the young compared with that o adults.

Age ratio is of practical value in ·ild-life management. A low ratio of immatures to adults indicates a poor reproductiv season and should caution against excessive take or yield, as th population is declining.

Adaptation to Biotope

In the stabilized population of any species, whatever th number of eggs or young produced per pair of adults, the averag number of offspring reaching reproductive status can never b greater than two in sexual forms, which is the number re quired to replace the parents on their death. With each nev generation there is, therefore, a population turnover, wit newly born individuals replacing the adults that die. In stabllized population, the rate of increase of population throug the course of several productive cycles must equal the deat rate, so that the value of one factor is also a measur of the other. Either factor is indicative both of the rate o population turnover and of the intensity of environmenta resistance.

There is a general inverse relation between growth rate an longevity, T. However, if growth rate were dependent only o the longevity of the species, then r_mT would be a constant. Obvi ously this is not true. It appears that different growth rates ma correlate with various intensities of environmental resistance i the different habitats occupied by different species. If ther

untrained observers (Carpenter 1940b). The number of rodents may be in the hundreds or thousands per hectare, of insects in the millions. Outbreaks are known to have occurred since the beginning of recorded history of Europe, Africa and North America, especially in insects and rodents, plagues of European meadow voles were recorded 18 times in France between 1792 and 1931.

The cause and control of plagues or outbreaks have concerned man since civilization begin outbreaks are commonly associated either with favourable weather and food conditions so that survival and reproduction is increased, or with a disturbance in predator-prey relations. Rapid increase in population size may thus occur with almost any kind of animal in any habitat. Irruptions of the bean clam occurred several times between 1894 and 1955 in the intertidal zone at La Jolla, California. The abrupt decline of the last outbreak in 1951-52 was the result of an epizootic associated with a minute unicellular organism of uncertain identity, found in the tissues of the clam.

Catastrophes

Catastrophes occur at more or less widely spaced intervals and bring marked depressions in the population level of a species. Information from observers indicated that low populations were widespread in eastern North America both in this species and in many other song and game birds during most of these years. There is considerable evidence that these conspicuous variations occur in abundance.

Among mammals, fluctuations in the population of the common hare in Denmark have been correlated with the varying effects of summer rainfall, spring temperatures, and the number of days of frost during the winter.

Catastrophes may occur with practically any type of animal life. The severe winter of 1917-18, for instance, produced a marked reduction in the number of many species of marine invertebrates in the region of Woods Hole, Mass. In seeking correlations between catastrophes and weather or other environmental conditions, one needs to determine the critical period in its life cycle during which the species is most vulnerable, and the weather conditions coincident with that period that exert greatest effect.

Cycles

Populations are said to be cyclic when they alternatively irrupt and subside in a more or less uniform manner between high and low levels of density.

Reality of Cycles

To designate fluctuations as cyclic implies considerable regularity for them. What ecologists call cycles are really oscillations because both phase and amplitude are inconstant. The variability that is evident, however, especially in phase, is less than is to be expected by chance, and reasonably accurate predictions can be made of the course of future variations in population size.

The short-term cycle is commonly 3, 4 or 5 years long, although it may be as short as 2 years, or as long as 6 years. The coefficient of variation standard deviation divided by the mean, for different species having the short cycle, varies from 30 to 50 per cent, and is of the same order of magnitude for the longer cycle. It is of interest that by drawing numbered cards from a well-shuffled deck or rolling dice or plotting random numbers short and long cycles may be obtained of about the same relative lengths and variation coefficients as animal population cycles.

There has been considerable controversy about the true nature of cycles that is, whether they are caused by a master factor or factors or by the random impact of a large number of factors acting more or less independently of each other. Before oscillations in the size of natural populations are considered cycles, they should be tested statistically for randomness. Even then, the cause of cyclic populations can be expected always to contain some random components inextricably mixed with any systematic ones. The reality of cycles may also be tested by the amount of synchrony that they exhibit in different localities and on different species. The presence of such synchrony would indicate the presence of broad coordinating influences or mechanisms rather than haphazard effects of many random factors.

Intrinsic Causes

Intrinsic factor affecting animal cycles are those inherent in the populations themselves and in the interrelations of different species. They are primarily biotic in nature. A number of general the ories of possible intrinsic mechanisms have been found inadequate, and still other theories have limited application. However, disease, predation, food and genetical or physiological explanations merit consideration.

Extrinsic Causes

Extrinsic factors that may affect cycles are largely abiotic and external to the populations, such as the weather. Weather is affected by variations in solar radiation.

Solar radiation may also affect animals and plants in other ways than through the weather. A little ozone in the atmosphere at the earth's surface is stimulating to animals, but a high amount is harmful. There is some experimental evidence that ionization of the air, that is, the conversion of neutral gas molecules into electrically charged ions, may affect the health and the vigour of animals. The height of the ionosphere and ozone layers above the earth is controlled by the intensity of solar radiation, and it is possible that cyclic variations in the height of these layers may affect organisms in ways that are little understood at the present time.

Solar radiation is not received in equal intensities in all parts of the world. Its intensity varies because of inclination of the earth's axis relative to the sun, differences in terrain, amount of cloudiness, and so forth. Solar radiation likewise does not have an identical effect on all species because of differences in their sensitivity to it and because critical periods in their life cycles come at different times of the year. These theories are conjectural at the present time but influence of extramundance forces on animal and plant life should not be dismissed until their potential effects are thoroughly explored.

Population Explosion of Man

The world population of man increased from only 2-3 million at the time of Christ to approximately 1.5 billion in 1960. It then began to explode, reaching 3.6 billion in 1970 and may exceed 6 billion by the year 2000. This irruption is due to reduction in mortality rates without compensating decrease in birth rates.

In his early history, man had to depend on solar energy for the production of both food and heat. With the discovery of life, he used the burning of wood to furnish heat when solar energy was deficient, to cook his food, and to expand his capacity to use machines for locomotion, farming and industry. This allowed his dispersal into parts of the world hitherto closed to him and the production of more extensive food supplies. The next step came with the mining of coal about eight centuries ago and production of petroleum just over one century ago. World consusption of energy from fossil has increased about 4 per cent per year during the last century. The increase in world population is doubtless a response in large part to this increase in energy resources, and if atomic energy becomes generally available, there will be no restraint on continued population expansion on this account.

Early man through development of superior intelligence and the use of tools soon eliminated predation as a significant mortality factor. With improved medical practice and expertise, many common diseases have been eliminated or at least greatly reduced in incidence or severity. Emigration has extended to nearly all parts of the world, although less extensively, into deserts, tundra, and thick tropical forests. The oceans have no resident human population, although they could conceivably be occupied if there were sufficient demand and engineering ingenuity.

Reproductivity as expressed both in fecundity and success of young in attaining maturity, generally decreases in lower animals as populations expand, but in many species the decrease, especially in fecundity, is insufficient to prevent the populations from reaching thresholds where starvation, disease or competitive struggling becomes disastrous. Such thresholds will be reached also by man unless a determined efforts is mede to reduced fecundity to a point where the population can be stabilized. Man is unique among animals in that his intelligence has evolved to an extent that he can anticipate the future and plan for his welfare, although it remains to be demonstrated whether he has the will and the discipline to do so. It is encouraging that European man's birth rate in 1800 was about 35 per thousand adults per year, while by 1970 it had dropped to about 17 per thousand. This however, is not enough, and annual birth rates in the developing countries remain high, around 45 per thousand.

The average number of offspring per family in 10 European countries averages 2.7; in the United States it has been 3.8. A stable population could be established if the number were reduced to 2.3 per family. However, even when this number is reached, the population will continue to rise for a while. A high proportion of the present-day world population is under 15 years of age. As these young people mature, the size of the child-bearing fraction of the population will increase and more than offset the decrease in birth rate for at least one generation.

To check the population explosion, the approach to parenthood is to recognize that it is a privilege rather than a right. If voluntary family planning is inadequate, then stringest government regulation may be required. Continued population growth will lead only to disaster. Involved necessarily with deliberate control over population numbers is control over its genetic make up. Possibilities exist not only to eliminate factors from the germplasm that induce physical and mental defects but also to breed individuals or races with special characteristics that seem desirable. Man has control of his fate in more ways than one.

The population explosion of man is due to elimination of predation, emigration into new habitats, increased survival of young, reduction of disease, increased productivity of food – all related to man's increasing knowledge and technical skill in the exploitation of natural resources. Mankind is headed towards starvation, epidemics and competitive struggling, unless there is determined reduction in fecundity, a reorientation of social and industrial objectives and maintenance of environmental quality. Although man can make more efficient use of the space he has available, the world is still a finite body. Sears (1958) states : Man's physical body occupies space, somewhere between two and four cubic feet. At his present rate of increase it would be less than 700 years, say 22 generations – until there is standing room only, with eachspace of 3 by 2 feet or 6 square feet occupied. On this basis there is a room for exactly 14,646,000 people in each square mile. I have perhaps been over generous in estimating the per capita area but I did wish to leave space enough to permit each individual to reach in his pocket for the rent money when it becomes due. A little after this the hypothetical human population would weigh more than the planet.

5

Evolutionary Ecology

Background

We have been mostly concerned with the factors as to how organisms are adapted morphologically, physiologically and behaviourally to their environment. We are now going to discuss how these adaptations have evolved because of their survival value to the species. It is the consideration of these factors that constitute evolutionary ecology.

Evolutionary ecology is concerned not only with speciation and its accompanying niche segregation but also with the manner in which all ecological structure and function has evolved. Part of this evolution is biological in that species characteristics are transmitted genetically by means of chromosomes and genes, and part of it is cultural in that behaviour and social systems or ethics are transmitted through the training and learning experiences of the young. Human culture, including ethics and religion, has its a evolutionary roots in the behaviour and social systems of lower animals. Natural selection affects behaviour and social patterns as it does structure and function, to render the organisms more fit and able to survive in specific environments and to live together in particular relations with one another. In this chapter we will consider ecological niches and taxonomic species, how they have evolved together to give structure to the community, and how behavioural interrelations between individuals have evolved social patterns, ethics anc culture, culminating with man.

The ecological niche is a particular combination of physical factors (microhabitant) and biotic relations (role) required by a species for its life activities and continued existence in a community.

Hutchinson (1957) suggested treating the environmental variables, both physical and biotic, affecting a species population as a set of coordinates. The ranges of these variables within which the species can survive and reproduce then define as a dimensional abstract volume or hypervolume, which he terms the fundamental niche, If two species in a biotope overlap in their fundamental niches, only one species may survive in stable interaion in the volume overlap where they are in direct competition. If the two fundamental niches differ in part, the two species can survive together and coexist in the biotope. The realized niche of each species is, therefore, restricted to that portion of its fundamental niche to which it is better adapted. A community can thus be received as an assemblage of species in a biotope that have divided up the 'niche space' (the full range of environmental variables affecting those speci s in the community) so that they occupy niche volumes that differ, though they may overlap.

The restriction of a species to particular niche depends on its structural adaptations, physiological adjustments and tolerances, and behaviour patterns. At the same time, living in a particular niche brings increasing evolutionary perfections for living there rather than somewhere else. The variety of species living together in different niches gives ecosystem species diversity. The concepts of species and niche can hardly be considered one without the other, and both concepts need to be approached from the point of view of evolution and adaptation.

Behaviour Adjustments

Somehow an animal must get into a favourable niche out of the multitude of niches available to it. Doubtless, some animals accomplish this passively by the random dispersal of spores, eggs or larval stages, some of which by chance reach favourable locations, there to mature and survive. When immature stages of two or more species find a particular location favourable, that species which settles there first, in the greatest numbers, or grows faster is likely to acquire dominance over the area and force the other species to take a marginal position of disappear.

Higher animals with their more complex nervous systems, greater intelligence, and sense organs more highly evolved, are equipped to search actively for and recognize niches suitable to them either by sight, smell, contact or other means. For instance the intricate migratory behaviour of birds and other animals is such that they seek out nearly constant climatic environments throughout the year. Migratory behaviour may have evolved as the result of groups being unable to solve the problem of niche isolation locally.

The apparent ease and speed with which a new generation of individuals discriminates a hospitable niche means that they must in some way recognize the merits of that niche by definite characteristics of it that may be in the nature of sign stimuli Such characteristics are usually prominent, though they need not necessarily be the most essential features of the niche.

Probably most animals exercise a deliberate, although not necessarily conscious, evaluation process in choosing one niche from those available. This has been tested experimentally by exposing the animals to gradients of environmental factors, either in the laboratory or field. A variety of apparatus and procedures is available for such tests. Usually there is a coincidence of the species experimentally ascertained preferendum and its natural preferendum.

Where two species with similar niche requirements come into competition, one species must possess better adaptation to it if it gains full possession of the niche to the exclusion of the other species. Preadaptations or possession of suitable adaptations also appears a necessary prerequisite for a species to invade a new niche or habitat and successfully displace another species already occupying it, as frequently occurs. The search for such microadaptations as might give one species an advantage over another in competition for a particular niche, is a real challenge to ecological research.

Inheritance

The fact that all individuals of a species behave almost identically in many of their activities indicates that these behaviour patterns or instincts are in some form passed on or inherited from one generation to the next. Behaviour that is learned after birth is much more variable between individuals. Behaviour patterns are rooted in the structural arrangements of neurons and synapses and in the functions of hormones and enzymes. Once a stimulus is received a definite action results. Predisposition for a species behaviour patterns could well be inherited genetically through chromosomes and genes like any other structural or functional characteristics. These inherited behaviour patterns are doubtless subject to evolutionary development as much as are structural and functional adaptations; indeed, the one may have developed synchronously with the others.

There may be transmission of behaviour patterns to succeeding generations by tradition rather than by genetic mechanisms, that is, training of young, young imitating parents, conditioning or imprinting. It has been demonstrated that some parasitoid wasps lay their eggs only in the same kind of larvae as they themselves became conditioned to during their early growth and feeding. Young birds and other animals become imprinted to their own parents, to their own species, and perhaps to their proper niche, at critical stages in their development.

It is very probable that the manner in which niche segregation is passed on to succeeding generations is not the same in all species. We may believe, however, that most behaviour of animal has a genetic basis, but may become highly modified through practice, imitation and experience.

Interspecific Competition

Segregating Effect. Charles Darwin stated in The Origin of Species the case for interspecific competition as an instrument segregating species into different niches as follows (Crombie 1947) ;

1. The reproductive capacity of organisms is greater than the carrying capacity of the environment.
2. The range of an organism's tolerance of physical conditions and choice of food is limited.
3. The failure of an organism to survive, or be born at all, may be a result of the direct action of unfavourable habitat, predators, parasites or competitors.
4. When competition occurs, it is the severest between organisms with the most similar requirements.
5. In general, the closer the taxonomic relationship between them, the more similarity there is in needs and habits of species.
6. When new forms appear in a given locality, either by evolvement there or by nvasion after evolutionary divergence elsewhere, they will either eliminate or be eliminated by their nearest relatives if they competewith them, unless.
7. Each form becomes adapted to a different niche, in which case competition between them will cease and they may occur in proximity.

Evidence that interspecific competition is the most critical factor confining a species to this or that niche is available with the expansion of the species beyond the usual limits of its niche when this competition is removed. This expansion is often evident in geographic difference in the niche characteristics of a species. In Scotland, the mountain hare occurs at high elevations, the common hare at lower ones. In Ireland, which was isolated as an island before the common hare could reach it, the mountain hare occurs at both high and low elevations and is differentiated into distinct subspecies.

The general effect of interspecific competition is restriction of a population more closely within its optimum niche. Intraspecific competition exerts pressure impelling individuals to disperse into less favourable situations. The relative pressure exerted by these two forces determines whether at any particular time the species is contracting or expanding its habitat range. Because of these interacting forces, the total population of all species in a region is constant, conditioned by the food and other resources, although the population of individual species may depend on the number of competing species composing the fauna.

Interspecific competition is reduced or eliminated altogether when the combined requirements of all species are less than the supply of materials available. Species of land snails, insects, aquatic clams, and copepods sometimes occur together in considerable profusion but with little evidence of competition between them. Voles, during upswings in population, may become superabundant for periods of 2 or 3 years. During such periods several species of hawks and owls may feed in the same field without competing because there are more than enough voles to supply all. During downswings in the rodent population, however, competition does occur and some or all the predator species are forced to turn to the other prey.

When two species with similar niche requirements meet in competition under natural conditions, one of three things will generally happen:

1. One of the two species will be so much better adapted that it will spread rapidly through the range of the other and exterminate it.
2. One species will be better adapted to a portion of the range in which it will eliminate the other species, but the other species will be better adapted to the remainder of the range and will occupy it exclusively. Thus the two

species will occupy adjacent geographic regions with perhaps a zone of overlapping occupancy.

3. Each species will be better adapted to a different portion of the niche, to which it will become restricted, with this separation each species will then spread through the range of the other.

Predominance. It is characteristic of the taxonomic structure of communities that a few species furnish the greatest bulk of the population entirely out of proportion to the rest of the species. Thus in stream riffles, two species make up 85 per cent of the total riffles population and another two species constitute a similar percentage of the mud-bottom-pool populations. The abundant species are ordinarily of small size, herbivorous in their food habits and at the bottom of the food chains and pyramids. Predominance is better expressed in communities occurring in "harsh" environments, that is, in environments where relatively few species find conditions favourable than in those occurring in mesic ones that are favourable for many species.

Succession. Succession is the processes whereby a series of communities and habitats replace one another, beginning in an extreme habitat and ending in a mature climax ecosystem. With succession on land there is generally a progressive increase in height of the dominants, stratal differentiation of the vegetation, elaboration of internal microclimates, diversity of plant life-forms, biomass, and development of soil structure and fertility. Since the original habitat is bare of all life, at least in a primary succession, there is necessarily a progressive invasion of new species as each community develops in the sere. Species diversity tends to be greater in each successive stage, but the highest diversity index both for richness of species and equitability may sometimes occur in an intermediate or subclimax rather than the final climax stage. Successional trends may likewise vary between strata within the community.

Speciation and Cultural Evolution

Individuals within the species establish pattern of behaviour to which all adhere, so that the species functions in an efficient organized manner. Much of this behaviour is learned in each generation and thus constitutes a mores or ethics, the prototype of culture. The evolution of culture in human society is an outgrowth of these intra and interspecific interrelations in the behaviour of lower animals and developed concurrently with speciation of man.

Speciation is the process of evolutionary differentiation between populations, which may result in one older species becoming split into two or more new ones. Speciation also entails divergence of the new species into different niches or geographic areas. An understanding of the basic principles of speciation is therefore a prerequisite to an understanding of both local and geographic distribution of organisms.

Taxonomy

As the term is used by taxonomists, a population is a local aggregation of individuals that differs slightly, but characteristically, from other local aggregations of the same species. Geneticists define a population as a reproductive community sharing a common, characteristic pool. Every local population is different from every other one; but they are not easily distinguishable from one another, and therefore are given no formal nomenclature. What a population includes may vary from the entirety of a species to but a few individuals, according as the rate and extent to which individuals interdisperse between localities to make a common genepool.

Among bisexual forms, a species is a group of populations capable of successfully interbreeding and is reproductively isolated under natural conditions from other such populations, These criteria are often difficult to apply however, so reliance is commonly based on superficial or morphological characters. Species are usually morphologically distinct, but the distinguishing characteristics are sometimes barely discernible. Hybridization may occasionally occur between clearly defined species under captive or unnatural conditions, but does not occur with any significant frequency under natural conditions.

A subspecies is a geographically defined group of populations which differs in colour, size or some other taxonomic characteristic from other populations within the same species but nonetheless interbreeds with them freely, regularly and successfully, where their ranges come into contact. Subspecies are commonly distinguished by a trinomial nomenclature, although the desirability of this has been questioned. The taxonomic differences between subspecies are usually less pronounced than those between even closely related species but they are genetically fostered differences nonetheless. Race and variety are terms sometimes used as synonyms for subspecies, or for populations even less well defined than a subspecies. It is difficult to determine whether populations that differ phenotypically are subspecies or species when their ranges do not verge for in the absence of opportunity for them to commingle, it remains uncertain whether interbreeding would or could occur. The assignment of taxonomic rank under these circumstances is dependent upon considerable subjective inference. Boundaries between adjacent subspecies are frequently arbitrary in that they fit differences between the populations for some characters but not others. Differences between subspecies or populations are often correlated with differences in topography, soil, climate, or vegetation, and at least in some instances, appear to be adaptive to these differences in the environment.

Two species the distributional ranges of which do not overlap (that is, there is geographic isolation between them) are said to be allopatric. Two species are said to be sympatric when their ranges overlap, even though they may locally be ecologically segregated into different habitats; for example, the situation in which one species of snail is limited to floodplain and another species to upland habitats, or one species of rodent confined to the foothills and a closely related species to a higher zone on mountains only a short distance away. Populations should be considered sympatric if they occur within the common dispersal ranges of their young, so that there is at least possible a continuous and appreciable interbreeding and flow of genes among them.

Variation in Population Characteristics

Observable difference in structure, function and behaviour between individuals belonging to the same species are common. Actually, no two individuals, except perhaps identical twins, have exactly similar characteristics. It is the gradual accumulation of many small variations over many generations which eventually gives a population reproductive isolation and, consequently, species identity.

Non-heritable Variations

Not all variations of organisms are of direct significance in speciation. Non-heritable changes in body structures, functions and behaviour are common. If muscles are used continually and used intensively, they become thicker and stronger; if one kidney is removed, the other becomes hypertrophied; skin subjected to frequent rubbing or pressure thickens and becomes horny; and so on through a lengthy catalog. Animals progressively exposed to even more severe temperatures or lower oxygen concentrations will tolerate extreme conditions, which, had they been suddenly presented, would have been fatal. Insect larvae transferred to a new type of food often become so conditioned to it that they produce a strain that prefers food to other more usual food of the species. Chimney swifts in wilderness. Many phenotypic adaptations persist from generation to generation, either as similar responses made by each generation to constant environmental conditions; as the result of imitation of parents, conditioning of young or imprinting; or because the particular genes responsible for these characteristics have been sorted out from the general gene pool of the species. One would expect a change in behaviour or function, arisen in consequence or exposure to the new conditions, usually to presage the evolutional development of a new structure for natural selection cannot bring about the structural adaptation of perfection of an organ unless the organ is being used for the new purpose.

Polymorphism

When individuals of a population occur in more or less discrete groupings of colour phases, body sizes or other character variations that are not due to sex, age, caste, or continuous mutation, the population is said to display polymorphism. Polymorphism of a character arises in species when heterozygotes persist in an environment in the face of natural selection and homozygotes are reduced or eliminated. For instance, individuals with certain characteristics may be better adapted to environmental conditions during the spring, while individuals with other characteristics are superior in summer or autumn. This results in a mixture of types in the population more or less segregated by seasons. Likewise populations may vary in characteristic as adaptations to local habitat conditions. An increase in the frequency of a given characteristic appears to be a result of variations in the selective pressure of the environment, permitting individuals with certain characteristics of gene combinations to survive at one time or place; other individuals, at other times and in other places. In an environment that presents the same set of selective pressure year after, there commonly occurs a stability in the ratios of the different forms in which a character is manifested.

Mutations

Mutations may be the result of chemical changes in the individual gene or of chromosomal aberrations, the latter in the form of changes in the number of chromosomes (haploidy, polyploidy) or of arrangements of genes on the chromosomes (deletion, duplication, translocation, inversion). The rate at which any one gene mutates varies greatly from one kind of gene to another, but the average rate is of the order once in every 100,000 or 1,000,000 individuals. Between 0.4 and 10 per cent of the individuals in each generation may possess mutated genes. Other estimates of frequency of gene mutation are even higher.

Natural populations may, therefore, be well supplied with small. mutations of differing potential values to the organism. Mutations of different sorts apparently occur haphazardly and are not influenced by environmental conditions; only accidentally do they give special advantages to an organism. Adaptation of a species to a particular habitat or niche is effected through natural selection of the favourable mutations out of the many that occur.

The size of the population and the rate at which a particular gene mutates affect the odds that a mutation of that gene will become established in a population. However, if the gene mutates repeatedly in different individuals, the high mutation rate will greatly increase the number of individuals carrying the gene and increases the chances that the mutant character will become expressed in the population.

Natural Selection

Natural selection is a continual force exerted on each successive generation. Before natural selection can take place, however, there must be phenotypic variations between individuals from which selection can be made In order for these selected variations to have significance in speciation, they must be genetically fixed and heritable.

In most species, many more offspring are produced than can possibly survive. Because of this overproduction, there is competition between the offspring for the necessities of life, which, together with the strife between predator and prey and between organisms and their physical environment creates a struggle for existence.

There is differential survival in this struggle for existence because some individuals have structural, functional or behavioural variations that give them advantages over individuals lacking those variations. The superior genotypes will make a relatively large contribution to the gene pool of the next generation. The result of differential survival and differential reproduction is popularly known as the survival of the fittest.

The accumulation of favourable variations in a population brings the species generally to a better adaptation to the physical conditions of the habitat, avoidance of predation, more efficient physiological functioning, and new behaviour patterns. Natural selection favours those variations that are adaptive, and thereby fosters the continued existence and improved reproduction of species. These adaptations may never give a perfect adjustment of the organism to its environment, since the environment seldom remains constant for any length of time. If the population undergoing the changes is geographically isolated so that the favoured changes do not spread throughout the species, then differentiation of characters leading to speciation occurs.

Natural selection is relatively less effective in small populations than in large ones. Small populations may be locally restricted, come into conflict with few competing species, and experience only a favourable habitate. Because of the low selection pressure, chance combination of genes (genetic drift) may produce characters of little or no adaptive value that yet have a good chance to persist while really adaptive characters may be lost. In populations that are increasing rapidly in size, in the upswing of cycles, say there is little selection, and non-adaptive variations may persist as well as adaptive ones. As populations come to saturate habited niches and disperse into new or less favourable habitats, competition, predation and parasitism increase, and individuals become exposed to even more severe physical and climatic conditions. Natural selection then functions and characteristics that are adaptively advantageous will tend to persist while those less favourable or even harmful are eliminated.

Mutations upon which natural selection works are often recessive. They do not become fully expressed in the phenotype except when the individual is homozygous for the character.

Nevertheless they are important, and in stabilized populations tend to persist indefinitely in constant proportion to the dominant alleles. A recessive character will become more prominent in a population if that particular gene continues to mutate toward the recessive, or if the homozygous recessive phenotype has adaptations that give it selective advantages. In this latter case, natural selection may ultimately result in complete suppression of the dominant allele so that the hitherto recessive allele becomes permanently fixed in the population as the only gene for the character.

Since emergent species usually entrain adaptations to new environments, it would appear that those characteristics by which we distinguish species and higher taxonomic categories gnenerally are such as serve some useful purpose to the organisms either structurally, functionally, or in point of behaviour. It is often very difficult to determine a useful function for all distinctive characteristics of a species, yet one can seldom be positive that a seemingly minor character does not serve, say as a releaser for some critically necessary behaviour or is not vitally important in other unsuspected ways. However, not all characters that distinguish species or higher taxonomic categories are necessarily adaptive. Some characters originally adaptive may have lost their usefulness, although they persist in the organism. With natural selection no longer acting on them, they usually retrogress and may eventually disappear, as have skin pigmentation and eyes in many cave animals, for instance. Other characters may even have a slightly unfavourable value, but be closely linked genetically with selectively preferred favourable characters and thus continue in the organism.

Cultural Evolution

Webster's dictionary defines culture as the complex of distinctive, attainments, beliefs and traditions that constitute the background of racial, religious or social groups. The culture of a species includes its social or ethical as well as its intellectual and aesthetic attainments, especially as they are symbolized or expressed in many different ways. By ethics, reference is to definite behaviour patterns concerned with interrelations between individuals that have become establishedtfor the welfare of a species. Morals refer to the ethical relations and conduct of individuals. Animals have ethics and culture in so far as they have evolved characteristics behaviour and social interrelations but culture is more complex in the higher vertebrates, particularly in man correlated with the development of a higher degree of intelligence.

Different levels of social organization may be recognized:

1. Those individuals that simply tolerate each other's presence when crowded into a restricted space.
2. Those that show some integration so that the group responds more or less as a unit, and
3. Those that have established division of labour and definite social organization.

Division of labour started evolving very early, first with differentiation of somatic and reproductive cells in colonial protozoans, then differentiation of tissues in coelenterates, differentiation of organs in vermes separation of the sexes, specialization of castes in social insects and finally segregation of species into niches in the community. Among the social insects, for instance, ants and termites have for millions of years been building elaborate dwelling places, domesticating other species, raising vegetables, storing food, practising sanitation and living under elaborate social hierarchies – analogous to what man was to evolve independently and in a different manner later on. With division of labour, each individual of species contributes to the whole, and the welfare of the social unit or community on this cooperation. These cooperating units develop high efficiency in using natural resources, with the result that the constituent individuals survive and reproduce more successfully and the species is more likely to continue to exist,

In lower animals social systems are poorly developed if one can say they occur at all. Species multiply to the limit of their food or other resources except as held in check by predators weather, disease, or decreased fecundity. With development of competition between individuals in higher species there becomes established degrees of dominance and subordination expressed in social hierarchies territorial system, and niche segregation. These form a basis for more effective community cooperation, so that food resources become divided and utilized on a continuous self-sustaining basis.

Behaviour patterns, characteristic of species, are passed from generation to generation partly through genetic mechanisms and partly by imprinting, conditioning, or training of the young to behave, in particular w ays. For instance, in birds the capacity for fear is inherited, but the young need to learn in most cases what to fear, feeding is instinctive, but the young must learn through limitation of their parents and experience what is the proper food for the species, the ability to sing is inherited, but the young must hear the song of their species to perform; and so on in the lower invertebrates, behaviour appears relatively stereotyped and automatic, but in higher vertebrates, such as fish, birds, mammals, and finally man behaviour becomes more complex with learning relatively more important. This complexity allows for more variability between the behaviour of individuals and innovations to develop faster in the mores of the species.

There can be no doubt that evolution of social systems has taken place. Probably this evolution has not been by group selection of whole systems as such, as has been suggested (Wynne-Edwards 1962), but more likely by selection of those individuals that show capacity for particular kinds of inteir-relations and cooperative use of resources which result in their greater survival and reproductive success. This is cultural evolution often contrasted to biological evolution concerned chiefly with structure and function rather than behaviour. Behavioural patterns, social system and community interrelations can be traced phylogenetically through animal groups in the same manner as can structural characteristics. By the same process, human ethics, culture and religious beliefs can be traced from earlier beginnings, going all the way back to lower animals. In certain respects, evolution of social systems or more among animals is comparable to evolution of vegetative systems of physiognomies plants.

We have shown how competition is of prime importance in regulation of population size segregation of species into different niches and speciation. The competitive drive for advantages is also evident in modern civilization in the development of new techniques in industry, procedures in economics, and policies in government. Nations and ideologies compete in the same manner and for the same purpose. There is selection of those techniques, procedures, policies and ideologies that are most fit and rewarding to the individual. It is difficult to imagine how competition can ever be eliminated without the advancement of civilization coming to a standstill. Man, unlike animals, may through his intelligence deliberately change the course and rate of cultural evolution and through ill-advised action reduce fitness as often as he increases it. Anthropological and historical record shows that civilizations have risen and fallen depending on the vigour and wisdom of their activities. Over the course of time the healthier. more ingenious and more intelligent civilizations have persisted longer, with successive steps leading to the complex highly developed culture of the present day.

6

Geographical Ecology

Introduction

Through the study of geographical ecology we attempt to understand how organisms are distributed over the world and what forces brought about this distribution. The explanation of why a particular species has come to occupy a particular niche in a particular part of the world requires a knowledge of where it originated, how it dispersed, and how it evolved its present adjustments and characteristics. When populations invade new regions and environments and become geographically isolated, they commonly differentiate into new species. Each species thus becomes the product or visible expression of a particular combination of environmental factors, interactions, and locally, and as such the species is the practical taxonomic unit with which the ecologist must deal.

The general laws governing the distribution of animals can be reduced to three simple propositions; a species of animal will be found in any part of the earth having conditions suitable for its maintenance unless (1) its individuals have been unable to reach this region because of barriers; or (2), having reached it, the species has been unable to maintain itself because of inability to adapt to the region or to compete with the other forms; or (3), having arrived and survived, it has subsequently so evolved in the process of adaptation as to have become a species distinct from the original type.

In this lesson we will be concerned both with dispersal dynamics and with the distributional patterns of floral and faunal units and ecological communities that result.

Dispersal Dynamics

Dispersal is the spread or individuals away from their home-sites. Dispersal movements are usually slow, and cover relatively short distances in the life time of an individual. The cumulative result of short dispersals by successive generations, however, may become conspicuous in the course of years, decades, or centuries, when it constitutes range expansion of the species into a new habitat or area.

Under entirely natural conditions he rates of dispersal of all species, once a barrier is passed, would be similar, varying only with respective reproductive potentials, characteristic speeds of locomotion, and relative abilities to find unoccupied niches, overcome competition, acclimatize to new climatic conditions, and acquire new behaviour patterns.

Manner and Means of Dispersal

A common method of achieving dispersal is the broadcasting of enormous numbers of eggs, spores, encysted states, or young so that they scatter into a wide variety of places in a more or less random manner. Those that come by chance into suitable environments persist and become established; those that enter into unfavourable locations are destroyed or never develop. Broadcasting is a wasteful procedure; it is especially characteristic of aquatic species. The fresh water clam annually produces hundreds, perhaps thousands, of eggs. Only two fertilized eggs need mature that the two parents be numerically replaced when they die, and thus the population of the species be maintained at a constant level. Constrastingly, in those forms, such as birds and mammals that have developed a high degree of parental care, the number of eggs or young annually produced is commonly a half-dozen or less, and the off-spring exercise considerable discrimination in their choice of suitable habitat.

Although the dispersal of broadcast offspring is not under the control of the parents or the young, it is not often truly random. Water and wind currents and other agents of dispersal many channel eggs and spores in restricted directions. Such dispersal is described as passive conveyance. In streams, all agents of passive conveyance, except some other animals, direct dispersal downstream. Upstream dispersal must be the result of active locomotion.

Eggs of insects, snails, fish, or other aquatic organisms will sink unless they are buoyed up by currents, possess flotation mechanisms, or are attached to some floating object. Logs, masses of vegetation, and other debris are sometimes torn loose from the banks of rivers and float out to sea carrying the smaller animals attached to or trapped on them. It is estimated that may become attached to mammals, resident birds, or insects for more local dispersal. A species of Succinea snail, native only to St. Croix and Puerto Rico, was found alive in the feathers of a bobolink shot in Cuba. Ferrisia snails have been found attached to the wing covers of aquatic beetles and sphaeriid clams clamped on their legs. These insects occassionally fly from one body of water to another Crustacean eggs have also been found viable after passing through the digestive tract of different kinds of birds. Plant Seeds may be dispersed thousands of kilometers in the digestive tracts of birds. Fruits become conspicuous, abundant, and nutritious, thus attracting birds at the proper time to aid in their seed dispersal.

Some dispersal movements are determined by the manner in which the animals respond to environmental factors. Such directed movements, or taxes, may induce the dispersal of animals upstream positive to current rather than downstream.

Trial and error is involved to a considerable extent in the dispersal movements of animals. If, initially, an individual progress into an unfavourable habitat, it may be able to withdraw and proceed in a new direction. This may continue many times until by chance it discovers an area that is favourable. Trial and error movements are manifested in all groups of animals. In higher types, and even in many lower types, the individual may learn by experience and reduce the number of false trials that it makes. Thus initially random movements may eventually evolve into directed behaviour patterns.

Barriers

Dispersal continues until a barrier that makes further movement difficult or prevents successful colonization is encountered. Barriers are of different sorts and are classified as physiographic, climatic, and biotic. The difficulty of surmounting or bypassing any type of barrier varies greatly among species. River valleys, for instance, may be barriers to animals frequenting mountains ; but to lowland species, river valleys are important dispersal highways.

To fresh-water organisms in river and lake habitats, intervening land masses are usually effective barriers. The headwaters of different river systems may lie only a few kilometers apart, yet contain quite different species because the continuous water route down to the mouth of one river and back up to the other may be a distanae of several hundreds or thousands of kilometers. A waterfall may be a barrier for all non-flying aquatic species, and even riffles or swift water may prevent upstream dispersal of pond or lake species. The salt water intervening adjacent rivers flowing into the sea is a barrier to most fresh-water forms that might otherwise invade one river from the other.

Terrestrial organisms are hemmed in by a great variety of barriers. The oceans constitute the major physiographic barriers since they separate the faunas of the several continents and isolate islands from each other. Lakes are not effective barriers because they can be readily skirted, but wide rivers coursing long distances between banks of dense vegetation as does the Amazon river through the tropical rain forests of South America, may limit the range of forest mammals, butterflies, flightless beetles, land snails, and even birds. The Grand Canyon of the Colorado River separates the ranges of the Kaibab and Abert squirrels and several other species, even in country that is semi-arid and open. Mountains are sometimes considered barriers to lowland species. This is only true if the change in climate and vegetation that such barriers produce are unfavourable to the species.

Deserts are important climatic barrier since they are hot and dry. Temperature affects animals directly, since species have definite limits of tolerance, comfort, and efficiency. Precipitation is important because it controls the type of vegetation that occurs in a region, and the animals adapted to that of vegetation. A low relative humidity may directly limit the dispersal of moist-skinned species. Excessive solar radiation limits some species to forest habitats, excluding them from open country; without recourse to shade, such animals would experience overheating and critical loss of water from the body. Short photoperiods may limit distribution north-ward during the winter, especially if low temperature obtains, and may influence southward distribution of normal animals in the summer. The length of the season between spring and autumn killing frosts, or between dates when its limits of temperature tolerance are reached in the spring and autumn, may determine whether or not a species can complete its life cycle at a given latitude. A corollary consideration is whether the accumulation of heat is sufficient to furnish cold-blooded animals and plants sufficient energy for growth and reproduction.

Biotic barriers consist of changes in vegetation, food, competitors, and predators. The adaptations and behaviour patterns of many animals fit them to niches in specific types of vegetation; should the previously amenable vegetation change, the animals may have great difficulty in adapting to it. Tree squirrels, for instance, are replaced by ground squirrels in prairies and deserts. Most animals have a great adaptibility to food, so food is not so limiting a factor. Some insects, however, such as certain aphids, are narrowly limited to particular species of plants as a source of food. Where their food plant is not present, they cannot exist.

Function of competition between the two species Either species can live in th range of the other, but in the North, the house were usually wins in competition for territory and nest-sites. In the South, the Bewick's wren is more successful. Predators, such as the great horned owl and the swifter hawks, tend to confine the bob-white to forest-edge habitat, where it is less vulnerable to attack. Trypanosome parasites carried by the tsetse fly are effective barriers against successful introduction of domestic ungulates into certain parts of Africa, and the rabbits introduced into Australia have limited the range and greatly reduced the abundance of several species of native marsupial.

Dispersal of Young

The dispersal of a species is primarily accomplished in the immature stages. This is obviously true of eggs and spores, but banding and marking studies have shown that among the higher animals birds and mammals it is also the young which dispersed the species. Once a bird has reached sexual maturity and nested. It has strong tendencies to return to the same area in following years. The distribution of young birds is not random, however, as they tend to return to the general vicinity of their birthplaces rather than uniformly over the range of the species. Thus only 0.5 per cent of 557 adult house wrens recovered a year after banding nested farther than 33 km (2 miles) from the site where they had nested the year of banding, but 15 per cent of the 182 birds banded as nestlings were recovered at greater distances, the longest of which were 32 km (20 miles) 56 km (35 miles), 80 km (50 miles) and 1120 km (700 miles). Dispersal distances for young of other species are proportionally comparable. Apparently the young are at a disadvantage in the competition for nesting territories with adult that have previously nested in the area, hence are the ones forced to go elsewhere.

Of small mammals, it is characteristic that once an individual has selected a homesite, it rarely leaves it for another. It was observed that in the months following the time at which they had been captured, marked, and released, 95 per cent of 133 adult woodland white-footed mice resumed habitation within 183 m (200 yd) of the site of capture.

Rate of Dispersal

If dispersal from birthplace were typically limited to one direction, then a simple mean of the distances to which the young disperse before they breed would give the dispersal rate per generation. It is the case, however, that dispersal proceeds peripherally in all available directions and may extend to surprising distances.

The area exposed to invasion and the average time required to saturate that area increase proportionally as the square of the linear distance (d^2) from the centre, since the total area within which the individual could settle is πd^2. A number or equations have been suggested for computing mean dispersal distance. Consider the data on house wren nestling recovery distances, presented above. Excluding the truly extraordinary distance of 1120 km, and observing that only about 93 per cent of young female wrens nest when they are 1 year of age, we compute an annual dispersal rate of approximately 8 km (5 miles) for this species. The mean dispersal distance of one group of 154 young woodland white-footed mice (Burt 1940) is about 176 m (192 yd). However, mice bron in the spring mature sexually very quickly and breed in late summer or autumn, so the annual dispersal rate must be somewhat greater than this figure indicates. The annual dispersal of marked snails resulting from random movements along a linear bank of uniform habitat averaged only 5.5 m (6 yd) (Goodhart 1962).

Our dispersal rate data so far have described the outward diffusion of a local population through an area already occupied by the species. Once it has surmounted a barrier the dispersal rate of a species into a suitable area previously unoccupied by it should be faster. The European starling was introduced into North America about 1890. From a central locus around New York City it spread at an accelerating pace until, in 1940, it had become established over 6,500,000 km^2 (2,500,000 mi^2), a mean rate of about 130,000 km^2 (50,000 mi^2) per year. With amelioration of the climate in Finland during recent years, the lapwing spread northward between 1899 and 1954 at a mead annual rate of 7 km (4.3 miles); the roe deer, between 1850 1945, at a mean annual rate of about 95 km (58 miles). The Norway rat invaded southwestern Georgia and virtually displaced the previously established black rat at a rate of about 430 km² (167 mi²) per year (Eckh 1954). By way of contrast, it has taken the fresh-water amphipod Gammarus pulex the last 6000 years to disperse across 12 river systems from southern England into Scotland.

Causes of Dispersal

The reproductive rate of any species is so great that if all offspring survived the world would be overrun with that species within relatively few generations. Because species produce a surplus of young in most years, there is continuous pressure on individuals to move into all suitable niches, and to seek out new areas in which to settle. The impact of large numbers of individuals struggling for survival is described as population pressure, and is doubtless the most potent force inducing dispersal. It should be recognized, however, that population pressure is not uniformly constant year after year. When because of poor breeding conditions or catastrophe there is a reduction in the over-all population of a species, that decimated species may withdraw into its optimum habitat and be less put upon to exploit new or less desirable areas. In years favourable to the production of large surpluses of young, a species will often be found in less favorable habitats, even regions it would not otherwise occupy at all. The broadcasting of eggs or offspring, or the passive conveyance of the them to other regions, varies directly with the size of population producing them, and is hence as much an expression of population pressure as the active search for new areas engaged in by individuals under their own locomotion. In addition to relieving population pressure, dispersal is advantageous to the species for faster spread of mutations through the entire population and for occupation of new or depopulated areas. The tendency of disperse may even have been selected for in evolution so that it has become innate in young animals,

Animals cannot disperse successfully, if at all, into new areas to the characteristics of which they are not structurally, functionally, and behaviourally adapted. If an areas, which has excluded a species, changes so that the species is then adaptive to it, that species can successfully invade. The American robin, song sparrow, chestnut-sided warbler, house wren, and prarie horned lark have invaded Georgia only in recent years as the logging of forests, initation of early seral, grassy, and shurbby stages, and extensive general cultivation of the land have produced habitats meeting the requirements of these birds. Likewise, if an area remains unchanged but the species acquires new structural, functional, or behavioural traits by which it can adapt to that area, the species can invade the area. If the food supply fails, homesites or vegetation are destroyed, or a pernicious change in climate occurs there animals may be forced to leave an area to which they were well adapted and disperse, more or less temporarily, into an area to which they are less well adapted. The snowy owl, for instance, depends heavily on lemmings and mice for food in its usual range, the Arctic tundra. In correlation with the cyclic decline of the lemming population there, the owls invade our northern states.

Dispersal Pathways

An understanding of how animal and plant groups have dispersed over the face of the earth involves the controversy as to whether all the continental land masses were at one time intimately connected and if so, how long ago they drifted apart to their present locations. Separated continents need at least temporary bridges if there is to be dispersal and interchange of fauna between them.

Continental Drift Theory

This theory postulates that throughout the Paleozoic and much of the Mesozoic the presently distributed continents were grouped into two great land masses : a northern Laurasia was separated from a southern Gondowana by the vast sea of Tethys, although narrow connections between the two might have occurred for short periods. In the Jurassic of Lower Cretaceous the land masses fragmented, and the fragments subsequently drifted apart Laurasia is supposed to have split into North America, Greenland, Europe, and most of Asia; Gondwana, into South America, Africa, Arabia, Madagascar, India, Australia, and Antractica. This theory is suggested by the shapes of the continents that could conceivably be fitted together; by the characteristics of the Atlantic Ocean basin that make it appear to have been formed by a rifting apart of land masses; by new paleomagnetic studies; by the similar geological stratigraphy shown by characteristic assemblages of rocks and similarity in their ages, and by invertebrate fossils in Africa and South America; by some similarities of present-day fauna and flora at the same latitudes on different continents; and by the difficulty tropical species have of dispersing over arctic land bridges.

Evidence in favour of the continental drift theory has been increasing in recent years in fact, drift is apparently still going on. Deep sea drilling shows that sea floor has been spreding outward from the mid-Atlantic ridge at a uniform rate of 2 cm yr. However, the continents have certainly been separated thoughout most of Tertiary time and perhaps also the latter part of the Mesozoic so that dispersal of modern organisms over the world must have taken place in other ways.

Land Bridges

An alternative theory is that the continents have been situated in their present positions throughout geological time. They have changed their boundaries with rise and fall of ocean levels and with upheaval and subsidence of land masses, but there has been no significant lateral drifting. This theory was generally held to be correct by most geologists and biogeographers up untill the last few years. Regardless of the ancient history, the continents have been largely separated during the last 50 to 100 million years, and movements of animal and plant groups during this latter period have been most important in determining their present distribution.

The separated continents, however, have been connected by at least temporary land bridges at various times during the period. A land bridge now connects North and South America at the isthmus of Panama, and Eurasia and Africa are connected by the isthmus of Suez. One of the best known and generally accepted land bridges of the past, but not now in existence, connected. Asia and North America cross the Bering Strait and Bering Sea. Possibly bridges or extensive archipelagos occurred in tne Atlantic Ocean area in the past that permitted dispersal of organisms not possible at the present time, but concepts of this sort are highly hypothetical at the present time.

Land bridges may exist for only a short time in the geological sense, but they serve as important dispersal routes for those land animals and plants able to cross them. When a land bridge allows free passage of most animals and plants in either or both directions, it is called a corridor. If the land bridge is narrow, has an unfavourable climate, a lack of suitable niches, or too many competitors, it is called a filter—only a few specially adapted species are able to pass over it. A sweepstakes route of dispersal is one over which only a few species pass, more or less by chance. Because of its general unfavorableness, it is generally only one-way. This island-hopping pan-tropical dispersal of organisms from southeastern Asia through the south Pacific has been accomplished by relatively few species; the more distant the island, the fewer the species that have reached it.

A glance at a globe quickly show that the continental land masses are concentrated mostly in the northern hemisphere. There are three broad land extensions southward below the equator : the Malay Peninsula, East Indies, and Australia; Africa; and South America. The southern hemisphere otherwise consists largely of vast expanses of oceans. On a land route, North America intervenes South America and Eurasia. Eurasia is the largest continent, is central to all the others, and always has had a great diversity of climate and terrain. The size, arrangement, and positions of the continents are of importance to interpretation of the past evolution and dispersal of animals.

Considerable evidence indicates that some large groups of animals, notably mammals, first evolved in Eurasia and then spread to other parts of the world. North America was a less important centre of origin and dispersal. Warm, moist, uniform climates prevailed in the early Carboniferous, Jurassic, mid-Cretaceous and Eocene. During periods of continental emergence, climates on the northern continents changed from moist and warm to more arid and cold. Periods of aridity and glaciation are known to have occurred in the northern continents during the Permian, at the end of the Tirassic, at the beginning and end of the Cretaceous, and during the Pleistocene. Animals adaptively limited to moist, warm environments were restricted to tropical regions or dispersed outward into the southern land extensions. New more advanced animal types adapted to new conditions in the north appeared. Monotremes entered Australia at an early date. Marsupials probably originated in Eurasia and dispersed into Australia and South America during the late Mesozoic or early Tertiary, although there is no fossil evidence for these supposition. An oppossum still occurs in North America, but no marsupial is now present in Eurasia. Marsupials did not get into Africa, but are present in South America and have adaptively radiated into a variety of forms is Australia. The predominance of marsupials in Australia is probably due to their coming over at an early date on a land bridge on their chance success in surmounting the sweepstakes route bridge was an important route of dispersal between Asia and North America considerably more study is required before we will satisfactorily understand how this was accomplished by various kinds of animals.

Centers of Origin

The tracing of dispersal routes presupposes a starting point where the taxonomic group, whatever its size, first evolved. These starting points are called centers or origin. Various criteria for determining centers of origin have been suggested, but caution must be exercised in applying them. Of the many criteria proposed, the following two are especially important, although neither one is infalliable.

1. Location of greatest differentiation of the type or the greatest variety of endemic races, species, and general including primitive forms or fossils. The elder a group is and the longer it has occurred at a particular location, the more chance it has had to radiate into different habitats, become isolated, and evolve into new varieties. However, a shift of climate or a drastic change in physiography may render an original locality unhabitable and the group moves elsewhere. Also, a group losing its vitality may contract its range into some area other than the one in which it originated.

2. Continuity and convergence of lines of dispersal. Dispersal from a centre radiates in all directions in which conditions are favourable and until insur-mountable barriers are encountered. Lines of dispersal may be readily distinguished where one or more taxonomic characters can be traced from primitive or generalized types through more and more specialized types the greater the distance involved. However, once a species filters through a break in a barrier and invades an extensive, new, and favourable habitat, there may be increased evolution of new types, and a secondary dispersal centre formed. This has happened repeatedly in the geological past, so there is often difficulty in distinguishing which centre is the original one for a group.

Island Dispersal

Dispersal of animals from continents to islands and from one island to another presents special problems. Many of the world's prominent islands occur on the continental self and are separated from the mainland only by shallow seas. At times of land emergence, as when glaciers look up quantities of water as snow and ice enough to lower the level of the seas, these islands becomes connected by land bridges to the mainland, and dispersal of forms occurs. The British Isles have been thus connected to Europe; Japan to Korea and Siberia; Sumatra, Java, and Borneo to Malaya; New Guinea and Tasmania to Australia; and Newfoundland to Labrador. On the other hand, the Bermudas, Azores, Hawaiian and other small Pacific island, and possibly New Zealand, the West Indies, and Madagascar, could not have mainland connections and thus must have received their present faunas by some means other than overland dispersal.

Islands adjacent to continents, unless very small or long separated, generally have faunas similar to that on the nearby mainland. Oceanic island are more difficult to colonize, however, and often have unique unbalanced faunas or chance assemblages of species. Mammals, amphibians,and fresh-water forms are often absent or scarce. Flying birds, bats and insects blown in by strong winds, and small invertebrate forms readily transported on rafts are better represented.

Since islands are colonized with difficulty, the number of species on them is generally small compared with the mainland, and interspecific competition is reduced. A single established genus or family on islands free of close competitions may adaptively radiate into new niches to form a variety of species or races. Animals on island are often smaller in size than their relatives on the mainland, for instance ponies on the Shetland Islands, Key deer on the Florida Keys, and the gray fox on Catalina Island. Islands are generally too small to support adequate populations of large animals requiring extenive home ranges. At the same time, there may be selective advantages operating in evolution for becoming smaller. Because of lack of competitors and predators, primitive animals isolated on islands may survive long after their relatives on the mainland have perished, as has, for instance, the reptilian Sphenodon on New Zealand. Confinement of a species to a limited range permits extensive inbreeding so that the population becomes more homozygous in its various genetic traits, hence much less adaptable to new situations than are larger heterozygouspopulations Traits that would be eliminated by predation pressure on mainland sometimes becomes established in populations on islands. All these conditions render oceanic island species liable to extinction by invasion of mainland forms, and renders them impotent to reinvade the mainland.

There is continual invasion of individuals of the same or new species into island; at the sametime there is continual extinction of individuals and species. A biotic equilibrium is reached only when the two rates are equal. The relation between rates of invasion and extinction varies with the distance from the mainland. Successful colonization after invasion is more likely the higher the ratio of birth rate to mortality rate The number of established species commonly varies.

A continuous land connection occured between North and South American in late Cretaceous time. Ancient types of mammals, birds reptiles, amphibians, insects and other groups got into South America over this land bridge from the north and differentiated into distinct families and other toxonomic categoeis. During most of the Tertiary, no land bridge existed between the two continents; although there were scattered islands separated by relatively narrow water gaps which some groups, particularly birds, may have been able to use. The land connection now in existence was apparently formed in the late Pliocene or early Pleistocene.

Previous to late Pliocene there were about 29 families of land mammals confined to South America and 27 confined to North America; the two continents had not more than one or two families in common. During the Pleistocene, after the land bridge had been in existence for some time, 22 families were represented on both continents; 7 had dispersed from South to North America, 14 from North to South America, one is of uncertain origin. Some families have become extinct. The present faunas of the two continents contain 14 families common, 15 families found only in South America, and 9 families found in North America. Thus there has been considerable dispersal over the land bridge in both directions, but not a complete exchange or unification of faunas (Simpson 1940).

The modern fauna of North America is thus derived principally from (1) Eurasia, (2) South America, and (3) by autochthonous development. By autochthonous we refer to species evolved from very old indigenous types that may or may not be represented by related forms on other continents. The proportion of any local fauna that is derived from one or the other of these sources varies with each geographic area and with each group of animals. Northward on the continent and in the western mountains, the Eurasian bird element is strongly represented. The South American element becomes greater southward, especially in the lowlands of California, Mexico and Central America. Many Eurasian forms have dispersed through North America into South America, but no modern forms, at least, of South American origin have been able to invade Asia through North America.

Migration

Migration, like dispersal, involves movements and the invasion of new areas. Migration, as here defined, differs from dispersal in that it is a periodic movement back and fourth between two areas. In contrast, dispersal is a one-way outside movement. Migratory invasions of areas are temporary and repetitive, but invasions resulting from dispersal may be permanent.

Migration is best known in birds, as an invasion of breeding area alternating with an invasion of wintering area, annually. Representatives of other groups of animals also migrate, particularly mammals, fish, and insects. Migration may be classified as annual, diurnal, or metamorphic. Annual and diurnal cycles are correlated with the two most pronounced time cycles in the physical environment. Metamorphic migrations are movements from one habitat to another in different stages of an animal's life cycles.

Annual Migrations

Annual migrations may involve a change of latitude, or altitude, or be more local in extent.

Latitudinal migrations may traverse only a few kilometers or may extend almost from pole to pole. In terms of their occurrance in an area bird species are described as permanent residents species represented in an area throughout the year even though, some individuals migrate; summer residents, species present only during the warmer part of the year, which includes a breeding season that may extend from early spring to late autumn; winter visitors, species present only during the winter or non-breeding period; transients, species present only during migration periods, neither breeding nor wintering in the area; and accidentals, species that are rare or irregular in occurrence. The bird population in most communities reaches peaks during the vernal and autumnal aspects, as transients arrive and temporarily swell populations. The autumnal peak usually the greater because adult populations are incremented by the larger number of young birds hatched during the breedings season Latitudinal bird migrations also occur in the southern hemisphere but are less conspicuous because of the small continental land masses there and small populations of birds involved.

Altitudinal migrations are movements of no more than a few kilometers up and down the slopes of mountains.

By descending to lower altitudes in the autumn, an organism obtains some of the same benefits secured by those species that undertake latitudinal migrations – less snow, higher temperatures, and more food. Birds restricted to alpine habitats in the summer are common winter residents of lowland areas. Some of the larger mammals, such as the mule deer and the American wapiti have very regular migration habits in respect to herding, timing of movements, and migration routes. They move to the high alpine meadows soon after the vegetation renews its growth in spring, and come back down to the valleys in the time avoid the deep winter snows of the higher slopes.

Local migrations do not necessarily involve a change of latitude or altitude and are often quite limited in distance covered. However, in tropical grasslands and savanas, where wet and dry seasons bring great changes in available water, vegetation, and food, there is a great exodus of both mammals and birds during the dry season and an influx during the rainy season.

The Atlantic salmon, after reaching sexual maturity, may ascend fresh-water streams in subsequent years to spawn and return each time to the sea. Many deep water fishes spawn annually in shallow waters and then return to deep water again. Turtles come onto the land to lay their eggs, snakes disperse from their winter dens with the advent of warm weather in the spring; tree frogs go to small pools to mate and spawn; and resident birds move onto their breeding territories.

Insects perform regular migrations both into hibernation and out of hibernation. In the autumn, forest species migrate downward and may be found in peak numbers first in the shrubs, then in the herbs, then in their hibernacula in the ground. Many insects of the forest-edge, meadows and agricultural crops also hibernate in the forest, usually a few meters in from the south-exposed edge, where they derive some heat from the winter sun. These insects usually migrate into the forest in the same stratum, herb or shrub, in which they occur during the summer, then downward into the soil. These flights into the forest occur with declining temperatures and sometimes include spectacularly large numbers of individuals. As they come out of hibernation in the spring the direction of movement the insects take is just the reverse that taken in the autumn, that is, upward into their proper stratum, then horizontally outward into open country.

Daily Migrations

Ascent of plankton toward the surface at night and descent to greater depths during the day occur both in the sea and in lakes. The lake-dwelling culicid larva chaoborus lies on the bottom during the day, but becomes pelagic at night. Snails, slugs, and millipedes in the deciduous forest lie quiescent under logs or litter during the day, but come out at night to crawl around on the forest floor or even climb up on the vegetation to a height of perhaps a few meters. Although these excursions are restricted in range, they are more or less regular and periodic, and may be thought of as migrations.

Metamorphic Migrations

Acquatic larvae and naiads of several orders of insects eventually change into adults that leave the male aquatic habitat and become aerial. The adult stage is often short-lived. The eggs are deposited in water, or the immature stages return again to water to begin the cycle over again. The length of the cycle may be part or the whole of a year, or a longer period. The seventeen-year cicada is a well known insect whose nymhps spend 17 years underground feeding on juices from the roots of trees. The adults appear above ground in large numbers in late May, mate lay their eggs on twigs of various trees, and then disappear, all in a few weeks. The eggs hatch in about 6 weeks, and the nymphs drop to the ground and bury themselves for another long period of years.

The anadromous Pacific salmon ascends freshwater streams but once to spawn. The breeders die; it is their nff-spring that return to the ocean to develop for a period of years before they make the migration. Many adults return to spawn in the same stream or even in the same headwaters where they were hatched. The Pacific salmon apparently uses the sun as a compass mechanism to find its way from the ocean to the stream, then swims upstream against the current, rejecting all odors until it arrives at the home tributory to which it was initially conditioned. The two species of catadromous eels, of the western hemisphere and of Europe, migrate to the open sea in the region northeast of the West Indies in order to reproduce. The immature stage of the eel not the adults, returns to the two continents.

Ecesis

Dispersal or migration of individuals into new areas is without great ecological significance unless those individuals become established and can build up significant populations. The process by which organisms become established in new areas is called ecesis. Ecesis will occur if a species disperses into a habitat favorable to it and can secure its proper niche or become adjutsed to a new niche, new competitors, predators, parasites, and disease organisms Ecesis, to go to completion, requires first the establishment of individuals in an area, then the growth of the population that they form, and finally or simultaneously, the maturing of community structure with the invasion of many other species

Temporary ecesis is the rule with migrant species, as the periodic change of location is normal in their life behaviour. Ecesis as range expansion is not surely permanent until the species demonstrates that it can survive critical years in weather cycles. For instance, northward dispersal of tropical forests is thwarted by frosts that occur only at rare intervals. Insects may continuously expand their ranges for a period of years, only to be forced back hundreds of kilometers by a severe drought or cold spell.

The total community is an aggregation of many species. With seasonal onset of favorable weather or when a new habitat becomes available with progressive development of substratum, microclimate, and vegetation, the coordinated invasion and ecesis of these many species also follows the sigmoid or logistic curve. When, however, the new habitat is fully developed and ready for colonization by a group of species, as for instance islands the lower self-accelerating portion of the sigmoid curve may be eliminated and only the upper of self-inhibiting portion expressed Furthermore, colonization curves for insects arriving on small mangrove islands in the Florida Keys, after removal of the original fauna by fumigation, show that a larger number of species invade than can become established Extinction of less aggressive or more poorly adapted species results from various causes, including competition and predation. The asymptote of the curve is reached when the species present establish interactions between them so as to form a closed or stabilized community that discourages further invasion. Thus the process of growth or ecesis is much the same whether it is at the level of the individval cell, organism, species population, or community.

With succession, a series of communities become established one after another. This is best represented by a sequence of sigmoid curves, each indicating a particular stage in the sere. As a new community invades, the previously established one disappears, probably at a rate indicated by a reverse sigmoid curve. Ordinarily a larger number of distinct plant communities can be recognized in a particular sere than animal communities. The ecesis of a single animal community may overlap several plant stages. Thus the deciduous forest biociation begins its development on the Mississippi River floodplain with the establishment of the first trees in the cottonwood-willow associes, continues to develop through the sugarberry-sweet gum, oak-hickory associes, and into the climax oak-tuliptree association. However as the climax reaches its full dominance and all remnants of earlier plant stages are eliminated, the decrease in diversity of plants is often correlate with a decrease in species diversity of animal. The ecesis of distinct animal communities does not correspond, therefore, with ecesis of distinct plant communities, but the ecesis of animal communities is correlated only with the development of distinct types of vegetation. A background of understanding of the processes of dispersal and ecesis and the problems encountered by species in their invasion and establishment in different parts of the world is desirable for proper evaluation of present-day distribution of organisms and the distributional units that are recognized.

Distributional Units

Only a few groups of animals are cosmopolitan or nearly so, in having a worldwide distribution, notably cyprinid fish, frogs of the genus, Rana, colubrid snakes, passerine birds, rodents, and man. Most organisms are restricted in distribution because of barriers of one sort or another or because of past history of origin and dispersal. An understanding of their present-day distribution rakes us into biogeography, climatology, and paleoecology. Major units of distribution are the zoological regions, areas defined largely by the past and present relations of the continents to each other. Each region may then be subdivided into faunal or ecological units, depending on the criteria used.

Zoological Regions

Zoological regions have long been separated one from another by water, mountain, or desert barriers so that they have evolved distinctive and characteristic orders and families of animals.

The Australo-Popuan region has been isolated the longest and is marked by the absence of most groups of placental mammals, although some have been introduced. The original mammalian fauna, derived from Eurasia when Australia and New Guinea (Papua) connected to that continent by a land bridge, consisted of monotremes and marsupials. Murid rats and mice and bats invaded later and in a different manner. It is of interest that woodpeckers among birds, true frogs among amphibians, and fresh-water fish are absent or scarce.

The Neotrophical region contains the largest number of unique endemic forms of birds and mammals perhaps because it is the farthest by a land route from the oriental tropics where the major taxonomic groups probably originated and because the region is almost completely surrounded by water. The occurence of many related forms in South America, Africa, and Australia, and the absence or poor representation of these forms. Even as fossils, in North America and Eurasia may be explained either by the continental drift theory or by connections in the geological past with a much warmer Antarctica continent. For instance, the ratite bird, rhea, and the lungfish, Lepidosiren, occur only in South America; the ostrich and the lungfish Protopterus only in Africa; and the emu and cassowary birds and the lungfish Epiceratodus only in Australia and adjacent islands.

The Ethiopean region is noted especially for giraffes, antelopes, zebras, elephants, and other ungulates that wander around in large herds; by the large hippopotamus and rhinoceros; by a number of cats and dogs; and by the primate lemurs, babcons, macaques, chimpanzees, and gorillas. There are several endemic families of birds and other forms. Absent are deer and bears among mammals and salamanders and tree frogs among amphibians.

The fauna of the Oriental regions shows relationship to that of the Ethiopean region in several respects. India was connected with Africa at various times.

Faunal Groups or Elements

In all the systems above described, faunal distribution has been analyzed in terms of geographic areas, and the chief criterion for determining the limits of an area has been that of distinctiveness between the fauna of different areas. The fauna or flora is analyzed in terms of groups of species with similar centers of origin, dispersal routes, geological histories, and habitat preferences. The objectives are to learn the place where and time at which these groups of species originated, how the groups came to occupy a particular part of the world, became adapted to live in new environments, and how they evolved into the present-day living species. A type of vegetation, life zone, or biotic province may contain species from several different faunal elements that have come to live together.

Biomes

A still different approach for the study of distribution is the ecological one, involving the recognition and classification of natural communities. A number of different ecological systems have been suggested but we will here be concerned only with the biome concept. The systems described earlier in this section are zoogeographical, in their attention to centers of origin, dispersal routes, and evolution. The biome system emphasizes community dynamics and environmental relations. This does not mean that zoogeographers do not consider the relations of climate and terrain in controlling dispersal and evolution, nor that ecologists are not concerned with the geological history of the forms with which they deal, but the viewpoints and objective are different.

Basic to the understanding of the biome system is the recognition of communities composed of characteristic combinations of animal and plant species, of successional relations between communities and that succession in all local habitats eventually converge into a climax community pattern, the most important ecological characteristic of geographic area.

A biome is a biotic community, characterized by distinctiveness in life-forms of the important climax species On land, the most important climax species are usually plant dominants that occur in distinctive vegetation and landscape types; in the ocean, the important organisms that define biomes are usually the predominant animals which are sometimes also dominants.

Seral communities are developmental stages and are consequently considered part of the biome. The animal and plant constituents of seral stages are, however, more widely distributed than are species belonging to climax communities, since the habitats in which they belong are more nearly alike in different parts of the world than are the habitats that contain the climax. Seral species are not generally useful, therefore, to defining the limits of biomes. The majority of animal constituents of the climax community however, are characteristic only of the climax vegetation or habitat and therefore of restricted distribution.

The principal biomes of the world, or biome-types of some investigators are the following :

Terrestrial	**Marine**
Temperate deciduous forest	Oceanic plankton and nekton
Coniferous forest	Balanoid-gastro-
Woodland	pod-thallophyte
Temperate shrubland	Pelecypod-annelid
Tundra	Coral reef
Grassland	
Desert	
Tropical savenna	
Tropical forest	

7

Marine Ecology and Oceanographic Environment

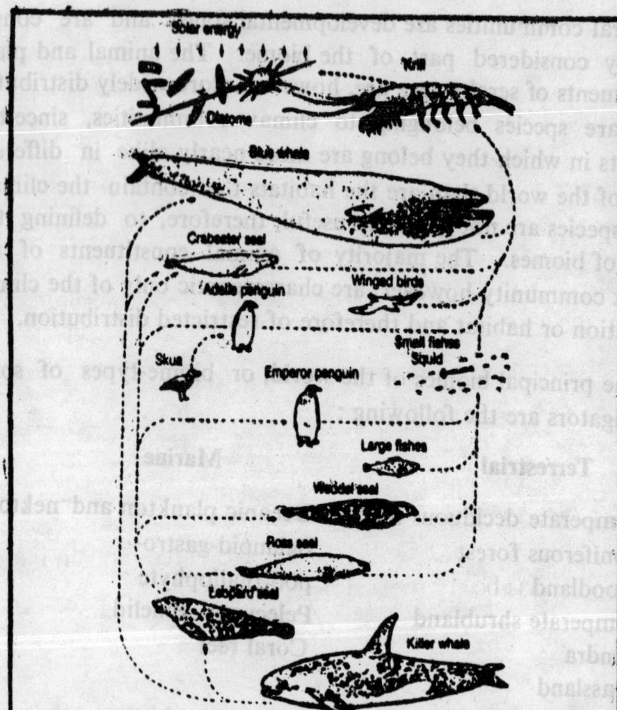

Food cycle of marine animals in the Antarctic (Murphy 1962 : from *Scientific American*).

Introduction

The geographic distribution of organisms in the sea depends on their responses to currents, temperatures, and physical barriers; local distribution is affected by waves and tides type of bottom, salinity, and depth. Marine ecology is concerned with environmental factors and problems of organic adjustments quite different from those in fresh water. Animals are relatively more conspicuous than plants. Succession is less evident, but such ecological processes represented by chemical cycles, co-operation and disoperation, food chains, productivity, population, dynamics, niche segregation, speciation, and dispersal are fully as import nts as on land.

Distinct self-contained community units are more difficult to recognize in the sea than on land because of the apparently greater interrelation of benethic species and the free movement with circulating currents of plankton and nekton. Plankton is everywhere a basic link in food chains, but the general distribution and importance of plankton species in the sea is no more remarkable than that of soil organisms in terrestrial biomes. To consider the entire ocean community as a single biome, as has been suggested by some investigators, is stretching the concept beyond its usefulness. Since we identify biomes by differences in the life-forms, and functional adjustments of the conspicuous dominant or predominant organisms, we may properly recognize biomes that occur in the open ocean, on eroding rocky shores, on muddy and sandy beaches, and composing the coral reefs and atolls. Each of these biomes may be subdivided by the taxonomic composition of the predominant organisms into secondary communities equivalent to the biociations that we have recognized on land. Much of the early literature on marine communities has been reviewed by Gislen.

Habitat

The marine biocycle is considered to have benthic (bottom) and pelagic (open water) divisions, the littoral zone of the ocean shore extends between the limits of high and low tides, the sublittoral zone covers the continental shelf to a depth of about

200 m. the approximate depth at which maximum wave action produces any effect. The average depth of the ocean is about 3800 m. but oceanic trenches (hadal zone) extend much deeper, the Marianas Trench in the Pacific Ocean to approximately 11 600 m. The neritic biochore is above the continental shelf and is commonly 16-240 km (10-150 ml) wide. The oceanic biochore is subdivided vertically with the boundary between the epi-and mesopelagic zones, depending on the extent of effective light penetration.

Tides

The level of water in the ocean rises and falls usually twice each day or at an interval of 12 hours and 26 minutes. In some parts of the world the tides are less regular or there may be but one daily. Flood-tide is the period in which the level is rising and covering more and more of the shoreline; ebb-tide is the period in which the water level is less than a meter, but the change may be much more than this on the shore, depending on its configuration, the Bay of Fundy opens broadly to the sea and tapers to a narrow head landward, and tides may be 6 to 10 or even 15 m. On the other hand, when bodies of water have only a relatively narrow connection with the sea, as does the Gulf of Mexico with the Atlantic, the range in water level is less than 30 cm. Even lakes have seiches, but they are hardly perceptible except in the larger lakes, where it may amount to a few centimeters.

Tides are caused by the attraction of the moon and, to a lesser extent, the sun. When the sun's attraction is added to that of the moon, as occurs twice each month at the time of full moon and new moon, the fluctuations of the tides are unusually high and unusually low, these are called spring tides. When the tidal influences of sun and moon are opposed, as happens twice each month, the tides have the least amount of flow and ebb and are called neap tides.

Tides have their greatest effect on animals on the seashore, because of the associated pounding of waves and the alternative submergence in water and exposure to the air. However, the organisms appear well adjusted to this rhythmic submergence and exposure (Flattely and Watson 1922, Korringa 1947). For instance, as the stones on which the chiton occurs become exposed, the animals react positively to gravity and negatively to strong light, and move downward. They travel at maximum speed while the stone is still moist and become aggregated on the damp lower sides of the stones. When the stone again becomes immersed by the returning tide, the animals lose their geotactic orientation, and, since, illumination becomes more or less equal on all sides of the stones, they move about at random until they reach the upper surfaces again.

Substratum

The pounding action of waves on rocky shore may have tremendous force, estimated in one instance at 15,000 kg/cm^2, animals occupying exposed rocky shores in the surf belt must be strongly protected and firmly attached (Flattely and Walton 1922). The conical shaped limpets present a minimum of surface to the waves. Large depressions in the rock retain water at ebb tide to form tidal pools and thus may contain the more delicate species because of the protection they afford. The various sea weeds absorb some of the wave shock for the animals living with them. The shape, form and size of corals, sponges, and other colonial type are affected by the amount of wave action to which the animals are exposed, a shall of the Mytilus mussel may weigh 58 g where the animal is exposed to a heavy surf but only 6.5 g in more quiet waters.

Sandy beaches occur only where the force of waves is reduced by being spread over a more gentle slope. Even here, especially during stroms the sand makes a very unstable substratum and not many animals except mollusks and some of the echinoderms can keep from being smothered or buried. Mud buttoms occur only in relatively quiet waters. Burrows made in mud hold their form better than in sand, so large populations of animals can occur in mud.

Organic deposits are either calcareous or siliceous, the former being derived from the shells of foraminiferans, small pelagic mollusks, or flagellate coccolithophorids, and the latter from skeletal material of diatoms and radiolarian protozoans.

Pressure

There is an enormous increase in the pressure of water upon the bodies of animals at great depths, and this may effect the vertical distribution of such species as crabs and mussels (Menzies and wilson 1961). This is not, however, an important limiting factor in the vertical distriburion of animals in general, as internal pressures closely counterbalance external pressures and life is known to exist at the greatest depths. Adjustments of internal pressures are not so rapid, however, to prevent injury in many species that are dredged at great depths and quickly hauled to the surface. Furthermore, individual species have different limits of pressure tolerance.

Temperature and Currents

The temperatures of surface waters vary between the freezing point (- 1.9°C) in the tropics. Seasonal variations are small in polar and tropical waters but somewhat greater in the temperate zones.

Temperature varies with depth, more so in the tropics than elsewhere. At 60°N lattitude in the Atlantic Ocean, the mean temperature of the warmest and coldest months beneath the surface is about 10°C. On the equator the temperature beneath the surface is approximately 26°C, at 200m 13°C, at 100m 4.51C, and at 2000m 2.3°C, (Ekman 1943). In temperate regions, seasonal thermocline develops near the surface in summer and is destroyed in autumn and winter when vertical mixing creates a layer of relatively uniform temperature in the upper 20 to 300m.

Currents moving toward the poles from the equator consist of warm water, and currents moving in the opposite direction of cold water. Surface currents make wide circular movements in opposite directions in the northern and southern hemisphere.

Light

Much of the solar radiation is reflected from the surface of absorved in the upper layers. Even in the clearest waters and at maximum radiation; the red, orange and ultraviolet are absorbed in the first 20m. Green, yellow, and blue wavelengths penetrate further, depending on the water colour. When the sun is not at the zenith, light penetration is reduced, and the maximum penetration in the winter at higher latitudes is much less than during the summer.

Salinity

The salinity of sea water varies from place to place depending largely on the amount that it is diluted by the yellow of fresh water from rivers or melting glaciers or the amount that it is concentrated by evaporation. The Red Sea, for instance, has a salinity of 40% (40 g of dry salts in 1000 g of seawater) while in some polar seas the salinity is less than 30%. The average salinity of the oceans as a whole is commonly given as 35%, of which the chloride ion constitutes about 19%, and the sodium ion little over 10%. The various major salts occur nearly everywhere in definite and constant proportions. As one would expect, the pH of seawater is high, averaging about 8, there is some similarity in relative proportions and concentrations of the variations in seawater and in the blood or body fluids of many invertebrate organisms. This may indicate that the sea is the habitat in which living forms first evolved.

The contrast in salinity between seawater (35,000 ppm) and fresh water 15-660 ppm) requires important difference in physiological adjustment of organisms to occupy these two habitats. The problem is one of osmotic regulation (Block 1951). Most marine invertebrates are poikilosmotic in that they are nearly isotonic with seawater; they are highly permeable to water; and gain or lose water according to concentration of the medium. A few marine segmented worms, flatworms, and crabs and all marine fish and mammals have at least some internal osmotic regulation and tend to be homoisomotic. All except the elasmobranch fishes maintain body fluids hypotonic to seawater in various ways. The skin has decreased its permeability to the free movement of water back and forth, the necessary water is obtained by swallowing, surplus salts are secreted outside the body, especially through the gills, and there is a general decreased function or atrophy of watersecreting organs such as the kidneys. The practical absence of insects and amphibians from the sea is largely due to their inability to secrete salts outwardly. The high asmotic concentration found in elasmobranchs is the result of huge quantities of urea retained in the body tissues and fluids.

These salts keep cycling through the ecosystem, but additions to the supply come continually from the land, being washed into the oceans by the rivers. Neritic waters are especially fertile and support a great mass and variety of animal life because of this land drainage and the pattern of water circulation on the continental shelf. Biological productivity decreases progressively from shallow waters over the continental shelf. to deeper waters, to the open ocean, but it also high over offshore banks and in areas of upwelling substantial amounts of nitrogen salts are also swept out of the air by precipitation, and there is nitrogen-fixation by bacteria.

Oxygen

The oxygen supply of seawater comes by diffusion from the air at the surface and from photosynthesis of green plants down to the compensation point. It is continuously used at all depths in respiration of animals and plants and in the decomposition of organic matter.

The oxygen content of seawater (Hedgpeth 1957) is seldom limiting for the occurance of animals, except in the deeper waters of the brackish Black and Caspian Seas, where it is practically absent. Oxygen concentration is especially high on shores where there is splashing of waves. Surface waters of the Atlantic Ocean commonly have 4.5 to 7.5 cc/litre and abyssal regions may run 5 cc/litre. Oxygen is somewhat less abundant in the Pacific and Indian oceans. Oxygen may be reduced to lower concentrations between 100 and 1500 m, because of its use in animal respiration and in decomposition, than at lesser depths, where there is photosynthesis, or at greater depths, where the abundance of animals is greatly diminished.

Marine animals have a variety of mechanism and adoptations for respiration (Flattely and Walton 1922). Greatest difficulties occur in shore animals at low tide when they are exposed to the air, but the need for oxygen at this time is decreased in many forms by curtailment of activity. Some crabs, barnacles, snails, and fish have become almost amphibious in being capable of respiring in air, although at reduced rates as well as in water. Pune mud bottoms may present anaerobic conditions a short distance below the surface, bud mud bottoms mixed with sand contain an abundant and diversified fauna.

PLANKTON

Composition

The plankton of the sea includes a great variety of forms, even more than in fresh water (Biglow 1926, Hardy 1956), Rotifers, however, are uncommon in marine plankton and cladocerans are much less important.

The nannoplankton consists mostly of protozoans, algae bacteria, fungi, and viruses. The bacteria are largely periphy.ic, in that they are attached to the surfaces of floating plants, animals, and to particles of detritus. Very few occur freely suspended in the water (Harvey 1955). Bacteria, including nitrifying forms, occur at all depths but are especially abundant in or close to the bottom. They are generally more numerous in the winter than in the summer Nitrogen-fixing organisms are chiefly blue-green algae and possibly yeasts (Oppenheimer 1962).

The green phytoplankton is composed primarily of diatoms, dinoflagellates, and small unarmored flagellates, but several other kinds of algae are present and occasionally important. The dinoflagellates Noctiluca and Ceratium are luminescent and in some regions may give a glow at night to the entire sea. Bioluminescence is not limited to these organisms, however, but occurs also in various forms of bacteria radiolarians, sponges, coelenterates, ctenophores, nemertineans, worms crustaceans, brittlestars, mollusks, balanoglossids, tunicates, and fish (Harvey 1952).

The most important groups of protozoan zoo-plankton, other than the green flagellates which are usually considered with the phytoplankton, are the rhizopod Foraminifera, the actinopod Radiolaria, and the ciliate tintinnids. They may be enormously abundant at times.

Abundance

The actual abundance of plankton varies greatly from place to place and from one season to the next. Smaller species tend to be more numerous than larger ones. The mean annual abundance of diatoms is commonly in the tens of thousands per litre and for shorter periods the year algal blooms may increase the population to hundreds of thousands of cell per litre (Ricketts and Calvin 1948), zooplankton is, however, much less abundant. Large populations of zooplankton generally follow large populations of phytoplankton and by their grazing maintain the standing crop of phytoplankton at the size suited to prevailing conditions (Nielsen 1958).

The total net zooplankton (cc per unit volume of water) is some 50 times more numerous in the neritic coastal waters of the Atlantic coast of North America than in the Sargasso Sea (Grice and Hart 1962). The abundance of plankton is generally higher in cold than in warm ocean waters correlated with the greater amount of phosphate present in colder water (Harvey 1955). Cold-water plankton tends to be of larger individuals size. There is generally in most taxonomic groups, however, a lower variety of species in cold than in warm waters. Annual productivity is also less in cold waters correlated with fewer generations per year. Productivity of zooplankton varies between 14 and 58 per cent of the net primary productivity.

Nekton

Mollusks, fishes, birds, and mammals make up the nekton of the sea. Mollusks are represented by the squids; fish by the sharks, flying fish herrings, meackerels as many others, including numerous varieties of small species; and mammals, by the seals, porpoises, and whales. The distribution of fish is irregular, but in general they occur more abundantly in neritic waters than in the open ocean. Likewise, they are much more numerous in the epipelagic than in lower strata. Most pelagic fish, except sharks, possess a swimbladder useful for maintaining hydrostatic equilibrium at the depth where there they occur; those fish that lack one are commonly bottom forms (Marshall 1954). In arctic waters fish are less abundant, and mammals relatively more important, than is the cease farther south.

Birds, like many other marine animals, are more numerous in the neritic biochore than in the open ocean. In the ocean far from land occur only penguins, albatrosses, hearwaters, and petrels, and even these species become more common shoreward. Other marine species in nerittic water are tropic-birds, pelicans, gannets, boobies, cormorants, frigate-birds, ducks, gulls, terns skimmeers, auks, and murres. These marine birds may spend many days or weeks feeding and travelling over the water, but all must search out some shore, cliff, or isolated islands on which to nest. Here they sometimes concentrate in enormous numbers during the nesting season because of the limiting number of suitable nesting locations available.

Benthos

Benthos is of much greater variety in marine than in fresh-water habitats. These animals are very abundant in the littoral zone and decrease in numbers with depth until only scattered individuals are found in the deep ocean trenches (Saunders et al 1965). Benthos consists of sessile forms, the sponges, barnacles, mussles, oysters, crinolds, corals, hydroids, bryozoans and some worms; creeping forms, such as crabs, lobsters, certain copepods, amphipods, other crustacoans, many protozoans, snails, echinoderms, some bivalves, and some fishes; and burrowing forms, including must clams, worms, and some crustaceans. Sessile and creeping forms are often grouped as ebifauna are burrowing forms as infauna. Epifauna in the littoral zone decreases in variety towards the poles since it is subjected composition of infauna remains about the same.

In food habsts, different species may be grouped as seston eaters, mostly sessile or semi-sessile forms that capture suspended food particles; sluggish motile forms collecting detritus or food particles from the bottom surface; sluggish forms that extract food particles from bottom material that they swalloos; carnivores, and scavengers.

Oceanic Plankton and Nekton Biome

This biome is characterized by the predominance of organisms possessing life-forms adopted to keep them afloat. Plankton and nekton predominate, although the deep-sea benthos may also be considered as belonging to this biome. Seasonal aspection may bring drastic changes in species composition, especially in plankton. Dominance, in the sense used for terrestrial communities, probably does not exist, except possibly in the Sargasso Sea, where the floating vegetation establishes the habitat. The ecosystem is self-contained, however, since energy is derived from the sun and nutrient material continues to recircuate with little or no dependence on terrestrial resources.

The sargassum community of the Atlantic Ocean is of special interest, the floating Sargassum alga accumulates and is held within a limited area by circular ocean currents. This plant belong to the intertidal zone of the Caribean islands but in turn lose in large amounts along with attached animals during the hurricane season. It continues to grow thereafter, but does not reproduce. The fauna that it contains to a truly littoral one rather than pelagic, but because the algae to cold and ice erosion, but the species accumulates in fresh amounts as fast as old plants die, the animals reproduce and maintain a continued existence far from any shore.

Food Chains

As in fresh-water and terrestrial communities, bacteria in the sea are largely responsible for the final decomposition of excreta and dead bodies to make their essential nutrients available for reabsorption by the green phytoplankton.

Nitrogen and phosphorus are least concentrated near the surface of the ocean, since this is the stratum in which they are most rapidly absorbed by the photoplankton. Excreta and dead organisms sink during the process of decomposition, so nitrogen regeneration is most evident at depths of 500 to 1500 m. The total non-living organic matter, either dissolved or in the form of particulate matter, is generally much larger than the biomass of living organisms at any one time. Organic matter makes up less than one-third of the total particulate matter, and all of it sinks slowly through the water, requiring months or even years to reach the bottom. Much of the particulate organic matter is still undissolved and accumulates on the ocean bottom. Numerous species of invertebrates depend on it for food, especially on the bacteria that it contains. The deep-sea fishes feed on these invertebrates or are carnivorous on other fish. Many of them have very wide mouths, distensible stomachs, and formidable teeth. In addition to these food coactions, it s also likely that many deep-sea fish and larger invertebrates undergo vertical migrations, so that they obtain food by preying on living organims at more moderate depths. Much that is known about the life histories of these deep benthic species has been summarized by Marshall.

Especially fertile regions of the open ocean occur when there is deep mixing of waters by turbulence and upwelling. Vertical water currents bring nutrients up to the surface from intermediate depths where they had accumulated. Prominent regions of more or less permanent upwelling occur around the Antarctic continent, off the coasts of California, Peru, and Somali, and Off the west coasts of both north and south Africa. In many areas, very small phytoplankton and bacteria form the first link in the food chain, which goes next to the microzooplankton before reaching the carnivorous net zooplankton of copepods, euphasiids, and other crustacean forms. Most animals depending on small organisms and organic detritus for food have various filter-feeding mechanisms for straining the material out of the water They do not actively search and catch individual items through directed actions. Invertebrate animals may also be able to absorb some essential salts and dissolved organic compounds to build their skeletal structures and for general metabolism, but there is considerable controversy on this point.

The baleen or whalebone whales (Mysticeti) are toothless but possess large plates in their mouths that strain out the plankton (especially copepods, euphausiaceans, mysidaceans) that they use as food. Only occasionally are small fish as food. Only occasionally are small fish or other invertebrates ingested. Some whales reach tremendous proportions, and the differential in size between these animals and their food is one of the most remarkable in the animal kingdom. Much more common is the feeding on plankton by squids, the young stages of most fishes, and such adult fishesas sardine, an chovy, meuaden, herring, and mackeral. The menhaden is unique in having such fine-mesh gill-rakers that can feed extensively on diatoms, which because of their smaller size cannot be readily secured by other large marine animals.

Balanoid-Gastropod-Thallophyte Biome

This community extends from high-to below low tide levels on rocky shores (Lewis 1964). Benthic animals and attached algal plants are conspicuous and important, the benthos is mostly epifauna, as too hard to permit development of extensive infauna. When the tide is out the organisms are subjected to drying, the occasional inflow of fresh water, higher temperatures, and greater light intensities. Organisms avoid desiccation when the tide is out by variously crawling under stones or thick algal growths, closing thick shells or operculae, retreating into crevices, or secreting a mucous seal. Most organisms are also faced with the pounding action of waves., various holdfast or anchoring devices have developed, and many species protect their more delicate structures with a hard shell. The adoptations for life on the seashore are many and varied (Yonge 1949). The plankton and nekton associated with the benthos include many species not common to the oceanic biome,

Zonation

Vertical zonation of species on rocky shores is usually conspicuous although individual species may extend widely into adjacent areas (Hewatt 1937, Yonge 1949, Stephenson 1949 Southward 1958).

Begining on the landward side there is a supralittoral zone mostly above the action of tides and inhabited as much by land as by marine animals. This is followed seaward by a supmalittoral or Litttorin fringe which is wetted by the highest tides and by the splashing of waves. Because of the presence of either Myxophyceae or lichens, this zone is often discovered; commonly black. The fringe is especially characterized by large numbers of small snails and sometimes isopods.

Next below this fringe is the midlittoral or balanoid zone. It is strictly intertidal being covered and uncovered every day, and is occupied characteristically by acorn barnacles. This zone is often divided into subzones with the barnacles predominate in the upper portion, while polybhaets, colonial hydroids, or other forms are relatively more important in the lower part. The subzonation of algae as often also well marked.

The lowest zone over exposed and then only at extreme low tides, is called the infralittoral fringe. It is a transition area. The entire area between extreme high and low tides, including the mide-littoral fringes, when considered as a unit may be referred to as littoral, eulittoral, or tidal zone to distinguish it from the infralittoral or sublittoral zone that extends from the lowest of low tides to the edge of the continental shelf.

Littoral Zone

Brown algae from thick masses and give protection to those animals that find shelter in or under them. A fauna of copepods, ostracods, water mites, and young littorinids inhabit these seaweeds. In England, the numbers of individual per 100 g of seaweed vary from about 44 on brown algae to over 13,000 on lichens (Colman 1940).

The animal life on rocky shores is varied and luxuriant. Several species of ocorn barnacles, snails, marine impets, marine mussles, goose barnacles, sea anemones, chitons, sponges, hydroids, bryozoans, flatworms, annelids, isopods, crabs, sea urchins, starfishes, tunicates, and insects are present. Total abundance of animals may run into tens of thousands of individuals per square meter

Sublittoral Zone

Th s community in not subjected to exposure by tides or to the pounding of surf, but is affected considerably by wave action and the complete circulation of water. Animals move around somewhat more freely and there is less need for strong holdfast structures. Most organisms lack physiological tolerance for long exposure to the air and hence differ fundamentally in structure and mores from the community described above.

Laminarias or kelps are the largest of the brown algae and occur commonly in this community. They have root-like holdfast attached to the bottom and their stalks, which are often several meters long. bear leaf-like branches that float of the surface in the larger species. A long list of animals finds shelter and food in the kelp beds and especially in the protection of the holdfasts (Andrews 1945), Polychaet worms are particularly abundant in these holdfasts (Colman 1940). Filamentous red algae (Rhodophyceae) are also prominent

Abundant characteristic animals on the Pacific coast are sea urchins, sea cucumbers, starfishes snails, rock oyster, chitons, limpets scallops, mussels, nudibranches, barhacles, crabs, hermit crabs, hydroids, tunicates, shrims, and various fish. Distribution of fish species correlates strongly with the type of bottom or benthos that is present,

Tidal Pools

Seawater is often retained as depression or pools in the littoral zone and hence organism here are never completely exposed to the air. They are however, subject to high light intensity an increase in the temperature between tides (Klugh 1924). Tidal pools are usually richan both plant and animal life and some species are largely restricted to them. Red algae and kelps prefer the more shaded, cooler pools; the green algae and some of the smaller brown algae prenominate in the well-insolated pools. Animals of both the sublittoral and littoral zones are found here.

PELECYPOD ANNELID BIOME

Habitat

This biome develops on depositing sand and mud bottoms in contrast to the biome just described that occurs on eroding rocky shores. There is still a good deal of wave action over sandy bottoms. Fine sand particles shift about almost continuously, and animals have difficulty in preventing their burrows from collapsing. In general, the water over muddy shores is shallower, quieter, and warmer. The mud forms a soft, compact bottom, but is also easily moved or shifted around by storms and wave action. Animal burrows in mud are more permanent. Species tend to segregate depending on the amount of organic matter present. Shores of high mud content may be low in oxygen because of decaying organic matter, so animal populations tend to be largest and most varied in a mixture of mud and sand. Tidal currents are weaker, and change in the level of water is less pronounced on sand and muddy shores than on rocky ones.

Composition and Characteristics

Important plants in this biome are the marine eelgrass, which is seed plant and green algae, particularly the sea-lettuce, which grows in sheets either attached to the substratum or lying fragmented overlarge areas, and Enteromorpha, which grows in tuffs or tangles. Occuring on eelgrass and sea-lettuce may be several kinds of epiphytic algae. These plants form extensive stands and are important to animals for attachment, shelter, and food. Eelgrass was almost eliminated from the Atlantic coast in 1931-32, possibly because of a protozoan disease. This disturbance had a profoundly deleterious effect on the abundance of many animals, including the brant; a bird that depended on it almost exclusively for food, and on scallop and other coastal fisheries. Twenty years later there was evidence that eelgrass was recovering much of its former abundance (Cottam and Munro 1954).

Predominant animals are pelecypods, polychaere worms, particularly Arenicola and Nerels; starfishes, brittle-stars, sea cucumbers, crabs, amphipods, and snails. Populations may run to several thousands of individuals per squre meter. A variety of small fish occur here. Birds include sandpipers, plovers and herons.

The biome is worldwide in distribution, but the characteristic life-forms are represented by different species locally. Thus a number of secondary communities (Biociations) may be recognized (Petersen 1914, Jones 1950, Thorson in Hedgpeth 1957).

Many of the animal constituents in this biome are burrowing forms. The sub-stratum of mud and sand holds considerable water and when the tide is out on exposed flats pelecypods, worms and other animal constituents retract their fleshy argans into their burrows shells and remain in a water saturated environment (Hesse, Allee and Schmidt 1951). They are thus not exposed to the atmosphere with changes in the tide Furthermore, most forms are generally tolerant of low oxygen and high carbon dioxide concentrations. In order to maintain respiration when retracted in their underground burrows these, animals have long siphones, sometimes longer than their bodies or long tubes or canals that extent to the surface. Through these they maintain a circulation of water, often by means of special pumping organs.

Coral Reef Biome

Coral reef are formed by the accumulation of the calcareous skeletons of myriads of organisms. They extend from the sea bottom at depths of 46 m or rarely 74 m. to slightly above low-tide level. The best formation of coral reefs is confined to warm waters above 10 C, although individual species may extend into colder regions (Vaughan 1919, Wells in Hedgpeth 1957).

Predominant organisms involved are commonly the anthozoan stony corals and organ corals and the hydrozoan milliporids. Some reefs however are formed principally by foraminifera and still other by calcareous algae. All massive coral structures employ calcareous algae as cement. These algae not only thrive in the pounding surf on the windward side of the reef, but by their growth are able to repair damage to the reef caused by storms. Most typical reef-building animals are colonial and of shapes varying from closely compact, globose, or encrusting, to loosely branched or dendritic, depending in part on their exposure to wave action. Each polyp in a colony secretes its own calcareous skeleton and when it dies the next generation builds on top of the old so that the accumulation of a lime structure is fairly rapid (Yonge 1963).

The bright yellow or red colors of corals near enough to ther water's surface for adequate light penentration are the result of algae, the zooxanthellae, which are either embedded in the body wall or free in the internal cavities. In addition, there are bands of green filamentous algae growing to a depth of 2 or 3 cm in the pores of the insert coral skeleton that may have a biomass 16 times that of the zooxanthelale. In their photosynthesis these algae may absorb dioxide and nutrients derived from animals or the coral reef and liberate oxygen of value to the animals (Odum and Odum 1965). Perhaps because of this symbiotic relationship which requires solar radiation, living corals are largely confined to the upper, shallower waters. Coral animals actively ingest zooplankton, but apparently not phytoplankton, from the surrounding water.

Succession to Land

The three great biocycles-ocean, freshwater, and land-come into contact with each other around the margins of the seas. The change in the physical nature of the habitat from salt water to fresh water is a drastic one, but not more drastic than the change from salt water to land : The transition of animal and plant life is brupt, and a zonation or physiographic succession of communities can be recognized. This transition from the ocean to fresh water and from the ocean to land as we see it today is of special interest since it parallels the probable evolution and dispersal of life in past ages.

Life is generally believed to have originated in the littoral region. Apparently no great group (phyla) of animals originated except in the ocean. The routes by which animals probably left the ocean and reached fresh-water and land have been various. Some animals probably migrated directly across sea branches; others probably ascended rivers, passed through marshes and swamps, or burrowed through soil. Some animals were transferred from the ocean by land elevations which isolated them in bodies of water which gradually become fresh .. Imigration from the sea did not take place at one time. It has occurred many times in the past and is slowly progressing on many shores today .. The most successful animal colonizers of the land have been :

1. The arthropods, which have in many cases developed book-lungs or tracheae for breathing air;
2. The vertebrates, with lungs and dry skins; and
3. The snails with slime and spirally coiled shells to prevent desiccation There are at present many examples of animals which are in the midst of their transformation from marine to fresh-water animals, or from marine of fresh-water into land animals. Not only have plants and animals emigrated from sea to land, but there are countless instances when migrations have taken and are taking place in the opposite direction.

Grasses, insects, reptiles, birds and mammals have left the land for the sea Fishes began in fresh water, but new range through the ocean at all depths.

On rocky shores and cliffs there is a splash or suprallittoral zone above high tide level. Green algae occur here and scattered individuals of marine snails, acorn barnacles, limpets, amphipod sandfleas and flatworms as well as insects, especially. Diptera and other forms that come from the land. Above the influence of splashing, the rocks may be covered with lichens and mosses, representing the initial stages in the terrestrial rock sere. However, salt spray is often blown inland a considerable distance to affect conspicuously the development of normal terrestrial vegetation and its accompanying animal life. Cliffs along the ocean, as well as sandy beaches and islands, are favourite nesting places for large numbers of pelagic birds.

Succession to Fresh Water

Where rivers flow into the ocean on low coastal plains and there are extensive embayments or estuaries as long the Atlantic Coast, there is very gradual change from salt water to brakish water (salinity : 0.5 30%0) to entirely fresh water. This habitat gradient fluctuates back and forth with the tides. Since fresh water is less dense and often warmer, it flows over the top of the salt water, with the result that strata with different physical characteristics are formed and these different strata are inhabited by different kinds of fish and other organisms, influx of fresh water is one of the principal sources of dissolved nutrients, and the gradient of salinity is dependent upon continuous inflow of fresh water. Any reduction of inflow, as by damming of streams, will allow salt water to penetrate further inland.

APPLIED ECOLOGY

Food Production

Although 71 per cent of the earth's surface is occupied by oceans and only 29 per cent by land, nearly all of the food

and raw materials used by man is derived from the latter. This is in spite of the fact that agricultural soil is only a few inches thick and must be cultivated, protected from erosion, and fertilized, while the ocean with its chemical fertility, its photosynthetic production of basic plant food, and its great fisheries is an entirely natural resource ready to be harvested. However, fish harvesting in the ocean has definite limits determined by photoplankton primary productivity. It is estimated that maximum sustained yield to man is of the order of 100-150 million metric tons per year and in 1967, 6000,000 tons were being harvested. The ocean is relatively infertile, yet it covers most of the area; much more productive are the coastal areas, particularly the areas of upwelling but these are of comparatively small size. A large part of the production is taken by predators other than man and part of it must be left to maintain the breeding stock. The possibility of the ocean serving as an abundant future source of food for man is probably exaggerated. Seafood is important mostly as a supplementary source of protein. Man must look to other resource for his main energy requirements.

The plankton of the sea represents a possible food supply for man (Davis 1955). Its energy value is approximately 4 kcal/g dry weight, and it is more or less palatable (Clarke and Bishop 1948). However the expense involved in filtering and concentrating it is too great to be economical; poisonous species sometime occur, it is not easily digested and assimilated, and consequently it has not as yet proved to be a feasible diet. The energy value of the plankton is used by man at the present time primarily as it is transferred into higher links of the food chain. Aside from the fishes, the chief marine organisms used as food are the oysters and other mollusks, shrimp, crabs, lobsters, and sea turtles. These are mostly animals of the continental shelf and estuaries.

There are a number of problems in the use and conservation of marine organisms. Natural beds of American oysters on the Atlantic coast have nearly all been exhausted through overfishing and pollution. Because of heavy erosion of the land, silt deposition has become excessive in most of the bays and estuaries and there is increasing difficulty for oyster spat to find clean hard surfaces on which to set Surfaces that are loose or covered with silt are not suitable since the spat is very small and easily smothered. A common practice is to return to suitable areas all shells of mollusks removed or to introduce other suitable hard objects to furnish the necessary substratum for oyster setting. Control of erosion over the watershed would greatly alleviate the problem. The trend is increasing to lease suitable areas of water and to farm oysters in the manner of an agricultural crop.

In spite of their position at or near the top of the food chain, the greatest utilizable food resource of the sea is its fishes.

Ocean "farming" is largely limited to estuaries and salt mashes. Yield in the Orient occasionally reach or even surpass 1000 kg/hectare-yr (880 lb/acre-yr). Principal species are milkfish, grey mullets, eels, salmon, sea trout, tilapia, oysters, and shirmp. Conversion of marsh into ponds where water level, fertility, and cropping can be controlled has met with varying success in different parts of the world, but may be expected to expand as fish harvesting in the ocean itself approaches its yield limit.

Summary

Geographic distribution of marine organisms depends on their responses to current temperature, and physical barriers; their local distribution is affected by waves and tides, type of bottom, salinity, and depth. Major division of the marine biocycle are pelagic (open water) and benthic (bottom). Major communities recognized are the oceanic plankton and nekton biome in the open sea, the balanoid-gastropod-thallophyte biome on rocky shores, the pelecypod-annelid biome on sand and mud bottoms, and the coral reef biome.

Organisms making up the oceanic biome are widely distributed around the world but may be divided into warm-water, arctic, and antarctic faunas. Coral reefs are found only in the tropics. The two biomes on the continental shelf subdivide into warm-water, temperate, arctic, and antarctic faunas and into more restricted regions and subregions. The warm-water faunas are richest in species, especially in the Indo-Malayan and West Indian subregions.

Marine plankton include a greater variety of forms than does fresh-water plankton, although rotifers are nearly absent; cladocerans, less important. They possess various unique mechanisms for flotation. Although abundance varies greatly from place to place and from season to season, plankton is generally much more numerous in neritic coastal waters than in the open sea, Diel movements between the surface at night and greater depths during the day are pronounced.

Mollusks (squids), fishes, birds, and mammals constitute the nekton. The taxonomic composition of the fish fauna varies with depth. Bioluminescence is exceptionally well developed among deep-sea nekton and benthos.

Benthos includes a great variety of sessile, creeping, and burrowing forms. It is very abundant in the littoral zone, and decreases in numbers with depth, although individuals are found in the deepest ocean trenches. There is considerable difference in the life-form and species composition of benthos occurring on rocky shores and on sand and muddy ones. Zonation of species is more prominent on rocky than on depositing shores. Succession and dominance occurs in some situations, but is less important than in terrestrial communities. Coral reefs have many special features, but are now exposed in the Pacific and Indian Oceans to damage by a species of starfish.

Food chains in the sea are similar to those in fresh water but the number of links and species composition varies in different communities. Productivity is especially high in regions where upwelling and turbulence bring nutrients from deeper levels up to the surface.

The three great biocycles of ocean, fresh-water, and land come into contact around the margins of the seas. The succession from the ocean to land is abrupt and from the ocean to fresh water through estuaries only slightly less so.

Effective use of the ocean resources as food for man is largely limited to the continental shelf, regions of upwelling, estuaries, and coastal marshes. Plankton is abundant and has high productivity in these regions but is likely to be useful to man only as its nutrient material and energy are transferred with considerable loss into the higher food chain links of mollusks, larger anthropods, fishes, and other vertebrates. Coastal regions need protection and management for good production and all waters kept clean from pollution of all sorts.

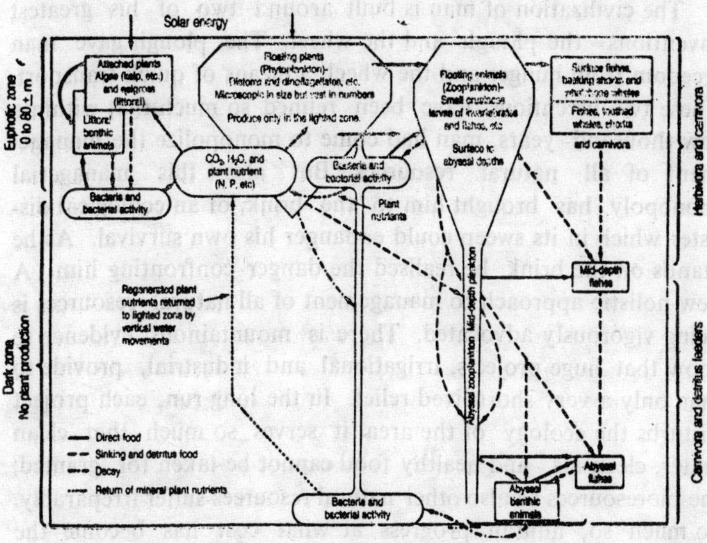

8

Development vs. Ecology

Introduction

The civilization of man is built around two of his greatest inventions—the plough and the wheel. The plough gave man freedom from hunger and the wheel, a means of quick transport. These two inventions have been refined so much that within a few thousand years, man had come to monopolize the management of all natural resources. But alas ! this managerial monopoly has brought him to the brink, of an ecological disaster which in its sweep could endanger his own survival. As he stands on the brink he realised the danger confronting him. A new holistic approach to management of all natural resources is being vigorously advocated. There is mountainous evidence to show that huge projects, irrigational and industrial, provide to man only a very short lived relief. In the long run, each project disturbs the ecology of the area it serves so much that clean water, clean air, and healthy food cannot be taken for granted; the bioresources as also other natural resources suffer irreparably. So much so, human progress at what cost has become the central theme of the debate raging through the whole informed world.

We in India are in an unenviable position. Our population is fast outstripping the country's carrying capacity. And we dont have the technology to improving the carrying capacity of our land without detriment to our ecology. No government dare impose restrictions of a compulsory nature on the family size; neither can it deny itself plans to press into service every available resource to meet the demands of the immediate present. In fact, we see a variety of contradictions rocking the government and people of our land; each one of them exemplifies this dilemma. The creation of several wild-life reserves through every conceivable ecosystem, the vociferous opposition to and the planning of a good many multipurpose projects, the creation of the Departments of Environment and Forests at the Centre and in the States, the acquisition of legislative powers to regulate human intrusions into natures arrangements reflect the growing concern of the government and the people of India over the likely damage to environment. On the other hand, politicians whose vision is necessarily for here and now are equally vociferous in demanding projects that benefit their vote banks through the next quarter century. Often, what we want directly goes against what we need and that is the crux of the problem. An understanding of this dilemma is absolutely essential for every practising ecologist. We will examine this dilemma under the following headings.

1. The Environment and Environmentalism

Inamdar, N.R. (1989) looks upon environmentalism as a social activity aimed at distinguishing between the factors impinging on the social system from outside and those inherent in the system itself. The quality of life in a local context engages the attention of every environmentalist. This is not altogether a modern idea. In Mahabharat. we come across passages in which the pandavas, while they lived in the forest, were advised no to stay in one forest area for too long. Dharma Raja was briefed on the ecological degradation, his party was causing and was advised to move to another forest. He was told in no unmistakable terms that a forest lives on the basis of a net-work of devourer-devoured relationship and too much of human presence would unsettle this relationship. And, the pandavas respected this advice.

In the totality of environment, the biological component goes through cyclical changes but the non-living component does not. And the all-important fact is that the living world sustains itself at the expense of the non-living. A portion of the non-living finds its way into the making of a living object but eventually whatever had thus moved from the non-living state of existence into the fabric of a living object has got to go back to its original state. Otherwise, ecological imbalance sets in. In such a situation, the threat posed to the ecology of an area comes mainly from man, for while the non human segments of the living world do not play any planned role aimed at sustaining their numbers, man had grown too clever to be outmanoeuvered by nature. On the other hand, he began to reorganize nature's order of things to suit his own ends. He had played this game for too long. Now, he finds that the game did not pay off. He realises that while he had been clever, he had not been wise. Hence, this reappraisal. Thus is born tne environmentalist school. It works towards generating a fresh awareness regarding the importance of maintaining environmental quality in mans own interest. Environmentalism is not against projects aimed at improving the "standard of life", a term usually understood in terms of human consumption. Environmentalism advocates development geared to improving quality of life for man. The quality of life which the school of environmentalists espouse is attended by a demand that man merge with nature, in an unnoticeable way. We should not understand this as a call for going back to a primitive style of life. It is a call for promoting human welfare in a manner that would not endanger other life-forms. Environmentalists stress thr need to examine a project not merely by working out the cost-benefit ratio through the next fifty years but in terms of thousands of years to come.

2. Meaning of Development

The Industrial Revolution that was ushered in through the last quarter of 18th century and progressed at an ever-increasing pace through 19th and 20th centuries proved to be a vital factor in the upgradation of consumerism and consequent degradation of environment. The two processes moved along a collision course, with almost identical speed. The economic and political fall outs of that revolution unleashed forces that could grapple with those problems somewhat successfully. But it took a very long time, almost not until into the 7th decade of this century, to understand the harm the industrial revolution did to the environment. In his hunt for procuring to the new rich articles of fancy, the new breed of industrial marketeer began to exploit the bioresources. This exploitation reached limits of lunacy when it led to killing bird and beast, not to improve living standards but to satisfy the ego and facny of man. When one learns of the extermination of a bird species for the only reason that it sad attractive plummage, attractive enough to make a wealthy women want to adorn her hat with it or when one learns that the rhino populations suffered heavily just because it was believed that its horn had approdisiac properties, one feels at once had and ashamed for being born a man. Added to these, the knowledge gained in regard to green-house effect, acid rain, ozone hole etc., have made man realise the danger that he is bringing upon himself. But above all, the science of genetic engineering has opened up new possibilities of transplanting non-native genes into microorganisms so as to induce them to synthesize chemicals of great significance to human welfare. In the light of all these things. Development has come to assume new meanings.

Development now has come to mean installation on pollution-regulating units side with industrial units especially those which produce or consume, chemicals of all kinds, so that the water and air are left untouched by contamination of any kind. Often, the effects of contamination are insidious. The examples of DDT and CFC-s come to mind. Not for years since they began to be employed, their harmful impact on the biological food chain and stratosphere had become visible. But in the meanwhile, they caused enormous damage. The Bhopal tragedy in our own country opened the eyes of government that was overanxious to produce chemicals basic to manufacture pesticides: if only there had been an appraisal of the possible consequences of an accidental leakage of Methyl isocyanide, that tragedy could have been possibly averted. Even in countries where technological achievement is of a very high order such as USSR, accidents of the type of Chernobyl could not be prevented.

The problem could be spelled out as follows. The modern way of life cannot be sustained without an industrial base. Yet the fact that industrial activity spells, doom to environmental quality cannot be igonred. The global debate that is going on is seized of this dilemma. Vested interests are not prepared to stop exploitation of natural resources not are they willing to order closure of industrial units with a potential for destruction of the environment. Yet, man worke up to the dangers inherent in such activity. The Gandhian thought seems to point a way out of this impasse. But who cares ? Saints are always applauded but never taken seriously.

All the same, the new awareness had stirred the conscience of man. Through the last twenty years, public opinion had asserted itself so much that governments the world over have started looking beyond the immediate future. Let us take a look at the environmentalist movement in India and the response of the Government to this strong and enlightened public opinion.

3. Environmental Movement in India

There had always been a concern for every form of life in the Indian mind. This Indian concern is projected through the doctrine of non-violence, preached by every religious school that flowered in our land. Indeed, it is enjoined on every Hindu to spend his last years of life in a forest, meditating upon the meaning of life (Vanaprasthasrama).

Violation of life for the sheer joy of it marked the early years of colonial rule. An Englishman who was here during those years wrote to his friend at home thus : "Wild pig, porcupine, wild fowl, game fowl and other animals, dear to the Sportsman, are to be met with in incredible numbers". It was almost an invitation to go to India to kill. The demand for railway sleepers led to a huge-scale felling of trees. Even after the advent of Independence, the Government of India went ahead with large irrigational and industrial projects, costing hundreds of crores of rupees, not taking into account the disastrous consequences to the ecology of the area. In fact, there was no ecology-consciousness at that time. This absence of any environmental awareness is best expressed by a local farmer on the occasion of the commissioning Nagarjuna Sagar project in Andhra Pradesh, when he said to Jawaharlal Nehru, our first Prime Minister. "You lighted a lamp in our homes", little knowing that his children were soon to suffer from a discease called Knock-knee; the discease had been traced to a change in the flouride content of the fields that came to be irrigated soon afterwards.

The dangers to which all forms of wild life got exposed were realized long before Shri Sunderlal Bahuguna launched his non-famous 'Chipko movement. Several wild-life sanctuaries have been carved out and poaching in these areas was made a cognizable offence. Indeed, laws which prohibited killing endangered species of animals and birds were passed through mid-sixty's. The Gir lion, the Bengal tiger, the great Indian Bustard and a few other species benefited a great deal through such laws For some species, they came too late : for example, the Indian cheetah and the musk-deer. But the questions raised by environmentalists are far more basic and it is in questions so raised that the quintessence of environmentalist movement lies. A close scrutiny of the demands made by the Narmada Bachao Andolan spotlights this Development *vs.* Ecology debate. The Sardar Sarovar project and the Narmada Sagar Project, as the projects underway on Narmada are christened, are designed to take water to the parched area of Kutch in Gujarat. Four States are involved in the execution of this project Madhya Pradesh, Maharashtra, Gujarat and Rajasthan. Gujarat is the beneficiary-in-chief and small wonder, that State is also the chief financing agency. By the time the project reached the take-off stage, it has become legally mandatory that every project obtain clearance from the Ministry of Environment and Forests, which is expected to go into the social and environmental costs of the project. Very recently, an environment status report had been prepared in respect of eight points touching upon this project : catchment area treatment, compensatory afforestation, command area development, rehabilitation, flora/fauna studies, archaeology, seismicity and rim stability and health aspects. Each of these titles is self-explanatory and so do not need elaboration. This status report points out that while "plans for all environmental and rehabilitation aspects should be ready by 1989 and that these should be implemented simultaneously with the main construction work", the construction work progressed so fast as to advance the submergence of inhabited area by four years, to 1994 instead of 1998" but the "plans for afforestation, rehabilitiation, relocation of three temples in Gujarat and studies on seismicity and flora/fauna studies are far behind schedule". In fact, the work in respect of these obligatory duties had been so slow that all the four States own up that the completion of these studies and subsequent formulation of plans will take at least two or three years from now: which is to say that 1993 is the earliest by which such studies would be completed. When the plans would be implemented is anybody's ,guess. Of the 1982 families from Maharashtra which are to be resettled in Gujarat, so far only 63 families are taken care of and for the rest, the land where such resettlement can be effected is yet to be identified. Studies on flora/fauna, identification of the site where the archaeologically significant temples are to be rebuilt, the question of health hazards such as appearance of malaria and spread of water-borne diseases are not just given due consideration as yet. It is against this background that the pleas of environmentalists like Baba Amte, Sunderal Bahuguna and Ms Medha Patkar have to be heeded. But there is always the other side of the coin. A commoner from Kutch wrote to Baba Amte that he should give due weightage to the perennial drought that the 30 lakhs of people in Kutch are exposed even as he is worrying over displacement of 3 lakhs of people in Madhya Pradesh. Of course, the Baba talks of a different technology which would meet everybody's needs without hurting anybody's interests.

The dilemma that India faces is really pathetic. Her population threatens to outstrip that of China by the turn of the century. No Government can afford to adopt a 'go to hell' attitude. It has to press the resources to the utmost limit possible. As it is, India uses only a tenth of the rainfall it receives annually and even with all the reservoir-building plans which are now on the anvil, it will be able to use only upto a quarter of that resource even after 40 years from now. Let one thing be made clear. The dams and the reservoirs are not an production becomes possible. They serve to save on water and also to generate more power. They check floods, create opportunities for developing inland pisciculture which could go a long way to remedy the festring evil of protein deficiency dogging the nutritional aspect of our national life, meeting the water needs of the Industry and so on. All this in turn means generation of more employment opportunities for our people. But it also means huge displacement, unaffordable financing, submergence of vast tracts of land etc.

4. Environment and the Law

Caught on the horns of such an intractable dilemma, the later day governments which came to rule the country found it necessary to embark upon legislation as a means of tackling it.

As early as 1969, New Delhi hosted a conference under the auspiecies of the International Union for Conservation of Nature. The then Prime Minister, Smt. Indira Gandhi inaugurated it. A participant from India, obviously an enthusiastic conservationist remarked in a mood of exultation "From being a catch-phrase, wild-life become a serious commitment of the Government. The Ministers got involved; all those who hindered now began to help." And in 1972, the Wild Life Protection Act was passed. This was followed by a series of enactments, all of which reflected the concern of the Government of India for environment. However, it should not be construed that there was no law whatever checking pollution. The Civil Procudure Code provides relief to those that seek it. This relief permeates every level of administration, from the Central Government to the local administration through State Governments. The Municipal Council of Ratlam had been arranged for not constructing enough number of public latrines in the town, thus leaving a vast scope for committal of public nuisance. Likewise, one Krishna Gopal sought remedy against noise and air pollution caused by a drug factory built in close proximity to a residential area : this pollution affected a heart-patient living in the area. The Company's argument that it was a private nuisance if it is any nuisance at all and not public nuisance was invalidated by the local magistrate and the invalidation was upheld by the High Court. Thus, remedies exist in civil procedure code and public spirited bodies and individuals can obtain redressal. In both the cases cited above, as Dr. Leela Krishnan put it, "the judiciary played activist role ; adding new dimensions to the rarely used provisions of Law.

The case of Mayur in Kerala illustrates what law can do. Mayur is a small dot on the map of Kerala. The Birlas built a Rayon factory there, taking advantage of abundant availability of bamboo in the area. Soon, the local population began to enjoy a level of prosperity that they had not known earlier. The factory management allowed the effluents to flow into the nearly river. Soon enough, the river water became non-potable. An agitation forced the State Government stop supply of bamboo to the factory and that forced its closure. The prosperous township went back to its earlier poverty. Now, a new agitation to get the factory reopened got underway. The State Government successfully pursuaded the management of the factory to reopen it, assuring it of the necessary supply of bamboo. But the new agreement ensured that the waters and the air did not suffer from any pollution.

Through the last fifteen years, there has been a great deal of legislative activism in India in respect of environmental protection. The Water (Prevention and Control of Pollution) Act, 1974, the Air (Prevention and Control of Pollution) Act, 1981 and the Environment (Protection) Act 1986 are all enacted with a view to curbing activities likely to affect the quality of environmental resources. In combination with the Constitution (42nd Amendment) Act, 1976 these laws give to the Government enough muscle power to prevent environmental degradation. Further, an amendment added Article 48A to the Directive principles. This made it mandatory on the part of every State Government to do all that it could to protect and improve the environment and to safeguard the forests and wild life. It also cast the fundamental duty of every citizen to protect and improve the natural environment. This amendment also transferred to the concurrent list some subjects which were in the State list earlier. Thus, now the Centre also has power to make laws on forests, wild life and population control. This new found power of the Centre was very much in evidence in the case of Telugu Ganga Project.

The awareness is there. The governmental agencies are armed with enough legal and executive powers to intervene when necessary. The duties of citizens are spelt out in unambiguous language. The Central Government, when satisfied that there is transgression of laws, can intervene on its own or through State Governments or local administration by advising action through cut-off of Power supply. Water supply or any other service to the offending industry. But unfortunately, political exigencies and clash of personalities often come in the way of the right type of intervention. Not only that. The Boards meant to supervise environmental protection are more often interest-infested than expert-oriented and do not enjoy a coersive power. Such Boards can only act as prosecuting agencies and a citizen seeking redressal under the provisions of the acts mentioned above can do so only under approval from the concerned Board. A new Environmental Impact Assessment (EIA) of developmental projects is the trying need of the hour. There are models to go by but the fact is that these models operate in the developed world. With our population pressure, illiteracy, shrinking resources and many other disadvantages, how far we can stretch our legal system to offer a desirable level of environmental protection is debatable.

The foregoing analysis presents a gloomy picture. It explains the problems but offers no solutions. The ecologist has a great responsibility. Can he rise to the occasion ?

5. Environmental Problems and Action in India*

The pressure on natural resources from an increasing population-800 million at present and projected at 1,000 million by 2000 AD.

- Of the 266 m ha considered productive, about 175 m ha are degraded in varying degress (arid, alkaline, saline water logged, ravines, etc.). About 90m. ha are acutely degraded, chiefly on account of a loss of tree cover and top soil, leading to floods and droughts.
- Depletion in forest cover to about 19 per cent of the total geographical area, instead of the desired 33 per cent. India has only 2 per cent of the forest land of the world but supports 16 per cent of the world population.
- Shortage of fuelwood and fodder for rural needs, leading to pressure on our forests.
- Threats to faunal and floral species and biological diversity because of disturbance of their habitat.
- Adverse impacts of developmental activities such as mining, power generation, industrialisation and irrigation.
- Degradation of fragile ecosystems, such as mangroves, wetlands, beaches and hill areas for reasons such as over-exploitation, lack of tree cover, ill-advised agricultural practices, tourism and indiscriminate building activities.
- Pollution of water from domestic and industrial waste.
- Pollution of coastal areas and seas.
- Air pollution due to emission from industries.
- Increased production, transportation and use of hazardous chemicals.

---

* Presented by Shri Maheswar Prasad, Secretary, Ministry of Environment and Forests, Government of India, at a Special Training Programme for IAS Officers on Policy Analysis conducted at IIPA, New Delhi from August 28 September 23, 1989.

—Degradation of the urban environment because of rapid expansion and inadequate basic services.

—Global warming and depletion of the ozone layer.

—Lack of sufficient environmental awareness. The consciousness began only in the 1970s, not only from concerns voiced around the world, but from our own immediate experience also.

ACTION TAKEN

Legal

—The wildlife Protection Act, 1972.

—The Water (Prevention and Control of Pollution) Act, 1974 amended in 1978 and 1988

—The Forest (Conservation) Act, 1980 amended in 1988.

—The Air (Prevention and Control of Pollution) Act, 1981 amended in 1988.

—The Environment (Protection) Act, 1986.

Institutions

—The following major institutions have been established, apart from the existing ones.

—The Department of Environment in 1980 and the integrated Ministry of Environment and Forests in 1985 at the Centre.

—Departments of Environment in 20 states and three Union Territories. The Central Pollution Control Board and State Pollution Control Boards in 21 states.

—Indian Council of Forestry Research and Education.

—National Land-Use and Wastelands Development Council 1985

—National Wastelands Development Board in 1985.

—Indian Board for Wildlife in 1952.

Land and Soil

—Surveys by the All India Soil and Land-use Survey Organisation.

—Treating catchment of river valley projects, integrated watershed management projects in catchment of flood prone rivers.

—Assistance to states to control shifting cultivation.

—Assistance for reclamation and development of ravine areas.

—Drought prone areas programme.

—Desert development programme.

Biosphere Reserves

—Thirteen biogeographical zones in the country have been identified for establishing biosphere reserves. Seven reservers have been notified so far. Management plans for these reserves under preparation. Central assistance provided to three States.

Islands

—Island Development Authority set up in 1996 and ecologically appropriate plans under implementation.

Wetlands Mangroves Forests

—A National Wetlands Committee has identified 16 wetlands for conservation, management and research. Management plans under preparation. A directory of wetlands in the country has been completed.

—A status report on mangrove areas prepared. The National Managroves Committee has chosen 15 areas and has taken up the preparation of management plans.

—The first Forest Policy of India was announced in 1894 and revised in 1952. The new Forest Policy of 1988 approved by the government proclaims the principal aim as ensuring environmental stability and maintenance of ecological balance including atmospheric pheric equilibrium. The deriviation of direct economic benefit from forests will be subordinated to this principal aim. The objectives are : maintenance of environmental stability, preservation of bilogical diversity, soil and water conservation, increase of tree cover, meeting the requirements of the rural and triabl population, increase in the productivity, efficient utilisation of forest produce, substitution of wood and peoples involvement for achieving these objectives. The goal is to have a minimum of one-third of the country under forest of tree cover.

—Afforestation taken up from the First Plan, but up to 1985 only 6 m ha covered. The loss was assessed at 9.17 m ha between 1972 and 1985 alone. The National Wastelands Development Board (NWDB) established in 1985. In three years 5.04 m ha planted. The Seventh Plan achievement will be over 9 m ha.

—The Forest Conservation Act of 1980 has stringent provisions for checking the diversion of forests land for any other purpose. Between 1951 and 1980, the rate of diversion of forest land was about 1,50,000 ha per annum. After 1980, it has been reduced to 6,500 ha per annum.

Wildlife

—National Wildlife Action Plan was adopted in 1983. The Indian Board for Wildlife is the apex statutory body, with Prime Minister as Chairman. Sixtythree national parks and 358 santuaries in the country covering an area of 1,33,000 sq. km about 4 per cent of the countrys area. Sixteen tiger reserves in the country in which Project Tiger launched in 1974 has had great success. The Wildlife Institute of India at Dehradun established in 1982.

Environmental Impact Assessment (EIA)

—EIA and approval of all projects which require approval by the Government of India and which require diversion of forest land, made obligatory. Guidelines for EIA issued by the ministry for mining projects, industrial sitting, river valley projects, development of ports and harbours, and development of beaches.

Pollution Monitoring and Control

—Standards notified for 26 industries and under enforcement by the Central and State Pollution Control Boards.

—Environment (Protection) Act, 1986 enforced in 80 cases to take stringent action against continuously erring industries.

— Over 3,500 prosecutions launched against erring industries throughout the country.

—3000 Water Quality Monitoring Station and 106 Air Quality Monitoring Stations established throughout the country.

— Of 4,056 large and medium industries in the country, over 50 per cent have installed effluent treatment plants.

—Use-based zone and classification of the 14 major rivers in the country has been completed.

Control Over hazardous Substances

— The Environment (Protection) Act of 1986 provides for control of hazardous substances.

— Draft rules prepared for notification of hazardous chemicals and control of manufacture, storage and transportation, and disposal of such substances and wastes.

—A crisis management plan for chemical accident prepared and a Central Crisis Group set up. Guidelines for similar action at State and district level issued. Forteen States have so far prepared on-site and off-site emergency plans and set up State level coordination committees.

Eco-Regeneration and Development

— Ganga Action Plan lanuched in 1985 at a total estimated cost of 292 crores to clean up the river and restore water quality. Two hundred sixty two schemes at a cost of

Rs. 259 crores sanctioned for treatment of waste water and for prevention of pollution of the river.

—Eco task forces of ex-servicemen deployed in Uttar Pradesh, Rajasthan and Jammu and Kashmir for ecological restoration through afforestation and soil conservation.

—Eco-development camps organised for youth through NGOs and public participation

Research

—Forestry research reorganised under the India Council of Forestry Research and Education, Dehradun. It would coordinate research through the institutions established at Coimbatore, Banglore and Jabalpur. Two more institutes to be established at Jodhpur and Jorhat.

—Wildlife research carried out by the Wildlife Institute of India, Dehradun.

—Gobind Ballabh Pant Himalaya Parayavaran Evam Vikas Sansthan established in 1988 near Almora (UP) to develop integrated sustainable development strategies by coordinated research.

—Funding of research projects, on all environment related subject. Analysis of results for possible implementation.

—Multi-institutional research projects such as the All India Coordinated Research Project on Heavy Metals effect of Air Pollutants on plants, Ethno-biology, etc.

—Environmental Education and Awareness.

—Non-formal education and awareness through seminars, workshops, training programmes, etc.

—National Museum of Natural History establishtd in 1978 at Delhi.

—National environmental awareness programme launched in 1986 as an annual feature in which more than 200 NGOs carry on the compaign aimed at different section of the population through all the media.

– Centre for Environment Education established at Ahmedabad in 1984 to devise education materials and curricula.

Global Warming and Ozone Depletion

– Reduction in pollution from thermal plants.

– Energy conservation.

– Promotion of non-conventional energy sources.

– Steps to check deforestation and promotion of afforestation.

– Coordinated study of ozone levels in the Indian Middle Atmosphere Programme.

– Initiation of studies on the impacts of possible rise in sea level.

– Participation in international flora and support for global action.

Need for Coordinated Scientific Research

All activities to protect the environment require considerable scientific imputs in the areas, such as land, soil and water conservation, pollution control, environmental impact assessment, preparation of environment management plans, ecological restoration of degraded areas, forestry, wildlife protection, assessing the impacts and finding solutions for global warming and ozone depletion.

9

Forest

Introduction

Of all the environmental problems facing the country, the problem of deforestation has received the maximum public attention. Ironically, no set of government policies in the field of environment has attracted greater public criticism than the policies for afforestation.

Social forestry programmes have been launched by several states to promote afforestation on essentially non-forest lands, that is, on private farms and on village commons. Several state governments have stepped up their own afforestation efforts and are also planning (in most cases, without much public discussion) to hand over large tracts of degraded government forest lands to industrial firms for afforestation.

All these programmes have one thing in common the planting of trees to meet the needs of urban and industrial markets while the glaring fuel and fodder crises facing the common person continue to grow. The species chosen-like eucalyptus, teak and pine-also do little for ecological restoration, for enhancing soil fertiliy or for soil and water conservation.

The very same process of commercialising the forest resources base that has led to widespread deforestation in the country is today the motive force behind the government's afforestation programmes being carried out with the full support of foreign aid agencies, India's afforestation is, thus, as anti-people as is its deforestation. Rural women, for instance, whose lives revolve around the collection of fuel and fodder, have almost nowhere been involved in these programmes.

People are responding with greater alacrity to voluntary afforestation efforts aimed primarily of meeting people's needs of fuel, fodder and small timber—and popular protests grow against both government sponsored deforestation and afforestation. Few in the government have the imagination to realise that afforestation of India's degraded public lands could form the core of India's largest reforms, anti-poverty and employment generation programmes.

Social Forestry

Social forestry is a term used by the National Commission on Agriculture in 1976 to denote tree-raising programmes to supply firewood, fodder, small timber and minor forest produce to rural populations. Nearly a decade later, it is emerging as the most controversial initiative of the Indian government.

Critics of the programme strongly contend that the wood produced from social forestry programmes is ending up in urban and industrial India instead of the poor in rural India, reducing rural employment and land under food production, and promoting absentee landlordism.

The Union Government and the World Bank are discussing a $600 million proposal for a nationwide social forestry programme which will cover 13 states including a number of those in the Northeast

Ambitious social forestry progr amms have already been launched by several state governments, most of which have been financed by foreign aid agencies like the World Bank, US Agency for International Development (USAID), Canadian International Development Agency (CIDA) and the Swedish International Development Authority (SIDA). Forest departments in most states have set up separate social forestry wings.

Growing Criticism

Despite these achievements, present and projected, doubts and criticism about the actual benefits of social forestry have grown into a shrill chorus.

Social forestry programmes have mainly three components : farm forestry, encouraging farmers to plant trees on their own farms by distributing free or subsidised seedings; woodlots planted by the forest departments for the needs of the community, essentially along roadside, canal banks and other such public lands; and community woodlots planted by the communities themselves on community lands, to be share equally by them.

Judging by the World Bank's own mid-term reviews of social forestry projects in Uttar Pradesh and Gujarat, big farmers are emerging as the primary beneficiaries. In Uttar Pradesh the World Bank assisted social forestry programme overshot its farm forestry target by 3,430 per cent a phenomenal and unprecedented degree of success for a government programme. However, efforts in the creation of community self-help woodlots achieved only 11 per cent of the target, a disastrous record even though this component of social forestry programmes is aimed most at meeting the needs of the neediest in the rural communities, the landless and small and marginal farmers. In Gujarat, too many forestry targets were exceeded by more than 200 per cent whereas community self-help woodlots met only 43 per cent of the target.

In communist-ruled West Bengal, the best state according to the World Bank, the creation of village woodlots by the forest department has fared well, achieving 109 per cent of the target. But these are not self-help community woodlots. Landless families have been given small plots of government wastelands for tree planting. The Marxist lead United Front government has also ensured involvement of its party cadres in the Programme, thus evoking a good grassroot response. Even then farm forestry in West Bengal has fared better, achieving 200 per cent of its target.

The Agriculture Ministry itself admits that the percentage of seedlings reaching farmers with landholdings of more than 2 hectares has been 61.5 per cent in Haryana and 57.6 per cent in Gujarat. Jammu and Kashmir and West Bengal have fared better : this percentage there is down to 35.3 per cent and 19.5 per cent respectively.

A significant result of the success of farm forestry -- the "spontaneous response of small farmers or communities to the commercial incentive of rising prises (for wood) and the perception that farming fastgrowing short rotation trees as a cash crop can be as profitable as farming some of the traditional cash crops", as a World Bank document puts it is the conversion of large tracts of prime agricultural land to tree farming. In Uttar Pradesh, the average area to be planted annually under farm forestry was targeted at 2,800 hectares over a four-year period between 1979 and 1983, but over 98,000 hectares have actually been planted. Seedlings sold by the social forestry wing have been cornered by farmers owning over 4 hectares. In eastern Uttar Pradesh where landholdings are extremely fragmented, peasants extremely poor, and the energy crisis extremely acute, farm forestry has fared poorly. This is yet another indication that it benefits better-off farmers and does little to meet the energy needs of the poor.

The fact that seedlings are distributed free or are heavily subsidised essentially means that the programmes, started with the professed aim of meeting the dire fuel and fodder needs of the poor, are becoming a scheme of subsidies to support lucrative cash cropping by the rich. Additionally the fact that seedings are often not available immediately after the rains means that small farmers with unirrigated lands cannot take as much advantage of subsidised seedlings as rich farmers with irrigated lands can. In many states, irrigated lands have come under tree farming and in at least one state there are now demands to can free farming on irrigated lands so as to affect food production seriously.

Farm Forestry

A major reason for the distortions in social forestry programmes is the acute shortage of building poles in the urban market and of pulpwood, required by the rayon and paper industries. The active propagation of farm forestry by the government and the financial incentives it holds out has come as a windfall for the big farmer. Labour needs go down when farmers switch from short-term agriculture to long-term tree farming and they can even contemplate becoming absentee landlords. If they plant a tree like eucalyptus which is not palatable to animals, supervisory problems are also reduced.

However, for the landless in the village this can be a real disater. For instance, in Punjab an influential farmer with over 100 hectares of land switched over form cotton to eucalyptus farming. The change in the type of biomass being produced created immediate energy problems for the farmer's labourers who earlier got cotton sticks regularly for fuel. Now they get nothing. After five to seven years, when the eucalyptus is harvested, they might get the lops and tops, as the trunk will go to the urban market and that only if the farmer is unable to sell the lops and tops of the tree. There is every reason to believe that with social forestry. India is getting into a new green revolution which will probably increase wood in the villages but will also make the poor even poorer in terms of energy.

A major reason why farm forestry has become such an overwhelningly dominant component of social forestry is what the World Bank calls, "Project economic justification". Government strip plantations cost about Rs. 4,200 per hectare and village woodlots Rs. 3,000 per hectare and the internal rates of return for such plantation range from 12 per cent to 15 per cent. By contrast, the cost of farm forestry is about Rs. 1,600 per hectare and the internal rates of return are of the order of 25 per cent to 30 per cent. Concludes a World Bank document : "This comparison clearly reinforces the conclusions now being drawn by the community forestry wings project management that farm forestry is likely to be the least costly and economically the most effective approach to afforestation of the rural areas and that a major thrust of project development in the future should be in this direction."

One of the best indicators of the insensitivity of the World Bank to the needs of the poor is the fact that all its documents harp on the importance of fuelwood almost without any mention of the fodder problem, while there is no reason to believe that the fodder crisis is any less severe than the fuel crisis, if not more. Even the weeds that take over degraded lands can provide some fuel but they provide no fodder. They spread only because they are not palatable to animals. The World Bank and foresters are happy to spread eucalyptus, a plant which specifically cannot provide fodder – a plant which socially speaking has all the characteristics of a weed – and yet make loud claims about the social relevance and success or social forestry.

The champions of "commercialised social forestry" assert that the gains of farm forestry far outweigh the losses. By increasing the wood available for urban and industrial purposes, pressure on natural forests will be reduced. Pro-eucalyptus economists argue that once wood supply overtakes industrial and urban demand, prices of wood will fall and farmers will diversify their plantations. But they do not answer why farmers will go on to produce wood for the poor who cannot pay.

That the commercial wood market is of key interest to the World Bank can be seen form the following paragraph from one of its own documents : "Marketing studies are seen as a critical issue in social forestry projects being funded by the Bank in India. It is becoming increasingly obvious that the rising prices for poles and fuel wood have been one of the main factors in stimulating farmers' interest in tree planting. Early definition of the size and geographic location of such cash markets can influence the siting of tree nurse ies and the design of appropriate extension programmes."

Revealing Survey

A model study prepared for the World Bank by the International Institute for Environment and Development on the wood market in Gujarat claims that World Bank officials are worried that if existing cash markets were to be satisfied in the conceivably foreseeable future, this market saturation could have an adverse effect on farmer and community response to farm forestry. The study, therefore, looks into the existing and potential market structure for future project planning. It concentrates entirely on the commercial wood demand, urban, firewood demand, demand for construction poles, sawlogs, wood and pulp, and paper, and other wood products like activated charcoal, catechu (a tanning extract) and 'katha' (a condiment used in the preparation of betal nut). But no where does the study talk about the non-commercial firewood demand of the poor, as if that was never an objective of the social forestry project in Gujarat.

The study, which is to be used by the World Bank as a model for planning social forestry programmes in other states, goes into great detal about how the commercial wood market can be expanded to continue to promote social forestry.

"In addition, enterprising and innovative individuals are actively considering potential markets that could be tapped, should any surpluses become apparent. The demand for wood products in India is exceptionally high and the lack of a sufficient raw material base has greatly hindered market development. If the raw materials become available through social forestry activities, it can be expected that new forest product industries will develop, particularly pulp and paper, wood-based panels, and saw milling industries.

"Because a relatively large amount of wood will become available over a short period of time, there is some concern that the current marketing structure may fail to protect vulnerable small farmers with little experience. A promising sign is that wood growers' cooperatives have been formed in a few areas (emphasis in original".

The study also talks of various new opportunities being discussed in Gujarat to expand the commercial wood market. For instance, it reports that there has been some discussion among farmers, particularly in Bhavnagar District, on establishing small-scale village mechanical pulping devices that would be capable of handling eucalyptus. They have also been discussing with the State Agro-Industry Boardand the Gujarat Energy Development Authority the idea of wood-based electricity generating power station whose power would be purchased by the State and fed into the central grid. Jyoti Ltd. has been working on such a dendro-thermal power plant. This company has also designed a wood gasilier-powered irrigation pump, which will extend the wood marked further. The economics to the Jyoti system appears to be quite favourable, depending on the availability of wood. Units of varying sizes are expected to be put on the market in the near future. Another company has approached the forest department for establishing a hardboard plant in Bhavnagar area. Good quality eucalyptus timber could also be used by the ship building industry along the coast. Essentially this adds up to a wood production, marketing and development scheme.

Admission of Failure

Access to fodder, fuelwood and building wood essentially means access to land and social forestry projects. If they are really to serve the neediest, must be directed to landless, small and marginal farmer households. The World Bank admits in its latest global review of World Bank-financed forestry activity that it has failed in this respect : "Ensuring adequate supplies of fuelwood, building poles and fodder for the landless, who in some developing countries number more than 25 per cent of the population, is probably the most difficult forest policy issue that most developing county governments face. Various approaches have been tried, such as government-sponsored establishment of road-side plantations or woodlots with preference in allocation of employment and output to landless families. The problem of creating viable employment for the landless is an issue which can only be resolved by general economic growth, by movement of people out of the rural areas into industrial employment, by resolution of inequities in land distribution, by growth of the service sector, by development of markets for processed primary products and by other development-oriented interventions. Forestry's role in this area will remain marginal. A second constraint is that of how to provide effective incentives to persuade small farmers to take up tree planting in situations where the main purpose of such tree planting is to create a supply of wood for domestic consumption."

The landless, in fact, are distrustful of social forestry programmes. The World Bank itself lists a number of reasons whose response to community forestry efforts has been poor: heterogeneous village communities; mistrust of the system for ensuring equitable distribution of the produce from community woodlots, disputes amongst farmers concerning availability of the village common land for the establishment of village woodlots, and shortage of panchayat funds for payment to labour and protection of plantations. "The 'commons' problems is particularly difficult issues as it affects the willingness of rural communities to cooperate in the establishment of village woodlots, and to protect them from grazing animals", admits the World Bank.

Surveys in Tamil Nadu show that the cattle-owning landless households are against woodlots on village commons because they are afraid this will further reduce the availability of grazing lands, block the ancient right of passage for carts and reduce the supply of brushwood (the poor person's fuel). Also experience has shown the poor that the produce from such commu ity efforts seldom reach them. Both Madhya Pradesh and Uttar Pradesh have tried to ensure equitable distribution of the usefruct by drawing up elaborate contracts betbeen the panchayats and the social forestry directorate. But as experience in a village in Sultanpur District shows despite a list of the "poorest of the poor" being made by the village pradhan the actual distribution of firewood was in favour of the landed: the landless got nothing. Uttar Pradesh and its donor World Bank, are not even pursuing such contracts in other villages because they consider it a futile exercise.

The World Bank's reaction to this failure exemplifies its limited commitment to get social forestry programmes serving community needs. The Gujarat social forestry appraisal report describes how project staff are resorting to management decisions to "increase the size of the harm forestry programme, to reduce expectations under the village self-help component and to increase the forestry department, supervised village woodlot component". In other words, if self-help woodlots have not met with success, forget the poor and switch to farm foresty. J. Bandyopadhyay, one of the authors of the Kolar study, strongly questions the propriety of international borrowing for financing social forestry programmes mainly executed by rich farmers, and whose benefits do not reach the poor If social forestry is so profitable why should there be any subsidies. especially those which make the country indebted to foreign bodies, he asks.

In many states additional resources for social forestry have been mobilised by linking if to the National Rural Employment Programme (NREP) and the Integrated Rural Development Programme (IRDP). In Delhi, Madhya Pradesh, Uttar Pradesh and Gujarat, trees are being planted on public lands with seedling provided by the forest department and the wages component funded under NREP. In Gujarat the selfhelp component of the community woodlots scheme picked up momentum after the forest department decided that the poor and the landless should not be expected to volunteer their labour but should be paid out to NREP funds. However, this has only led to some token social forestry with an eye to obtaining more funds. In Pudukottai District of Tamil Nadu, for instance, it was found that because of the stipulation for receiving grants under IRDP or NREP, most of the panchayat unions had reluctantly raised plantations, but these were neglected.

The World Bank and the forest departments do not admit that tree farming is eroding rural employment but they do not produce any hard data or studies either to prove otherwise. Figures produced show that social forestry is creating employment but no comparative studies have been published to show whether his employment is less or more than cultivation would generate in irrigated and non-irrigated conditions. The Kolar study estimates that for each hectare of land shifted from food crops to eucalyptus, there is a loss of about 250 person days of employment per year. Other experts, however, argue that forest plantations can increase employment for the rural landless when compared to food crops.

A major failing of the social forestry projects is the lack of involvement of poor women who ought to be the main beneficiaries. It is women who collect fuel and fodder every day for the family and the male orientation of the projects has, not surprisingly, turned them into cash-generating rather than basic needs generating exercises.

Shifting Priorities

Whatever propaganda the World Bank officials may continue to use for their ignorant admirers in Washington, in India they have kept a low profile. In private, World Bank Officials admit that, having learnt their lesson, they are now stressing afforestation of degraded forest lands as an important component in social forestry programmes to meet primarily the fuel needs of the poor. They insist that the World Bank is now increasingly emphasising individualisation or privatisation of trees grown on degraded forest lands. Right to the usufruct of the planted trees will be given to poor households living near such lands.

In West Bengal, for instance, small plots of government-owned wastelands are being allocated to landless families for tree planting, the landless do not get the title to the land, but they are given full right to the usufruct of the trees they plant. Both wages and seedlings are given at the planting stage. Stage States like Gujarat and Maharashtra have also developed schemes to involve the landless in tree planting but it is too early to evaluate these schemes. Certain reports indicate that the forest departments show little interest in these schemes and tribal families are reluctant to take advantage of them for fear of getting involved in something that may only harm them in the long run. Voluntary agencies may have to play an important role in organising the poor to take full advantage of these schemes.

In any case only those landless families can be involved in the schemes which live close to the degraded forest lands. Moreover, any major move to settle the landless on degraded forest and will face strong opposition from wood-based industries which are pushing for allotment of such lands to them, to meet their own wood requirements.

A fundamental reason for the failure of the social forestry programme to help the most needy is that it expects an unrealistic degree of collaboration between the villagers who use the forests and the forest department personnel who police them. Many villagers in Gujarat, Madhya Pradesh, Uttar Pradesh and elsewhere are reluctant to let the forest department afforest their common lands for fear that they may claim the afforested area as theirs. Poor rural families have even been reluctant to accept saplings distributed free for homestead plantations through social forestry projects for fear that planting them will somehow give the forest department a claim to the land.

Foresters too, do not relish the new role thrust upon them of wooing the rural public as partners in afforestation. A young divisional forest officer in Madhya Pradesh sums up : "We used to be kings (of the forests), now we have to go as beggars (to persuade people to plant trees)." It is clear that there has to be a fundamental change in objectives and attitudes if the social forestry programme is to make a lasting impact in areas where such an impact is of prime importance.

Government Forest Lands

Attention in the last two years has mostly focussed on social forestry programmes, programmes mainly promoted on private farmers and panchayat lands. Little public attention has focussed on the management of 75 million hectares of forest lands under the control of the forest department. Inside government committees, the paper industry. facing an unprecedented raw materials crisis is arguing for control over large tracts of degraded forest lands which it claims it could afforest and generate its own raw materials. Until now the industry has only dipped into the vast natural forests, in most cases at throwaway prices, and caused massive deforestation, often in collusion with forest officials and money-hungry politicians.

Paper is a vital commodity. The per capita consumption of paper—often regarded as an index of literacy and standard of living—is around 2 kg per person per year, a pitiful amount when compared to the 268 kg in the US, 124 kg in the UK and 115 kg in France and even 11 kg in Thailand and 10 kg in Egypt. The paper industry in India has steadily increased its capacity from 0.14 million tonnes in 1951 to 2.16 million tonnes at the beginning of 1983. There were then a total of 179 paper mills. About 20 per cent of the present paper-making capacity was in existence before the 1950s. Another 50 per cent came into existence between 1950 and 1970. The remaining 30 per cent was added during the 1970s. This period was marked by a drastic reduction in the growth rate compared to the 1950-70 period. The capacity added during the 1970s mainly came through schemes involving addition of balancing equipments in large mills and through the proliferation of small mills based on agricultural residues.

The Indian paper industry, like the Chinese, is unique for its wide range of mill sizes. Mill capacities range from 5 tonnes to 10 tonnes per day in the smallest units to about 250 tonnes per day in the largest mill. The latest newsprint mills are of 300 tonnes per day capacity. But even though there are many small paper mills in the country, about 65 per cent of the total installed capacity is represented by the large integrated mills with plant sizes of over 20,000 tonnes per year, which are only 16 per cent of the total number of mills in the country.

The growth of large mills was greatly inhibited during the 1970s because of steeply escalating energy and investment costs and even more because of growing shortages of raw materials like bomboo. A "redeeming feature" of the paper industry during the 1970s, according to the Development Council of the Pulp and Paper and Allied Industry (DCPPAI), constituted by the government, was the emergence of many small paper mills with an annual capacity of less than 10,000 tonnes based on unconventional materials like agricultural resides (rise and wheat straw etc.), cotton rags, hemp, jute cuttings, and grass. Most capacities created in the past decade are based on agricultural residuces, helped by tax concessions offered by the government for use of unconventional raw materials. Small paper mills use on an average 55 per cent agricultural residues 40 per cent waste paper, and 5 per cent purchased pulp and other long fibre materials like cotton linters and rags.

Waste paper recycling is also increasing in India, even though it is still low (only 20 per cent) as compared to many European countries and Japan where it is as high as 50 per cent. Raw material and energy shortages, particularly electricity; have also become major constraints in existing paper mills based on bamboo. Capacity utilisation in the paper industry has steadily declined from a peak of 98.9 per cent in 1970 to 68 per cent in 1981. According to the Indian Paper Makers Association (IPMA), capacity utilisation in 1983 was only 58.1 per cent. Thus even with this large capacity, the paper and paperboard industry could achieve a production of only 1.11 million tonnes in 1983, valued at Rs. 1 200 crore. To achieve this production the mills used about 3.09 million tonnes of forest raw materials.

Large paper and paperboard mills are fully integrated to pulp production which is now based mostly on bamboo and hardwoods. These mills are generally located close to forests The general policy is to be self reliant in fibre supply and us market pulp only as a "buffer" material. Bamboo, essentially medium fibre material, was then available in the country abundance. But now with bamboo resources shrinking larg mills have no other option but to mix hardwoods with bambo even though there is a decline in pulp quality. Use of soft woo is still very limited. The 28 large integrated mills, with a installed capacity of 1.4 million tonnes, use on an average 60 to 65 per cent bamboo, 30 to 35 per cent hardwoods, and very minimally other fibrous materials. The paper industry claims that it today accounts for about 2 per cent of the country's annual consumption of wood and 51 per cent of bamboo.

Bamboo Crisis

The growing bamboo shortage poses production and ecological problems : over-exploitation of existing forests and increased exploitation of remote forests.

According to the Pre-Investment Survey of Forest Resources, annual production of bamboo varies from 2.5 million tonnes to 3.0 million tonnes. About 1.6 million tonnes is supplied to existing paper and pulp mills and the remaining is used for construction by poor rural and urban people. Bamboo is popularly known as the poor person's timber and is already in short supply because of the paper mills. No sizeable bamboo resources remain to sustain new paper and pulp mills or even to feed the expansion of the existing ones.

Forests in Danger

In such a raw material crisis, all companies, private and public sector, want to head towards the last forest frontiers like Bastar, the Northeast and the Andamans. The government expects that the three paper mills of the public sector Hindustan Paper Corporation at Tuli in Nagaland and at Nowgong and Cachar in Assam will soon increase the country's paper production capacity by 233,000 tonnes of paper and paperboard. These plants will use bamboo as their main raw material. Lack of transport infrastructure is the main stumbling block in the exploitation of the 'green mines' of the Northeast but, once better roads are built this area is certain to be devastated. Meanwhile, a public sector paper mill is being considered by the Andhra Pradesh Industrial Development Corporation for Kakinada. This paper mill will use wood chips from the virgin Andaman Islands as its main raw material.

State governments in their bid to attract industry and politicians in their bid to make big money have often given away bamboo forests on lease at throwaway prices. The Orient Paper Mills of Amlai was initially given bamboo at dirt cheap prices, 37 paise per tonne by the Madhya Pradesh government. The Seshasayee Paper Mills in Tamil Nadu was paying Rs. 6.89 per tonne until 1970, Rs. 11 from 1870 to 1975, and Rs. 22 plus 5 per cent for administrative expenses since 1975. Paper mills in Karnataka were paying Rs. 15 per tonne for bamboo when the poor could purchase it in the open market only at Rs. 1,200 per tonne.

In recent years, however, many state governments have begun to revise royalty rates steeply upwards. Inevitably, all kinds of cheap raw materials. When the Orient Paper Mills closed down its Brajrajnagar plant in Orissa, ostensibly because of the workers' agitation for more bonus, it was widely rumoured that the management had done so deliberately to bring in cheaper bamboo from Orissa and pressurise the Madhya Pradesh government which had suddenly jacked up bamboo rates to Rs. 102 per tonne in 1981. There is also the case of a West Bengal paper mill which had approached Bihar government for allocating forest areas for its raw material requirements after its own sources in West Bengal ran short. But when the Bihar government refused to do so unless the mill was set up in Bihar, all that the company had to do was to send its truck to the Bihar-West Bengal border.

Thousand of headloaders would illegally deliver wood to it every day.

In 1980, the Public Accounts Committee of the Karnataka Legislature had studied the sale of eucalyptus plantations to the Gwalior Rayon Silk Mills - the rayon industry is another major user of pulp at Rs. 24 at tonne which was lower than the cost of production of Rs. 44 a tonne. The Committee suggested that the price was very low and in February 1981 the price of eucalyptus was fixed at Rs. 81 a tonne. But five days later the then Chief Minister R. Gundu Rao, gave oral instructions that the forest department should continue to sell forest produce at the old rates to industries after securing a bank guarantee representing the difference between the old and new rates. In April 1931 fresh instructions were issued to continue supplies at old rates against a bank guarantee of only Rs. 5 lakh. The Committee indicated the Chief Minister for staying the operation of the new rates, declaring it illegal and staying that it had led to a loss of Rs. 22 crore. The Committee also expressed dismay over the disappearance of the file in which Gundu Rao's oral instructions, staying the revised rates, were recorded.

But many state governments have not learnt the lesson in charging low royalties in the first place. The new 80,000 tonne per year newsprint plant of the public sector, Hindustan Paper Corporation, will pay the state of Kerala only Rs. 11 per tonne of green wood of eucalyptus, far below the cost of plantation. The Kerala government will supply eucalyptus and reeds on a sustained basis in adequate quantities. The management, however, does not rule out acute raw material shortages unless a massive plantation programme is taken up. Part of the mill's requirements are already being met through imported pulp.

Importance of the Forest

Forest is needed to conserve soil, water, air; to maintain the required level of oxygen, carbon dioxide, etc.; to resist pollution of air, water and soil; to maintain nature's equilibrium between living things; to supply medicines; to have a biotic check on pest and diseases; to screen irritating smoke, dust, moving sand, etc.; to absorb sound; to cut off gloves and divide lands for directing traffic; to plant barren hills to check erosion; to ensure privacy by planting small species of trees; to act as store-house of food; to check floods and droughts; to induce rain; to increase further agricultural yields; to provide organic chemicals, employment, income and profits; to conserve gene-pools and over all for holistic management of the ecosystems. It is the base camp of life's journey to the planet. Yet, the allocation of funds in Central and State budgets for forest is a pittance compared to the allocations to irrigation, energy and other sectors.

Depletion of the Forest

The depletion of the forest is due to its exhaustible use as fuel wood without consideration of its regrowth; its use for industrialisation, towns, living habitat for exploding population; refugee settlements; commercial exploitation for housing, household goods, agricultural uses, construction of roads, railways, airports, dams, mining; destruction by army and illegal felling due to corrupt officials and politicians, etc. Vanmahotsva, Social forestry, reclamation of land, Chipko and Appiko movements, forest week, etc., are nothing but apology of the mankind to the nature for the age long holocaust done by him on the fragile forest ecosystems.

Problem of Forestry in Few Indian States

To draw some of the fundamental conclusions regarding the forest problems of India it will be better to discuss some of the appealing problems of the forest of the states and their efforts in the field of conservation.

Nagaland

The deforestation of Nagaland is taking place at a frightening pace. Nagaland government report of 1983 shows that total forest area were 91,525 ha of which 29,255 ha were reserved and rest were protected. Private forests covered 1,81,150 ha of land. The State has 10 veneer mills, 100 large and medium industries, 200 small and 50 mini saw mills with the annual deforestation of 18,000 ha. This way, Nagaland would run out its exploitable forests within 10 years. The only forests which will survive are, those which cannot be exploited because of difficult terrains or inadequacy of communication. Besides, the smuggling, here instead of felling trees for stretches, forests are thinned out and which can on every hectare of land.

Himachal Pradesh

About 50,000 trees mostly spruce and fir, measuring 1,80,000 cu m are cut down every year to meet the requirements of the orchardists in the State. It is an apple state and due to the pressure fro apple lobby it is very difficult for the Chief Minister to take any stringent decisions. But due to pressure from Central Government, the Chief Minister has decided to switch over from wooden boxes to cartons for packing fruits especially apples. The cartons are now supplied on subsidised rate and the government has decided to set up one cardboard factory. The popular trees are also planted in captivity to ease the pressure on packing woods.

Madhya Pradesh

In 1986, the Chief Minister of Madhya Pradesh has banned felling of trees in 11 districts which does not include Bastar district. Unofficially, it is reported that more than 500 truckload of forest logs are smuggled out of the district every month with the collusion of the officials. What is more alarming is that not only the dry or dead woods were cut, and even the green trees were not spared by these unscrupulous elements. They exploit the tribals who have free access to protected forest areas and are allowed to carry a headload of fuel wood. This headload is often converted into truckload, and these trucks cross the state border unchecked.

The new onslaught is now on the sal forest of Baster (spread over 22,703 sq km area) setting up the hydel power project or for the craze of opening industry. It is estimated that Bodhghat Hydel Project will damage about 45 lakh green trees, including at least 17 per cent sal forest. Here is an important fact that sal trees cannot be grown by planting saplings, but can grow naturally when the sal seed falls on forest ground.

Gujarat

Gujarat forest department has drawn a special scheme for forest protection and employment for the tribal population. Although this is on meagre scale but it can be model for 14 per cent tribal population of the State and also to the other States of the country.

In order to pay the special attention on the development of the 194 forest village colonies of the State, the chief conservator has adopted a new model. The social defence scheme provides employment to the Adivasi families on a permanent basis in forestry. Each year beneficiary family is allotted 1.5 ha land on which they undertake afforestation according to the instructions of the forest department. By this way, they are assured 25 days employment in a month. These beneficiaries are considered as participants of this programme. They will be given 20 per cent of the net profit from tree cutting. This scheme provides them a livelihood means for maintenance. The scheme has benefited 311 families in 1985 86. A scheme for awarding prizes to these individuals and Panchayats helping in nabbing the forests criminals is also in force.

West Bengal

In this State small plots of government owned wasteland are being allotted to landless families for tree planting. The landless do not get the tillership of the land but they are given full right to use the trees they plant. They are provided seedlings at the planting stage. The same philosophy has been adopted by Maharashtra and Bihar also.

Tamil Nadu

Tamil Nadu shows an interesting report on social forestry. In Chengalpattu district, a eucalyptus hybrid plantation was begun in 1980. However, immediately after planting, 5,000 saplings were uprooted by villagers. An evaluation of 20 years of farm forestry of the same district (1960-80) in which 69 species were listed for plantation only two, babul and eucalptus hybrid dominate and the forest department persisted with eucalyptus even when the villagers demanding casuarina because it was more suitable for fuel.

The Mangrove of Different States

India has about 7,000 km² of its area under mangroves forest. These consist of deltaic mangroves along the eastern coast, estuarine mangroves along the western coast and insular mangroves on the Andaman and Nicobar islands. These forests are very important because it has got very low floristic diversity. They are under serious demographic pressure and creating ecological hazards due to their reckless cutting and extinction due to human interference.

Measures for Improvement of Forest Ecology in India

It is very difficult to give some concrete suggestions for improvement of the forest ecology of India. However, few measures can be adopted for maintaining the biological treasure of India.

(a) The Hill development plan should be both ecologically sustainable and economically viable for the rural community. There should be harmonious relationship between man and forest. Natural mixed forests should be allowed to grow and mono-culture is to be given up. Cultivation of agricultural crops in areas above 15° slopes should be stopped and planted with five "Fs" (food, fodder, fuel, fertilizer and fibre) trees.

(b) Industry producing dust and fumes should not be encouraged in the hilly areas. Household industries like spinning, weaving, knitting, embroidery, carving, watch making, etc., are best suited for the same area.

(c) Polyculture, instead of monoculture can be encouraged. Besides this, cooperation and collaboration of the local people must be sought at each and every level of policy formulation and implementation undertaken to save the areas from landslides and floods.

(d) The National Forest Policy of India has given emphasis to restore ecological balance, optimum utilization of land, water, livestock and human resources. It is having the three priorities *i.e.*, priorities of social forestry for meeting the domestic needs of rural and semi-urban communities specially for economically poor; production forestry, to cater to the needs of the forest based industries and environmental forestry. which would encompass both in restoring the ecological balance as at present impaired.

(e) The National Wasteland Development Board had been set up for dealing with matters relating to development of wasteland in the country, with special emphasis on social forestry-oriented to increase the fuelwood and fodder resources by massive plantation. The function of this Council is policy planning and coordination of all matters relating to the health and scientific management of the land resources of the country.

(f) The life of tribal population is intimately connected with forest and forest economy. For example, in Santhal Pargana forests has been cut recklessly due to stone-breaking work. The trees like Mahua, Sekhua, Herbs, Asan are reduced to minimum. Formerly, they were doing 'Pattal', Datwan making work Now, they are dependent on contractors for stone-breaking works which has also accelerated the soil erosion, and forced them to work on their own wish. The Government must check these illegal activities and special attention is to be paid to the activities that provide higher incomes to tribals from forest-based activities like production of lac, tassar, silk and other minor forest produce and improve their quality of life.

(g) The National Rural employment programme can be associated with employment-oriented forestry and environmental development programmes as most of the States are doing. In this process degraded and marginal forest areas should receive special attention together with integrated soil and water management.

(h) At the interval of every commissionaries a Regional Forest Research Centre is to be set up. In order to involve people in afforestation by educating them on the importance of trees in the economy and preservation of ecosystem, an extension organisation can also be creased.

(i) The uncontrolled grazing is also playing havoc on naturally regenerated forest crop. In order to contain this problem exclusive pasture can be established everywhere.

(j) There should be an aggressive production programme on all wastelands, marginal and submarginal agricultural lands to meet the needs of fuelwood and small timber. The marginal farmers whose lands are not economically viable can be motivated for tree plantation on the place of traditional cropping.

(k) Smokeless stoves biogas, solar power, wind mills, etc., can be encouraged to ease the pressure on forest.

(l) Universities, colleges can play a vital role in tree planting. In community forestry programme, active participation of school children is must.

(m) Officials should be careful in exotic varieties of tree planting as it can displace indigenous varieties.

(n) The scientific establishments and research agencies should direct their energies towards result-oriented research on improving the ecology like tissue culture for quick multiplication of seedlings, work on location-specific species like Jojoba on coastal saline wastelands, redevelopment of the lost mangrove system and aerial seedling in inaccessible areas.

(o) There should be complete moratorium on commercial felling in the Himalayas and other ecologically hilly areas.

(p) Successful use of multi crop pattern in agriculture, the Madhya Pradesh Forest Development Corporation has introduced plantation of more than one species of timber trees, such as bamboo and sheesham in the teak forest in Bari (Hoshangabad). While the teak tree takes about 60 years to mature, the bamboo and sheesham get ready much earlier. Besides, the felling of teak trees too is now planned to be done at three stages – one at the age of 10 and the other at the age of 30 – to get better returns

(q) There should be complete banning of the burning of waste wood in the forest because it creates fire hazards in the summer.

(r) Selection of trees is one of the most important phenomena. If we try to go against nature by planting trees not meant for the areas, nature will revolt and retaliate. In all the degraded area, after a cursery survey of the field, interviews with the local populance about the depleted species, one can decide which species will be better for the area.

(s) To encourage plantation, the people subsisting on selling fuelwood by destroying forest, can be employed to plant trees. The wage should be paid in kind, preferably wheat. It can be stored, if possible, in villages itself to inspire confidence among the people that food is available easily.

(t) To support the massive social forestry programme, adequate research input are needed in respect of fuelwood, biomass, fodder, trees, agro-forestry and genetic engineering. In the field of agro-forestry indepth study is to be done for intercropping relationship between fuelwood and agricultural crops as well as soil/moisture/nutrient/mycorrhizal relationship.

(u) The peoples' participation is must in forestry programme. The meaningful involvement of the people would lead to social fencing resulting the reduction of the unit cost of afforestation.

(v) A huge amount of bio mass is burnt all over the country for the religious and social purposes such as for dead bodies burning and different festivals like Holi and other local rituals, etc. The electric crematoria can be constructed on mass scale in all over the country and for this people's mentality can be changed through persuasion.

10

Dams

Introduction

Large dams are no more revered unthinkingly as the temples of India's progress. They are not, of course, white Sepulcheres but their destructive consequences are now being increasingly recognised. Even so, bigger and bigger investments are being made in them All major rivers today are dammed or are in the process of being so. Only a fraction of the country's hydropower potential has been realised. This indicates that in the next two or three decades, dam construction is going to be a major development activity.

In the light of recent experience, questions are now being raised about the viability of large dams which have often proved to be ecologically disasters. Can there be ecologically sound and economically viable engineering alternatives to large dams ? If dams are inevitable for a large country faced with the task of feeding its growing millions andfor meeting its energy requirements, can the destruction they cause be minimised and adequately managed ?

Hardly a day passes without a dignitary taking up cudgels on behalf of the construction of a big dam somewhere in the country or against one. Things have now reached such a pass that ministers are trading charges, without naming each other. A Congress leader has accused Ms Maneka Gandhi, minister of state for environment, of "subversion" by lobbying a West German minister to vote against support to the Sardar Sarovar project at the World Bank.

If nothing else, this demolishes a myth : that environment is a "non-political" issue. As should be apparent, it is entirely political for the simple reason that it concerns the use of natural resources to which competing classes and communities lay claim. Specifically in relation to "megaprojects" like those on the Narmada or at Tehri, thereare people who stand to grin : immediately, contractors and their employees, as well as the funders; ultimately, the beneficiaries. In the "opposition" are those who will be displaced, together with environmentalists and human rights activitists.

While the Sardar Sarovar in Gujarat and Narmada Sagar upstream in MP are probably the biggest among dams being built in the world, and Tehri among the highest, disputes are also simmering over Bodhghat in Bastar, Koel-karo, near Ranchi and Pooyamkutty in Idukki district, among others.

Clear Policy

Clearly, the government will have to make up its mind on what its policy towards big dams should be if its does not want to confirm the impression that it is ruling in different directions. This is not a matter of public relations but hard economics. Already, Rs. 400 crores has been spent over 16 years on Tehri and half as much on Sardar Sarovar; every day's delay costs lakhs. Not that money spent should be the clincher : long-term losses, if a project is completed, could outweight such expenditure.

The time-honoured path will be to plump for "development", even while protecting the environment. However, as the heated exchanges between protagonists of big dams (like Mr Sanat Mehta of the Sardar Sarovar Nigam) and opponents (like Baba Amte) show, the debate has transcended such platitudes. The crux of the matter is : who is benefiting from such projects, and who is paying the cost in social terms as well ?

There is no question that big irrigation and power schemes have since the original "temple or today". Bhakra-Nangal, been an integral part of the agricultural and industrial growth of the country. At the same time, there is no getting away from the fact that for all the euphoria over the Green Revolution, the present annual output of cereals, a total of 170 million tonnes, is on the Planning Commission's own norm of 4 tonnes a hectare from irrigated land – 70 million tonnes less than what should be the contribution of irrigated area alone. As much as 70 per cent of arable land today is rain-fed and even when the total irrigation potential is exhausted, half will always depend on the monsoon. If the present productivity of farming is tremendously low, it is due to poor land and water management.

Two-Fold Benefits

The two-fold benefits of big dams can be subject to scrutiny. Take power, which may have helped the country figure among the dozen largest producers of goods in the world. But this does camouflage the fact that electricity is not quite the "general good" that it is made out to be, given its present use. Two-thirds of it is consumed by industry. While no one can deny the massive strides made by industry since independence, on one count-employment-its record is absymal. In 1988, the entire organised sector, which includes both the public and private sector, accounted for only 26 million jobs. Whatever the other benefits of technological advances, increased employment is certainly not one of them. Factories, between them, hired less than 8 million hands in 1987.

Five-year plans have allocated one-fifth of the resources to the power sector alone and this is considered a since qua non of progress. Unfortunately, planners mistake electricity for energy. For that matter, only "commercial" energy is accounted for, the rest being generated at no money cost. Half the total energy consumed in the country is by way of fuelwood for cooking, and most of it is collected "free" that is, at great expense of human labour. If one looks at the needs of the neediest, therefore, it is futile pointing to electricity. It is the hearth at home that is cold.

Only Luddites would argue against the need to generate power. As energy experts like Dr. Amulya Reddy (now a Planning Commission member) have long observed, people's homes should be the first to be electricified; in practice, rural electrification is the bane of all state electricity boards. The social and economic spin-offs of extending the working day would be far greater than those of lopsided industrial growth. Even though the country has been able to electrify nearly eight out of every ten villages, only 14 per cent of the households in a typical village on the grid can afford to instal connections to their homes.

This is the context of the debate over big dams. On both counts, irrigation and power, the benefits can be questioned. Even that votary of a 21st century India, Mr. Rajiv Gandhi, is on record as saying : "We need some definite thrusts from the (big surface irrigation) projects that we started after 1970. Perhaps we can safely say that almost no benefit has come to the people from these projects. For 16 years, we have poured money out. The people have got nothing back : no irrigation, no water, no increase in production, no help in their daily life".

The opponents of big dams are also often barking up the wrong tree. It is the entire development process that is being debated, not just a "big versus small' slanging match If the policies followed in the last 20 years of encouraging capital and resource-intensive inputs in agriculture are to continue, together with a highly selective pattern of industrialisation (mirrored in uneven urbanisation), there is no way of reducing the nation's dependence on such projects.

Anti-Dam Campaigns

Meanwhile, Chipko Movement leader Chandi Prasad Bhatt has started a compaign against the Vishauprayag Hydroelectric Project because of the dangers if poses to the unique ecosystem of the Valley of Flowers and the general geological havoc it could create. Bhatt, as he has shown with the Chipko Movement, is capable of mobilising strong support among the local population though, a section of the workers who have found jobs on the project oppose the compaign.

Another instance of intervention at the highest levels resulting of a large dam is Lalpur in Gujarat. Gandhian social worker, Harivallabh Parikh, an ardent supporter of Mrs. Gandhi and a close associate of Jinabhai Darji, an Influential Congress (I) leader in the state, runs an ashram at Rangpur, which would have been submerged if the project had materialised. He raised two issues-destruction of forests and displacement of tribals-in a letter to Mrs. Gandhi. A committee is preparing an alternative plan.

These very issues are cited by Baba Amte, while mobilising protest rallies against the Sardar Sarovar. Bhopalpatnam and Inchampalli Dams. He is also questioning the very reisond etre big dams by asking: whose development do these dams represent ? Amte's strategy is to mobilise people through mass rallies and public opinion by getting in touch with environmentalists, political leaders, planners, journalists and even the Prime Minister himself through meetings, discussions and personal letters.

The protest against the Tehri dam in Garhwal is another, where along with rehabilitation, environmental issues relating to large dams have also been highlighted. The Bandh Virodhi Sangharsh Samiti led by advocate V.D. Saklani has underlined the threat to the ecological balance of this sensitive mountain region through rallies, seminars and even wall writing. The movement has gained momentum and local support because of the strident protests by citizens of Tehri town, which will be submerged by the hydro project.

Organised Tribals

Tribal populations who are affected by alarge dam are fully aware that they will lose their trees, land and houses. The Munda tribals of Chhota Nagpur, who celebrate their festivals by planting saplings, are today fighting against the massive Koel Karo Project. As one tribal from the area, questioned about the compensation he will get, replied: "Can you give back our trees planted by our forefathers ? The tribals of Chhota Nagpur likely to be affected by the Koel Karo

Project are well organised, with a coordination committee and a representative from each village. Local leaders of various political parties are also part of the organisation making it probably the strongest popular movement in India against a dam.

The movement has a single demand : comprehensive rehabilitation-social, cultural and economic. The Koel Karo Jana Sangathan is involved in negotiations with the government for getting as much compensation as possible. In Koel Karo, people succeeded in stopping work by their Kam Roko (Stop Work) Movement in 1978 and want the authorities to involve the affected people in fixing rehabilitation terms. In mid-1984, the new Chief Minister of the state had ordered armed police into the area to put down the tribals.

By contrast, in the absence of organised protests or support from any voluntary groups, people's houses were bulldozed for the Srisalaim Project in Andhra Pradesh and the villagers were hounded out by demolition squads.

As far as rehabilitation is concerned the movement in the villages affected by the Sardar Sarovar dam on the Narmada is setting an example. The Narmada Tribunal Award spelt out terms for villagers displaced from Madhya Pradesh and Maharashtra and, under pressure from various voluntary groups, the Gujarat government had to agree to similar terms. To ensure that the rehabilitation programme is carried out properly volunteers of the Gujarat Chhatra Yuva Sangharsh Vahini have gone into the villages and are both educating and organising the villagers. Researchers from the Centre for social studies at Surat have conducted a detailed house to-house survey in all the villages and presented detailed reports on their needs. At the state level, there are other voluntary groups which have been instrumental in bringing pressure to bear on the government.

Future Issues

Large dams are generally being constructed in areas which have so far not been touched by modern development. Many people in these areas look at such projects as symbols of hope. They even accept the argument that someone has to pay the cost for the benefits to society, even if they are the ones who have to pay. As things are today, people are not likely to oppose dams en mase. But they are definitely getting ready to agitate for better rehabilitation terms. But also as more and more dams are chosen on environmentally sensitive sites, as in the Himalayas, or in heavily forested areas, protests against dams are likely to grow. Meanwhile, a number of experts have even begun to question the very usefulness of the large surface irrigation systems that dams promised, which means that a thorough analysis of India's current water management plans-and especially of dams-is urgently needed.

In most cases, as in Koel Karo, these groups are fighting mainly for better terms of rehabilitation. But in a few cases, citizens' groups have even begun to demand that a dam or a hydroelectric project not be built at all either because of the numbers of people it would displace, as with Bedthi or with Bhopalpatnam and Inchampalli, or because of the unique ecosystem it would destroy, as in the case of Silent Valley or the Vishnuprayag Hydroelectric Project. Some of the antidam campaigns have even been successful in stopping these massive projects : Silent valley and Bedthi have already been given up by the Kerala and Karnataka governments respectively. And government committees are investigating the environmental impact of the Lalpur dam, the Vishnuprayag hydroelectric project and the Bhopalpatnam and Inchampalli dams.

Dams and Forests

A large dam inevitably alters the course of nature and, because of the play of complex and largely unknown forces, the ecological consequences are usually much more serious than were bargained. Silient Valley has been dropped but the government has decided to go ahead with Tehri, Vishnuprayag and many other projects. Also being proposed is the Manibhadra dam in Orissa, in one of the prime forest areas remaining in the country.

The forest cover bears the brunt of dam construction in more than one way. In the Himalayan valleys, contractors who come to clear the forest which is going to be submerged greedily cut trees even in unaffected areas. People who are being displaced from Tehri town in Garhwal to make way for the dam will be rehabilitated on forest land that has been acquired and cleared near Hardwar and Rishikesh. The forest is also being cleared for approach roads, offices, residential quarters and for storage of construction material. With the reduction in forest cover and the entry of people, the pressure on the remaining forest increases. Their needs for firewood leads to further danudation. The construction of a dam, therefore, has a multiplier effect.

The 1983-84 annual report of the Department of Environment reveals that the construction of the Idukki Dam over the Periyar river in Kerala has hastened the degradation of vegetation and sharply reduced the forest cover. It is estimated that during the period 1951-1976, 0.49 million hectares of forest has been lost due to major river valley project. This is roughly 10 per cent of the area that came under canal irrigation – 5.5 million hectares.

Despite the increased awareness about the need to project forest cover, little attention is paid to this factor when a site is selected for a dam, basically because all forests are state property and no one has to be compensated for them. Even in cost benefit calculations, forest loss is grossly undervalued and the recurring loss in terms of timber and firewood yield per year and social costs of degradation of environment are never taken into account. Under valuation of standing forest helps in boosting the cost-benefit ratio for obtaining Planning Commission approval and finance from international agencies like the World Bank. Moreover, the contract for clearing the forest is also undervalued, thus leaving a bigger margin to be shared between the contractors and the forest department.

Because of public protest, some provision is now made for reforestation or compensatory afforestation. In Tehri, for instance, Rs. 10 crore have been provided for this. With the best will in the world, it will take at least 10 to 12 years for forest cover to appear. Compensatory afforestation in catchment areas should, therefore start with the planning of the project, since it will also help in reducing flow of silt into the reservoir.

Dams and People

It is unfortunate that, despite the massive dams built over three decades, there is no rehabilitation policy worth the name. Rehabilitation becomes much more important and difficult when most of the displaced are tribals. Tribal life and culture is so uniquely interwoven and integrated with the forests that once uprooted, they are totally disoriented.

The main problems regarding rehabilitation have been well evoked by the remarkable studies on the gigantic Sardar Sarovar (Narmada) Project in Gujarat conducted by the Centre of Social Studies (CSS) in Surat and the Lokayan study on Srisailam. The issues can be summarised as follows: rehabilitation has always been considered an obstacle in implementing a project and an extrafinancial burden; the authorities always try and get away with minimum; no guidelines exist to include even the barest arrangements for rehabilitation in project plans; and facilities provided differ from project to project.

Provision of measures like "land for land" and basic amenities like drinking water, schools, dispensaries and so on were made obligatory in the case of Sardar Sarovar. That had been made possible because of a dispute between the three states - Maharashtra, Madhya Pradesh and Gujarat– which was handed over to a tribunal. The tribunal outlined the rehabilitation arrangement to be carried out by the Gujarat government. As Gujarat was to bear the costs of rehabilitation of the affected villagers of Maharashtra and Madhya Pradesh these two states saw to it that fairly elaborate measures were given in the tribunal's award.

The award was silent about the rehabilitation of the 'oustees' from affected villages in Gujarat. It was only after a tough stand by voluntary agencies that the Gujarat government agreed to similar provisions for its own oustees. When Gujarat got down to implementing these measures, it was reported to be shirking its promises. In any event, the Narmada Award has become a precedent for subsequent hydroprojects.

Meagre Compensation

The only legal power with the government to handle rehabilitation is the Land Acquisition Act, enacted by the British in 1894. Apart from cash compensation, any effort to resettle people is considered charity. Compensation for land and houses is usually grossly inadequate, and none has ever been paid for loss of employment or disruption of livelihood other than from land. Landless labourers, marginal farmers, sharecroppers, tenant cultivators and those who live on collecting and selling forest produce have suffered most. Village artisans whose base is the village also get uprooted. Small farmers who supplement their income by rearing cattle are hit because new settlements never have enough grazing land. Villagers are forced to sell off their cattle at throwaway prices before shifting.

People engaged in non-agricultural activity, like fishing are also not entitled to compensation. For example, there are 750 families along a 150 km river stretch downstream of Broach town who make a living out of catching hilsa fish. The construction of Sardar Sarovar will interfere with hilsa migration and 750 families will be rendered unemployed.

No one in authoritty even pauses to ponder over the fate of the people involved in so-called marginal occupations.

To add insult to injury, the meagre rehabilitation measures are implemental half-heartedly. The entire process, beginning with the notices issued to the villagers for acquisition of land to the final settlement, is a stoıy of misery. Poor people are harassed, hounded out of their villages and handicapped for the rest of their lives. The whole operation is conducted in a most unimaginative and bureaucratic manner as the story of oustees from the Srisailam Project demonstrates.

Nightmarish Experience

The law requires that written notice be issued individually to the villagers for acquisition of their land and houses. The villagers, mostly tribals, are illietrate and ignorant of the legalities and get confused. As the experience of the Koel Karo project in Bihar shows, panic spreads from much earlier when the project is being planned. Rumours abound about the rate of compensation and resettlement sites.

In the case of Srisailam, the notice specified terms of compensation with a clause that appeals about the rate that had to be submitted in writing by stipulated date. The rate of compensation was very low and depended upon the type of land, house and so on thus leaving a lot of elbow room to the land acquisition officers. Though the law provided for compensation at the market rate according to the latest land deals in the area, the compensation was arbitrarily fixed by the authorities.

It has been the experience everywhere that the rich farmers bribe the officers and manage to get higher compensation and better alternative lands, whereas the poor keep running from pillar to post. Vidut Joshi of the CSS in Surat points out that the terms of acquisition and other procedures were never orally explained to the people in the Sardar Sarover project and in the process illiterate people were cheated by the officers and their intermediaries, except in cases where activists of voluntary organisation came to their help.

Even when compensation is paid to villagers through their bank accounts, it has been found that part of it goes in sustaining them during the shift to a new site and much of the rest is frittered away in non-productive expenditure. Another problem is that government takes only that land which is to be submerged or is required for other purposes, leaving many partly affected people in the lurch. When part of a village is shifted elsewhere, the families left behined face a life of isolation, in some cases, when the major portion of a family's land is taken over, it can neither survive on the remaining area nor get any compensation for the remaining land or for its house if it decides to shift along with fellow villagers.

Such cases were never considered until the Narmada Tribunal specified that if 75 per cent of a family's land was affected, the entire land should be taken over and it should also be paid compensation for the house and transport for shifting. Even now, there is no provision for families who are isolated in partly affected villages. Almost all the farmers left in the vicinity of Shivsagar (Koyna) reservoir live in scattered basis, below the poverty line as they have no means of subsistence except for a few quintals of paddy grown by primitive methods.

Resettlement Problems

"Land for land" is a better policy than cash settlement and the Narmada Tribunal has in fact specified that a minimum of 5 acres should be given to each affected family. Even in implementing this policy, the land is not given in the command area in most cases, forest land is either cleared on waste fallow land given without any provision for developing the land or for the supply of necessary inputs; a village is broken up and families dispersed; villagers are usually left to buy private land, take loans from the government, which puts poor villagers at a disadvantage land prices in neighbouring villages shoot up steeply if the government takes up resettlement, the villagers are resettled in distant places. sometimes in a totally alien environment and culture, thus creating insurmountable adjustment problems.

Oustees from Pong Dam in Himachal Pradesh were settled in Anupgarh in Rajastan, bordering on Pakisthan. Though the land falls in the command area of the Rajasthan canal the villagers, who were used to hill cultivation, found in difficult to adjust to the rigours of the desert.

Tribals affected by Inchampalli and Bhopalpatnam in Maharashtra do not even speak Marathi and if they are resettled away from forests and in an alien Marathi-speaking area, they will find it difficult to survive and shifting them to any other area would amount to 'ethnocide', in the words of Baba Amte, the social activist who has been leading a mass movement against the two dams.

When the construction schedule of a dam is planned, residential quarters for officers and other staff are always given priority. In the Sardar Sarovar project, kevadia colony; a well-developed residental area with markets, schools, parks and other facilities was kept ready before the staff moved in. On the other hand, when villagers wer served legal notices for acquisition of lands, no preparation was made for their shift to alternative sites. Even if some sites were located, no effort was made to develop the sites. The people were generally left to fend for themselves. Arrangements for drinking water dispensaries, schools, village roads or drainage at the rehabilitaion sites are only completed years later.

Helping Hands

In the case of the Ukai Dam in Gujarat, resettlement work was undertaken by the Ukai Nav Nirman Samiti, a voluntary agency. What is more, the then Chief Minister of Gujarat belonged to the same district. Even so, out of a total of 18,500 affected families, only 3,500 families could be resettled. Though Rs. 7 crore was paid as compensation, only Rs. one crore was deposited in banks, the rest being squandered or used for day-to-day needs. In the case of the Koyna Dam, when a survey of displaced people was conducted by Deccan College in Pune in 1969, it was found that around 12 families from practically every village had returned to their neighbourhood.

Since those who are displaced are the illiterate and unorganised rural poor they do not have only political clout. Whennever they have been organised with outside help they have succeeded in getting a better deal. The Ukai Nav Nirman Samiti handled the rehabilitation with some success with the support of the then Chief Minister of Gujarat Baba Amte in Gadchiroli District. Maharashtra and the Koel-Karo Jana Sangathan near Ranchi, Bihar, are raising issues which are fundamental to a rehabilitation policy.

The Koel Karo people are demanding that the government spell out comprehensive rehabilitation policy -- economic, social and cultural – to the satisfication of the people affected by the project. The struggle committee consists of heads of all the villages involved and they have succeeded in stalling the construction of even the approach roads to the dam site. The state government hag been forced to enter into negotiations with the people's representatives. The proposal put forward by the state government includes reservations for the displaced people in Class III and Class IV jobs and setting up of vocational training facilities for the youth but provision of land-for-land is not part of the deal.

In Gujarat, four voluntary agencies have been involved in rehabilitating the villagers displaced by Sardar Sarovar on the Narmada. Chhatra Yuva Sangharsh Vahini volunteers have gone into the villages to educate villagers about compensation and rehabilitation and mobilise them against corruption and for a better deal in accordance with the Narmada Tribunal Award. The action Research in Community Health and Development group of Mangrol and the Rajpipla Social Service Society are working in cooperation with the Sangharsh Vahini and are liaising between the people and state government. The CCS in Surat has done commendable work in studying each affected village in detail, and indicating the rehabilitation requirements of the people.

Rehabilitation Policy

A comprehensive rehabilitation policy can be spelt out only when development is understood as a process which benefits the people. If the completion of a project of the minimum cost becomes the overriding objective, rehabilitation simply amounts to evacuation. People who could previously barely manage to survive in their traditional environment are uprooted as a result. If people are the majo rfocus, any project should seek to improve the conditions of those living in object poverty.

According to the CCS, the objectives of rehabilitation should be :

- The people displaced should get an appropriate share in the fruits of development.
- They should be rehabilitated by creating new settlements within their own environment.
- Removal of poverty should also be an objective of the rehabilitation policy and therefore some land to all.
- Oustees (even the landless) should be given assurance of employment.
- While dealing with tribals one should also keep in mind the following five principles of tribal-development accepted during Jawaharlal Nehru's-era as tribal panchsheel'.
- Tribals should develop along the lines of their own genius and we should avoid imposing anything on them.
- We should try to encourage their own traditional arts and culture in every way.
- Tribal rights in land and forests should be preserved.
- We should try to train a team of their own people to do the work of administration and development.
- We should not overadminister these areas of overwhelm them with a multiplicity of schemes but should rather work through, and not in rivalry with, their own social and cultural institutions.

Keeping these in view, the following general guidelines can be specified for any rehabilitation programme :

— Land for land should be provided with an economically viable minimum of 5 acres even to the landless.

— Resettlement should be in the neighbourhood of their own environment.

If resettlement is not possible in the command area, top priority should be given to the development of irrigation facilities and supply of basic inputs for agriculture; drinking water, wells, grazing grounds for cattle, schools for the children, primary health care units and other amenities should be arranged.

In partly affected village, villegers should be given the option of shifting out with other with the same compensation as available to evacuees.

—Training facilities should be set up to upgrade the skills of affected people and reservation in jobs should be made for the willing adults among the evacuees.

—Special attention should be given to the rehabilitation of artisans and village crafts people.

— Villagers should be taken into confidence at every stage of implementation and they should be educated, through open meetings and discussion, about the legalties of the Land Acquisition Act and other rehabilitation provisions.

—The aid of voluntary agencies should be taken in the planning and implementation of the rehabilitation programme.

It can be argued that it is inevitable that any development project will benefit some citizens more than others and invariably some citizens may have to bear the additional cost. In the case of big dams, like most other development projects, the direct beneficiaries are generally the rural rich and the educated urban elites, whereas the direct social costs are borne by the rural poor, mainly tribals. If people who pay the social costs start to count in the development strategy, the entire orientation of the planning process will have to change.

Alternatives to Big Dams

Storage of water by building large dams is regarded by irrigation engineers as a crucial component of India's water management system. The Central Water Commission argues that benefits like hydroelectricity, irrigation, flood control, industrial and municipal water supply are not possible "unless the water is stored in reservoirs created by dams". Water storage is necessary because rainfall is very unevenly distributed, varying from 10,000 mm in Cherapunji to about 100 mm in western parts of Rajasthan. There are also considerable seasonal variations. Over 75 per cent of the rainfall is received during the monsoon period of June to October and even during this period there are many erratic spells of heavy rain and dry weather. On one hand as many as 99 districts are classified as draught prone; one the other, 85 per cent to 90 per cent of the mansoon water flows as waste to the sea.

Since Independence, about 700 storage dams have been constructed. The total storage capacity created so far, including those of small tanks, is assessed to be about 16 million hectare-metre. Out of a total of 178 mham surface water available in the country, only 27 mham are now being used and the utilisable potential is about 67 mham. About 90 per cent of the water being used is for irrigation alone.

By the year 2000, the government hopes to provide irrigation to 113 million hectares, the ultimate irrigation potential out of a gross cropped area of 200 million hectares. Major and medium irrigation projects would irrigate about 584 million hectares and minor irrigation projects 54.6 million hectares. Between 1980 and 2000, some Rs. 60,000 crore (at 1980 prices) will be required for the development of this irrigation potential. At the beginning of the Sixth Plan, a total potential of 56.51 million hectares had been created of which 26.61 million hectares was through major and medium schemes and the rest through minor irrigation.

By June 1984, about 65.6 million hectares of irrigation potential had been created.

Potential Capacity

As even by the year 2000 nearly 40 per cent of cropped area will be unirrigated, the government is preparing a national perspective plan for water resources development, which will allow for large scale inter-basin transfer of water from water surplus river basins to water scarce river basins. This will mean that an additional 22 mham can be utilised. But these inter basin transfers of water will again depend primarily on the creation of storage reservoirs : the ultimate storage capacity in that case has been assessed at 37.5 mham.

The irrigation potential will thus increase from 113 million hectares to 148 million hectares with about 25 million hectares coming under surface schemes and about 10 million hectares benefiting from increased use of ground water. This ambitious plan will require an investment of about Rs. 50,000 crore. The investigation and planning work of the National Water Development Agency charged with the development of the perspective plan will cost Rs. 107 crore spread over 7 to 10 years.

The Commission argues that if water resources have to be harnessed by building storage reservoirs then either developed areas with a large population, agricultural areas of forest areas with a tribal population, have to be submerged "If none of these are to be affected . we may have to stop all further progress in water resources development."

Thus, if government programmes go ahead as scheduled, there will be hardly any free-flowing rivers left in the country. Already the Cauvery is fully exploited. The Krishna and Godavari rivers, with large dams like Tungabhadra, Srisailam, Nagar junasagar, Mula, Jayakwadi completed or near completion and several new rivers in the offing like Inchampalli and Polavaram should be fully exploited soon. With the award of the Narmaada Tribunal, the Narmada valley development programme has also been launched and by the early part of the next century, the Narmada will be a fully controlled river. The Tapti with Ukai and Upper Tapti dams will soon become a fully used river. Mahi with Bajaj sagar and Kanada dams faces the same prospect, and so does the Betwa with Paricha, Dhukwan. Orchha, Matatila and Rajghat dams.

The waters of the Sone with Bamraha Bansagar, Sukhar Churhat, Hurma, Rihand and Kanhar will have been bottled up by 1995. The Mahanadi with Dudhawa, Ravishankarsagar, Hasdeo, Jonk, Arpa, Sheonath, Pairi, Hirakund and Tikkarpara dams will have been mostly used up in the next 20 years. The rivers in the Indus Basin, Suttlej. Beas and Ravi are also being controlled and the northern tributaries of the Ganga (Bhagirathi, Sharda, Ghagra and Kosi) too are likely to be fully exploited by the turn of the century.

Emerging Problems

But whatever the Central Water Commission may argue, more and more experts are beginning to question the very utility of big dams, and especially of the large surface irrigation schemes they offer to the country. In a study on canal irrigation and its impact on crop production and environment, D.R. Bhumbla, former Agricultural Commissioner and Vice-Chancellor of Haryana Agriculture University, argues that "from the experience of major and medium irrigation works in India, it is evident that its benefits in arid areas though spectacular for the first 10 to 20 years, gradually get reduced and a considerable portion of the land gets deteriorated because of water logging and salinity. In the humid areas, the benefits are doubtful from the beginning and in many cases negative. The programmes of major and medium irrigation works as has been envisaged for future in humid areas, in my opinion, would not only lead to disastrous consequences in degradation of soil and environment but would also result in reduced agricultural production. It is time to halt the expansion of these programmes".

In low rainfall areas of Rajasthan and similar areas of Punjab and Haryana (300 mm to 400 mm), canal irrigation is necessary because raising crops without irrigation is either not possible or uneconomical. Moreover, the ground water is saline in many areas and hence the potential for tubewell irrigation is limited. The introduction of canals in these areas has resulted in a significant increase in yields of crops like cotton in the kharif season and wheat in the rabi. In semi-arid regions of Tamil Nadu Andhra Pradesh and Karnataka, where rice and sugarcane are the main irrigated crops, increases in yields have been significant.

But in all these areas waterlogging and soil salinisation have emerged as serious problems. As far as unlined canals and distributaries are concerned, a study by the Central Water and Power Commission in 1967 revealed that about 71 per cent of the water is lost in transit from the reservoir to the field as highlighted in Table 1.

Table 1

Loss of water from unlined canals due to seepage (per cent)

Canals	15
Distributaries	7
Watercourses	22
Field Losses	27
Total Loss	71

Source : Interim Report of National Committee on the Use of Plastics in Agriculture 1982.

Increasing Waterlogging

The Water table in the Bhakra canal command area has been rising at the rate of about one metre per year. The managing director of Haryana State Minor Irrigation and Tubewell Corporation, in a press conference in late 1985, expressed serious concern over the rising ground water table in the state. He said that 65 per cent of the area in the state covered by canal commands had saline groundwater aquifers underneath and in 4 lakh hectares, the water table had entered the critical zone of upto 3 metres depth. This was creating water logging accompanied by salinisation.

The rising water table will affect the production of many crops adversely. The story of canal irrigated are as of pakistan where cotton has vanished from many fields, is likely to be repeated, says Bhumbla. The high water table, apart from affecting agricultural productivity, has many other adverse effects. Because of the loss of moisture storing capacity of the soil, even slight rains cause floods. The maintenance of road and buildings becomes difficult and diseases increase because of waterlogging. In Hissar in Haryana, the load bearing strength of the soil has become less than 50 per cent in 15 years.

In the sub-humid areas of the country (rainfall more than 1,000 mm), Bhumbla argues that the benefits of canal irrigation are questionable even in the short run. In fact, in many areas, introduction of canals has resulted in reduction in productive of crops. In Bihar, Madhya Pradesh and Orissa, which together have 2.5 million hectares of canal irrigated area and where rice is the main irrigated crop, the yield per hectare is not only less than one tonne, but there has been perceptible increase for the last 20 years. In Uttar Pradesh, which has the maximum canal irrigated area a part, there is a general complaint that from pre-sowing irrigation, hardly one irrigation is possible and the wheat crop frequently suffers from moisture stress.

During the drought of 1979-.0, the farmers in the Sarda Sahayak command area complained there was no water in the canal but in the wet year of 19 8, the canal water aggravated the flood problem created by excessive rain. In the rice growing area, water seldom available for raising rice nurseries before the start of the monsoon rains. Thus, in the non-rainy season, winter or summer, when irrigation could result in significant increases in yield enough water is not available.

Further, in the humid areas, it becomes difficult to message canal water during the rainy season and the problem of waterlogging becomes serious in a much shorter time. In fact, this may aggravate the problem of drainage and result in loss of production.

The situation is likely to be much worse as canal irrigation is introduced in high rainfall areas particularly if the soils have low permeability as has been demonstrated by the situation in Tawa and Barna command areas, where yields of all crops have declined since the introduction of canals. Bhumbia therefore concludes : The effect of canal irrigation on increasing cropping intensity is marginal". He argues that the irrigation departments do not even produce relevant data to show the cropping intensity in lands irrigated by canals and tubewells. All that we know is that in 1976-77, the net irrigated area was 34.8 million hectares, and the gross irrigated area was 43.1 million hectares, which gives a cropping intensity of 1.24 which is very low and a good reason for the low productivity of irrigated lands in India.

Bhumbla argues that in most canal irrigated areas cropping intensity is less than that in tubewell irrigated areas, generally two or more. In other words, a substantial increase in food production in the last decade or so has actually come from tubewell irrigated areas. Bhumbla argues that the government should concentrate on ground water development and management of water by storing it in catchment areas in small reservoirs of canal irrigation schemes.

Small is Beautiful

Groundwater development, minor irrigation schemes and micro hydel projects are often cited as alternatives to big dams, but are looked down upon by power engineers and planners. Small schemes cannot be considered as alternative to big dams for water management and power generation in all cases. But there is definitely a strong case for ground water development and minor schemes for irrigation, while power generation needs can be partly met through small hydel projects.

Like surface water, ground water resources in the country are being depleted because of the lower water retention in catchment areas which have been deprived of forest cover. The development of catchment areas, therefore, should become an integral part of an irrigation programme. This will also check the flow of silt into big reservoirs. There have been calls for the creation of a catchment area development authority.

Net Irrigated Area by Sources

(million hectares)

	Canals	*Tubewells*	*Total*
1950-51	83	NA	20.9
1960-61	10.4	0.13	24.7
1970-71	12.9	4.46	31.1
1978-79	150	8.2	38.0

Source : Indian Agriculture in Brief, 18th edition.

Wherever water resources have been harnessed through small reservoirs by constructing bunds or small dams, the lives and environment of the people in the area have been transformed without anyone having to pay the cost of development. Whether it is through the voluntary efforts of Prembhai, recipient of the Jamnalal Bajaj Award for Rural Development, who has been instrumental in the construction of over 750 small dams in Mirzapur district of Uttar Pradesh, or the initiative of P.R. Mishra, a soil conservation expert at Sukhomajri near Chandigarh, small dams have generated far more benefits than large irrigation schemes in terms of utilisation of water, improvement of agricultural productivity and ecological enhancement.

Successful Efforts

Prembhai claims that by constructing small bandhis (check dams), land belonging to the entire village has improved. A bandhi is constructed at an average cost of Rs. 60,000 and irrigates 40 acres. As a result of well-planned bandhis, the cropping intensity in their project area, has increased by 60 per cent and agricultural production by 160 per cent.

Because of improved water retention, even the forest has grown better. All the bandhis have been constructed entirely by local effort, thus giving employment to local people. In a bad monsoon, when the water dries up, desiltation is taken up and the fertile silt put back into the fields. Thus a farmer is helped even in the lean years.

At Sukhomajri, a small dam two more were added later was constructed at the foothill of a highly eroded ravine. The water provided irrigation to 80 acres of village land. Farm production improved and the farmers levelled and bunded their own land. Because of the increased yield the availability of fodder from crop residue improved, reducing the grazing pressure of non-agricultural land. The improved micro-climate in the watershed saw the spectacular growth of trees, bushed and grasses in the forest. Villagers who used to be driven out of Sukhomajri near-desert conditions during the lean season have now turned to growing fruit trees and vegetables to tide over the lean season.

Three rain-fed reservoirs, varying between 200 feet and 400 feet in diameter, were created by constructing small earthern dams in the water shed area of the Sukhana river. This has brought about a total transformation in the life and environment in Sukhomajri village, 35 km of north of Chandigarh. The implications of the project, basically a soil conservation effort to begin with, now embrace so many facets of development that Mishra, the project director, insists on calling a comprehensive growth model and not just a model for tank technology, water harvesting, recycling, or tree planting. The results are outstanding. The grass yield in the hill has shot up from less than 40 kg per year to more than 2,000 kg, Milk alone accounts for an increase of Rs. 5 lakh in annual income. Now the village society is planning to establish hill product-based village industries where the members of the society will get productive and more remunerative employment. Sukhomajri has shown that watershed management at the village level by construction of small dams has the potential of overall enhancement of the environment, without causing any destruction or displacement.

Viable Options

Very seldom are all technological options considered by irrigation planners. In a report submitted to the Department of Science and Technology on the environmental planning of Shahdol district, the Shahdol Group points out that in the context of the geography of the area there are five technical options : wells, ponds small dams, cascade dams and major dams to harness the water in the district. Of these four are dependent on surface water management while the fifth (wells) is based on ground water management. In an attempt to determine which is the optimal technology for the region the group has made a comparative assessment of the five technologies against several technical economic and parametres. On the basis of this matrix, they argue, that the system of cascade dams fits into the geographical and social characteristics of the region. But this is the one option of the five that the stateg overnment has totally ignored. On the contrary, major World Bank-aided investments have been allocated for the development of wells and the construction of a large multipurpose dam on the Sone river. Both of which are unsuited to local developmental needs in terms of topography and available resaurces.

The government is building a major multipurpose dam at Deolong on the Sone called the Bansagar Hydel Project, to be completed by 1984 at an estimated cost of Rs. 20 crore. Some 1,27,000 people will be displaced and 51,648 hectares of land (two thirds of which is cultivated and a fifth is forest) will be submerged. This has created enormous tension among the local population. The proposed ten townships for resettlement may never materialise and alternative employment may never be available as was the case with the nearby Rihand dam built on a tributory of the Sone. Most of the villagers are entirely ignorant of the schedules of evacuation, submergence and completion. Some farmers are hoping to get land in the command area of the dam and ask for immediate imposition of ceilings in the proposed command area before the beneficiaries of the region, mainly around Rewa and Idhi districts of M.P. and some parts of U.P. rearrange their legal records to retain the irrigated land. The changed status of land from unirrigated to irrigated will make one-third to two-thirds of the land in the command areas available for distribution.

Wells and pumps are the other technical options in which banks have been asked to invest. But this option has neither looked into the technical features of availability of subsoil water and its management nor into the social aspects of land distribution and repayment capacities. Consequently, wells have been dug where no water is available, the size of the wells has been determined by standard specifications set by the banks rather than by water availability, in order to meet targets the loans have been given to larger farmers rather than to the small farmers for whom the money was originally intended; further as the number of wells dug increases, the groundwater level recedes and thus negatives the whole exercise of water development.

Apart from wells and large dam ponds (which were mainly built in the days of the princely states) have been increasingly under pressure to change from domestic use to irrigation, pisciculture, and hyroponic cultivation; thus resulting in mismatch between design and utilisation and further increasing tensions within the village between ordinary villagers, rich, farmers and contractors. Small dams have never been seen as part of a river valley system and so their location has been technically faulty, subject to the whims of local leaders and a source of major revenue to petty contractors. As the government's own records show, these minor irrigation projects, designed for aiding kharif cultivation, have been used almost excessively for rabi and, there too, actual coverage has reached only 30 per cent of targeted irrigation. And when all on-going government schemes are taken into account, only about 5 per cent of the cropped area in the district will be irrigated

Given the undulating topography, the system of cascade dams based on an integrated view of the total water resources available in river valley, provides a distinct alternative to the present scheme Situated in the Vindhya range, a substantial amount of water can be stored by building small reservoirs on the nullahs (drains) which surround almost all villages. Several goods sites exist wherein runaway flows can be checked with little effort.

Flow irrigation is also possible in many sites. A number of villages can be served by shallow lifting devices. Others will need powered pumps. Existing tanks can be used or expanded, where necessary. Several small storage schemes in the foothills are possible. Low stone masonry weirs could also be constructed across river beds to store monsoon waters in the river bed itself. The area to be irrigated being only 30 per cent of the total area of the catchment, it should be possible to meet the entire irrigation needs by properly harnessing as much precipitation as possible.

Similarly, the setting up of mini and micro hydel projects will prove to be a boon for the electrification and overall development of remote rural and hilly areas where power from the central grid can never reach them or will entail heavy transmission losses. A considerable portion of the hydropotential in fast flowing hill streams, canals and river slopes can be tapped through small hydel schemes. Self contained turbine generators can utilise even the normal flow of rivers in some cases.

These small schemes will probably receive their due only when the overall development of far flung rural areas becomes the focus of the country's planning. China is an off-quoted example where thousands of small dams and mini hydel projects have been constructed which have provided water and power right at the point where they are needed and thus helped in overall development, ensuring basic amenities to the entire population. A much greater decentralised organisational effort is required to plan and implement schemes on the lines of China or of Sukhomajri or Mirzapur. On the other hand, large dams are constructed by centralised organisations and big contractors with huge funds borrowed from the World Bank and with imported machinery.

If irrigation alone is considered, small schemes are far better than large ones. But small hydel projects cannot meet the hydro requirements of the central power grid. The country will have to search for other sources of renewable energy because even so-called "clean" hydel is not so spotless if the destruction caused by big dams is analysed.

Ramchandra Singh Dao, former Madhya Pradesh Irrigation Minister has suggested that the construction of major reservoirs should be rescheduled. The building of minor or medium dams should start at the source of the river and proceed downwards. Major dams should be built only afrer the water power resources upstream have been fully harnessed. Such a system combines the advantages of big and small dams. Deo argues that no new major projects should be launched for the next 20 years; only medium and minor irrigation schemes should be taken up and benefits of the irrigation capacity already created should be consolidated.

FLASH POINTS

The Koel Karo Struggle

The potentially violent confrontation seems to be on the cards in Bihar over the Rs. 600 crore Koel Karo project, whose 710 MW capacity equals the entire state's current power generation capability. For years, tribals organised into the Koel Karo Jana Sangathan, made of people who will be displaced, have stalled construction, charging that the state government is deliberately underplaying the extent of inundation. According to their estimates, one lakh people, 15,000 families in about 100 villages, will be affected. According to government estimates, around 50,000 acres of land will have to be acquired, out of which half will be agricultural, affecting 7,573 families in 50 villages. Both estimates are just guess estimates since even the government department has not been able to survey the area so far.

The project is to be constructed by the National Hydroelectric Power Corporation (NHPC) and envisages a 44 metre earthen dam across the South Koel river and a 55 meter dam on the North Karo river. The two reservoirs are to be linked by a 4.7 km channel. Preliminary surveys were done in 1954-55 and the project was approved by the Planning Commission in 1973. Almost simultaneously struggle committees came up in some of the villages likely to be affected. In 1974, these committees were coordinated into the Koel Karo Jana Sangathan by political activists of the Jharkhand Mukti Morcha and the Communist Party of India.

The Jana Sangathan fired a volley of demands to the state government. However, no concrete proposals of rehabilitation came back and the state government decided to go ahead, early in 1978, with the construction of approach roads to the dam site. The tribals responded with a Roko (stop work) movement; a barricade put up by the villages in 1978 on the approach road even today stands testimony to their resolve not to allow any government vehicle to reach the dam site. As a result, a tripartite meeting in 1978 conducted by the Deputy Commissioner of Ranchi decided to survey the affected families to prepare a rehabilitation scheme.

This plan was also soon buried. In October 1981, the government sent a formal proposal to the people, mentioning only compensation without any rehabilitation. Two technical training institutes were set up in 1981, with 25 per cent of the seats reserved for affected people. At least one person from each family would be absorbed in government jobs.

The Jana Sangathan was not satisfied. They demanded land for land before construction, traditional rights and customs of tribals to be legally protected in the course of rehabilitation; and a concrete scheme for total rehabilitation.

The District Collector sat down with Sangathan representatives in July 1983, and the government came out with a fresh proposal which included : cash compensation for land varying from Rs. 8,000 to Rs. 20,000 per acre; 0.25 acre of free land as house site for each displaced family; reservation in Class II and Class IV jobs in Chhota Nagpur for the displaced; employment in NHPC and with its contractors for affected people—three quarters of Class III and all the Class IV jobs reserved for them; people displaced to be rehabilitated in labouring villages no more than a quarter of land in any village to be thus acquired; and full compensation paid into bank accounts and all amenities, drinking water, school, health centre and so on, to be provided before evacuation.

By December 1983 the government and people had still not reached an agreement. Nearly 200 NHPC staff, including 35 qualified engineers, are twiddling their thumbs at Torpa, the site headquarters 60 km from Ranchi. The NHPC has spent Rs. 4 crore on infrastructural facilities and chairman Pizada Ghulam Nabi admits that it has all been a colossal waste of human power and resources. About the only break through so far is that the road blockade is lifted for NHPC officials-but for no one else-whenever they undertake a periodic four of the countryside on yet another mission of persuasion. Only 3,000 acres of the 50,000 acres have been acquired s far.

While former Chief Minister Jagannath Mishra had dithered over the project, the new Chief Minister, has sent in armed constabulary to dishearten the villagers. Now rumours abound that the army will be called in and the villages thrown out without compensation. Rumours also say that the government is offering jobs and if the tribals keep fighting, they will get neither compensation nor jobs. But people are sceptical about government promises. They point to the technical institutes, which have no machines or facilities and the first batch of students did not get jobs.

At the moment the tribals are standing firm. They insist that they do not want cash compensation but new land for the land taken away from them, a principle now accepted in law in Maharashtra and by the M P. and Gujarat governments for the settlement of the people displaced by the Narmada Dam. The Ranchi tribals believe that if they waver and the government somehow succeeds in starting work with or without use of force-there will be a scramble. In the ensuing chaos, the poor will suffer, while middlemen, government officials and rich farmers will once again make a fast buck. The worst sufferers will be all those tribals who do not have any legal deeds to their land, as is common in many tribal areas.

The Story of Sardar Sarovar Dam

Work on the Sardar Sarovar project in Gujarat is in full swing. Whilst the dam is in Bharuch district of Gujarat, the lake which will form behind it is largely located in Madhya Pradesh. One hundred and eighty two villages in this State will be affected wholly or in part. Between 60,000 and 80,000 people will be displaced (the exact figure is still in some doubt), thousands of hectares of fertile agricultural land, continuously cultivated for centuries, and vast tracts of forests will be submerged and whole villages will disappear from the map of India.

It is believed that large areas of land in Gujarat will be irrigated and many more thousands of people benefited. For this some sacrifices will have to be made. Unfortunately, i is the people of Madhya Pradesh who will make the sacrifices. Who are these people ? They inhabit the fertile tract of Nimar, which has deep black cotton soil. Some of them live in Malwa, which is an equally rich area. Around them are the Satpudas, amongst the oldest mountain ranges in the world. Many of them are ancient tribals, Bhils, Bhilalas and Gonds.

Ancient Civilisation

Going by the cave paintings which have been discovered throughout the Narmada valley, these tribals represent a civilisation which goes back to when homo-sapiens was still evolving from ape to man. As the direct descendants of these ancestors the narmada Valley tribals are the oldest civilised inhabitants of this region. The land they till, the forests in which they dwell, the water that they drink have been in their ownership long before Gujarat developed as a settled region. If anyone should have rights in India, it is these tribals.

The justification for the dams on the Narmada is that the parched land of Gujarat can only prosper if the waters now flowing into the sea are impounded and diverted to the deserts, the dry lands and the cities of Gujarat. One cannot fault this argument because water is, in a semi-desert, a source of life. One wishes upon Gujarat the blessings of the waters of the Narmada. But Gujarat goes beyond this argument. In impounding the waters of the Narmada it argues that so high a dam should be built that not a drop of water reaches the ocean without first being utilised by the people of Gujarat.

Such impounding will certainly inundate the lands lying upstream of the dam. But these lands are after all inhabited by non-Gujaratis, largely wild tribals, whose habitat is surrounded by worthless jungle. If there is displacement of these people and if the jungles are felled, it is only wi derness which is lost. Therefore, the interests of Gujarat become paramount.

Displaced People

The Chief Minister of Gujarat is on record to have stated that the dam will be built, if necessary by calling in the army. Against whom will the army be used ? The displaced people crying for justice ?

The areas being submerged may appear wilderness to an urban dwelling Gujarati, but to the tribals of Madhya Pradesh this is their home. How do they view the forests of the Satpudas and the plains of Nimar ? Peter Matthiessen, in his book entitled "Indian Country", states what the American Indians felt about lands considered wild by the white man. He says : "The wilderness was merely uninhabited, a place where we do not live, a further part of home; it was only unknown when ones communion with it had not yet been revealed". The Gonds, Bhils and Bhilalas are in communion with their lands and to them it is not wilderness.

The award of the tribunal on Narmada dispute categorically states that the displaced people will be rehabilitated on land, with the responsibility vesting in the Gujarat Government The Gujarat Government has been unable to offer a square inch of public land which is arable to the Government of Madhya Pradesh for resettlement of the oustees. Instead an ingenious offer has been made, which states that the oustees may come to Gujarat, select private lands, strike deals for purchase with the land owners and claim reimbursement of cost from the Gujarat Government. Over which eyes is the wool sought to be pulled by making this offer ? Our cultivators have neither the time nor the means to go land hunting in Gujarat, nor do they have the skills to negotiate with their more cany counterparts in that State, nor are they likely to find any real absence of red tape in Gujarat obtaining reimbursement of cost. The Government of Madhya Pradesh cannot even ask the displaced people to express a desire to move to Gujarat unless they can be shown the land on which it is intended to settle them. By making this offer Gujarat probably hopes to duck its responsibility of resettlement, whilst expressing to the world its willingness to accommodate all the oustees. One would like to ask the Gujarat Government whether this is a fair and honourable thing to do.

What will happen to the forests when Sardar Sarovar is completed ? Once again Peter Matthiessen answers the question beautifully by quoting the American Indians, "When the buffaloes are all slaughtered the wild horses all tamed, the secret corners of the forests heavy with the scent of many men, and the view of the ripe hills blotted by taking wires, where is the thicket ? Gone, Where is the eagle ? Gone, And what is it to say good-bye to the swift and the hunt, the end of living and the beginning of survival ?" That is what will happen to the forests of the Satpudas. One does not know whether the prosperity that Sardar Sarovar will bring to Gujarat will last for ever, but certainly the forests will be gone for ever. Has Gujarat ever sat back and thought of what will happen to the Narmada when the forests are gone ?

Legal Position

Gujarat is proceeding with the construction of the dam on an assumption that it has the right to block the river and oust the people of Madhya Pradesh regardless of whether or not lands are acquired and the people resettled. Gujarat however cannot submerge an inch of land anywhere unless it has been legally acquired and transferred to State ownership. Even if the land is so acquired Gujarat cannot legally submerge any land unless the last oustee is properly rehabilitated on land to his satisfaction. The tribunal award itself makes this clear and the award is binding. It is for the Government of Madhya Pradesh to bring home this reality to the Government of Gujarat so that the plans for rehabilitation are quickly formulated and implemented.

Encroachment

If the people are not rehabilitated they will encroach on forest areas in Madhya Pradesh and further destroy the environment. If Gujarat and its people are to benefit from the project at the cost of the people of Madhya Pradesh, that State must also share the sufferings of our people and the benefits that will flow to the Gujaratis. Why should Gujarat not acquire from people in the command a sufficiency of land to rehabilitate all the oustees from Madhya Pradesh and Maharashtra ? If people in Gujarat can look forward to higher productivity because of irrigation they should be prepared to sacrifice a portion of the land of each beneficiary so that it can be pooled and distributed to the oustees. This is but fair and just. In a society wedded to socialism there has to be real social justice. For the people of Madhya Pradesh Sardar Sarovar personifies social injustice incarnate.

The Madhya Pradesh tribals are peaceful and docile, but they are not without rights. The right to land, their own land, is as fundamental as the right of the people of Gujarat to development. They cannot be treated like animals reserved for slaughter, to be herded out of their homes and discarded on payment of a pittance by way of compensation. There is a grounds well of resistance already building up against the injustice about to be perpetrated on the potential oustees. Gujarat cast the first stone when it declared several hundred kilometers of area around Sardar Sarovar as prohibited under the Official Secrets Act. What is secret about a dam ? One can understand prohibitory orders under Section 30 of the Police Act or Section 144, Criminal Procedure Code. But a notification under the Official Secrets Act ? There must be a strain of insanity in the officials who issued this notification.

11

Habitat

Introduction

The cities are booming. In 1981 India's urban population of 156 million was the world' fourth largest behind only USA, USSR and China. By 2000 A.D., India's towns and cities will have anywhere between 350 million and 400 million inhabitants. There will be 20 cities with more than a milion each, and Calcutta, Bombay, Delhi and Madras will have more than 10 million each.

Planning for this urban colossus will be a Herculean task. At the moment it looks a losing battle. Despite 60 odd statutoty development authorities in various cities and regions, slums continue to grow. The Planning Commission estimates that over 30 million people, one-fifth of the urban population, live in slums. Other estimates make it 30 per cent.

Some 3-5 million slum dwellers live on about 8,000 acres of Bombay's land; over 400 persons per acre. By current trends, 75 per cent of Bombay's population will be living in slums by the turn of the century. Some 100,000 people already live on the pavements, which is better than in Calcutta, where an estimated 600,000 people go to sleep every night in the streets. Bombay can boast of having the world's biggest slum, a sprawling conglomeration of shanties called Dharavi.

Conditions in the slum are more than shocking. A survey of about 4,000 households in nine slums in Bombay reveals that nearly 40 per cent have two to four persons packed into one room, another 35 per cent households have five to nine persons crrmmed into one living room, and one per cent have 10 or more persons living in the room.

No house has a private toilet. A quarter of the households do not even have access to community toilets and use the open spaces around the slum, for defecation. Over a third have no drainage facilities and another 40 per cent have uncovered drains. This lack of amenities makes the slum environment extremely dirty and prone to sickness.

Planning and legal systems, marked by upper class values, have turned slum dwellers into second class citizens. Leadin, urban planner M.N. Buch says that the "real tragedy of urban India" is that the texture of the built environment of the city is judged not in terms of what it provides to the citizen but rather what visual picture it presents to a (foreign) visitor".

For years local and national government agencies reacted by bulldozing the shacks. Sometimes efforts were made to resettle the slum-dwellers in 'pucca' (cement) houses in so called slum

Growth in Population of Major Indian Cities ('000)

	1901	*1951*	*1961*	*1971*	*1981*
Calcutta U.A.	1488	4589	5737	7081	9166
Greater Bombay	813	2967	4152	5971	8203
Delhi U.A.	214	1437	2359	3637	5752
Madras U.A.	594	1542	1945	3170	4277
Hyderabad U.A.	448	1128	1249	1796	2566
Ahmedabad U.A.	186	877	1206	1742	2515
Bangalore	159	779	1200	1654	2914
Kanpur U.A.	203	705	971	1275	1685
Pune U.A.	164	608	791	1135	1685
Nagpur	128	449	644	930	1298

clearance and resettlement schemes. By end-1976, the government had built about 1.16 lakh houses to accommodate displaced slum-dwellers. But the houses, built with construction standards and regulations meant for the middle classes, proved prohibitively expensive. Moreover, the cheapest tenement could not be afforded by 50 to 70 per cent of the urban population. As a result, houses built for low income groups have been acquired by middle income groups and houses for middle income groups by upper income groups.

The new strategy now accepted widely, is to let slums stay, and to provide water, drainage, community latrines and paved streets. Both, the Draft Five-Year Plan (1978-83) prepared under the Janata government and the Sixth Five-Year Plan (1980-1985) prepared under the Congress (I) government, had stated that substantially increased investments will have to be made in slum improvement.

Unemployment Myths

In a country like India, urban poverty and rural urban migration are closely related. When cities grow at yearly rates of 5 per cent or more, as Bombay, Calcutta, Madras, Delhi, Hyderabad, Bangalore, Ahmedabad have over the last three decades, at least half the growth is caused by net 'in-migration', concentrated very heavily in prime working areas, swelling the urban labour force, holding down wages and incomes in the urban economy, straining housing and related urban services to their limit.

But the belief that such migrants are unemployed parasites is absolutely incorrect. Survey results from Calcutta, Bombay, Delhi and Madras clearly show that most migrants find jobs fair quicly, suggesting prearranged employment. A recent study of unemployment in Calcutta, for instance found that unemployment was high among urban-born, educated youth but very low among the poorer sections.

Muslim migrants from Bihar and eastern Uttar Pradesh were largely absorbed in the city's jute industry. Hindus from rural Bengal were concentrated in the engineering, chemicals, rubber and printing industries. Calcutta's rickshaw pullers are largely migrants from Bihar and Orissa. The occupations of plumbers, domestic servants and paanwallas are dominated by people from Orissa, and tanning and leather works are largely a Bihar Muslim activity, Vast occupational networks are at work among the poor to find work for the newcomer.

In Bombay, textile mill workers are mainly from the Maharashta districts of Ratnagiri, Sangli and Satara. Construction workers are largely from Andhra Pradesh. Restaurant and catering industry jobs are cornered by young migrants from Udupi, Karnataka. Paanwallas and roadside food vendors are usually 'bhaiyas' from Uttar Pradesh.

A study of Bombay's labour market points out that less than 15 per cent of the poor migrants studied were employed, and that most of them regarded themselves as better off for having moved to the city. This study shows that the duration for which a migrant remains unemployed is related to the resourcefulness and "influence" of friends and relatives: "Majority of the migrants (77.7 per cent) have some support during the period o waiting (for a job). Family members and relatives from the largest single source which helped 45 per cent of the migrants. Caste fellows and friends form the native place each helped 11 per cent of the migrants get jobs. Of those with none to support them 61 per cent took up employment (usually of the casual variety) immediately.

Migration as a Strategy

The cities, compared with the villages, promise lighter work and higher and steadier wages. A study from Delhi states that the migrants' average earnings over the year were two and a half times more than what they could earn in the rural area. Many or most migrants see urban earnings as supplementing the rural, their urban stay as temporary and their places of origin as their permanent homes. Thus, the migrants range from the seasonal or shuttle migrant coming for a few months to the permanent settler seeking a new way of life.

Seasonal or short-term migration is usually resorted to by agricultural labourers, tenant farmers and small landholders during slack cultivation seasons. Many rural workers rely party on farming and party on some trade, craft or service skill, and their nonagricultural occupations may take them periodically to neighbouring twons and cities.

Then there are 'target migrants' : those who come with a speciffc purpose in mind, usually to earn a lumpsum of money for marriages or agricultural investment, and plan to return home once that is achieved. Target migrants may sometimes stay in the city for several years.

The most substantial section, however, are the "working life migrants", who move to the cities for the rest of their working lives. But they keep in close touch with their places of origin, often leave part of their nuclear families there, and plan to retire at home.

The bhaiyas of Uttar Pradesh, as they are called in Bombay, epitomise the working life migrant. William Rowe has provided a very vivid portrait of the migrant bhaiya as seen from the vantage point of an Uttar Pradesh village called Senapur :

"It is 'recurrent' rather than 'permanent' because these men traditionally go to the city without their wives and families and return home annually for visits of a month of six weeks. With few exceptions these men view migration to the city as a temporary solution to economic problems and maintain their involvement in the network of personal relationships in the villages. In severel cases they have worked for 25-30 years in Bombay which is 900 miles from the village... migrants to the city either remit or bring home with them all of their income beyond their minimum living expenses. This income has a differential function for the various families and caste communities who have deputed migrants to the city. Some men are pulled to the city because it is here that money can be earned to improve the family position at home (through purchase of additional land or farm animals or other capital expenditure)."

Slums as Solutions

Given that kind of attitude of life, how do the slum dwellers themselves look at their unstable huts ? The starting answer turn up by studies is that they do not really think it is such a terrible problem. The poor city bound migrants come from mud huts in the villages and expect little more, at least upon arrival. Village water supplies are neither more accessible nor much purer than those of the city. Lavatories and waste disposal systems are usually nonexistent back home.

A second factor is the correlation between caste religious status and housing quality. In Hyderabad, for instance, 59 per cent of slum dwellers are estimated to be from the scheduled castes and tribes, and 34 per cent are Muslims. In many parts of India, especially in the villages, prior to Independence Harijans were forbidden to build solid dwellings. They were forced to live in huts and forbidden to use common village wells. In that context the Harijans in cities may not feel so deprived today as outside observers might imagine.

The urban poor can therefore tolerate slum conditions better than the urban rich. On occasions, when government housing programmes have allowed other housing choices to slum dwellers, many of them have sold out their official houses and returned to their original slums. Research and interviews with groups of squatters conducted by Meera Bapat in Pune's slum clearly indicated that their principal concern is employment, housing is second. They seek the crowded, inner-city locations because the chances of finding jobs, especially the unskilled, casual jobs for which they are most eligible are so much higher in the central business district than anywhere else.

In a national survey of settlement patterns of the urban poor, Rakesh Mohan (housing expert and constultant Planning Commission) has concisely summarised the underlying logic and evolving priorities. "To poorest (for example, fresh migrants) are mainly interested in location. Being near job markets saves on transport costs. In their highly uncertain situation the only security they are interested in is job security. Their meagre income only allows for food consumption and other bare essentials in a kind of lexicographic prdering. The only amenity they need is space for sleeping. The next group, with a reasonably stable income put still not well-off, is interested in security of tenure. A temporary jobless does not then mean displacement of residence as well. They are also more interested in space than amenities and are willing to pay for it. Finally the richest, low income group is more interested in amenties, having a stable income and subsistence essentials. Electricity, plumbing, well-designed housing and recreation then become important and will be demanded by this group."

But central land is most likely to be owned either privately or by the government, squeezing the squatter out. The slum dweller sees the makeshift hut, often on encorached land, as a solution rather than a problem. It may be ugly, insanitary, illegal, but it does nevertheless serve as a shelter.

The Housing Crisis

There are thus two major aspects of the housing crisis. On the one hand, the failure of the government to view urban housing as a multidimensional process which involves making provision for several other needs apart from physical shelter as a result, after 35 years of governmental planning. There is a colossal housing and infrastructural shortage which has steadily accumulated over the years and is now, as the Planning Commission itself admits, beyond solution. On the other hand, the urban poor have solved this problem in their own desperate manner, by building clusters of hutments and shanties wherever and however they can. The resulting slums, barring a South African type authoritarian 'solution', are here to stay.

Given this grim scenario, Indians should perhaps learn to live with slums. Planners are also beginning to realise that squatters are economically valuable citizens who add to the gross national product by constructing their own shelter, no matter how make-shift, which saves the government a considerable amount of money. That squatters are upwardly mobile citizens in search of economic opportunity and have demonstrated high levels of enterprise, tenacity, and ability to suffer acute hardships; that the informal sector in which a majority of the slum dwellers are economically active contributes significantly to the city's overall economic growth; and that they should be helped and not hindered.

What can be done is to upgrade, not demolish, most existing slums, and move the remaining squatters and pavement dwellers to reasonably well-located vacant lands as part of a site-and-services scheme. In at least a few cities, local governments are attempting to act on these views and help the urban poor settle down. Ahmedabad, Calcutta, Madras and Hyderabad are experimenting with slum upgrading programmes. There are two basic elements; security of tenure and improvement of the physical conditions and basic amenities. The experience of these cities merit attention both as examples of what is possible with all the constraints, and as alternative to the constraints, and as alternative to the essentially demolition and eviction-oriented policies of Bombay and Delhi.

Ahmedabad

The squatter relocation project in Vasna suburb combines self-help with government assistance. With about 2.25 million people in the metropolitan region. Ahmedabad has about 7000 slums and squatter settlements, housing nearly 415,000 people. After the 1973 floods in the city's Sabarmati river which washed away thousands of squatter homes, the Ahmedabad Study Action Group (ASAG), a voluntary interdiciplinary organisation of various professionals, submitted to the municipal authorities an intergrated plan for moving the displaced squatters to a new suburban settlement. Their plan involved building for them new homes and providing social and economic welfare assistance to raise their living standards and their ability to participate in the city's economic activity. A constant refrain of the ASAG proposals was that 'slums are people, not places'.

The state government was eventually pushed into allocating 43 acres for a new settlement located about 7 km. from the city centre. The Ahmedabad municipality agreed to extend piped water, sewerage, street lighting, roads, and community facilities such as schools, shops and health centres. HUDCO (Housing and Urban Development Corporation) approved low interest 20 years loans to the occupants which was followed by state subsidy of Rs. 700 per family and another Rs. 400 per family was provided by Oxfam. A majority of the flood dislocated squatters agreed to move to Vasna.

The results, nine years after the first bricks were laid in May 1974, have, however, been mixed. Problems of quality of shelter construction, distance from the city and employment sites, an artificial juxtaposition of diverse ethnic/caste groups, incomplete city drainage facilites, differing economic and social priorities among the squatters themselves, and political corruption have marred the project. Many of the original squatters have returned to locations closer to the city centre. Other people have moved in and the neighbourhood now is ethnically more homogenous in Muslims now forming 75 per cent of its residents. Despite these problems and changes, however, Vasna remains a very interesting experiment.

Calcutta

In 1971 the Calcutta Metropolitan Development Authority (CMDA) started the country's most ambitious slum upgrading programme to cover in stages, by 1984, the city's slum population of approximately three million. By 1980 the total slum population had risen to 2.7 million crammed into some 3,000 reparate slums. Calcutta's planners have tried to improve the existing slum environment and make it more livable.

The CMDA has focussed on six areas : conversion of existing service latrines to sanitary latrines and the construction of a new toilet for every 25 persons; potable water tap connection to individual huts or for every 100 persons with an average water suyply of 90 litres per head; surface drainge facilities through underground conduits connected to an underground sewer system wherever available, or open surface drain discharging into the nearest open drainage channels; paving of roads, streets and pathways; and miscellaneous facilities such as garbage vats, dustbins and batching facilities at suitable locations.

These basic civic services cost Rs. 120 per person between 1971 and 1973, rising to Rs. 150 later. As the programme spread away from areas with basic infrastructure, the cost rose to Rs. 350 as totally new facilities had to be installed. By 1981 soms 1.9 million slum inhabitants had been serviced at a total cost of Rs. 25 crore.

"Implementing the programme was a huge technological challenge", says M.S. Moitra, Director General (Operations) of the CMDA and the brain behind the project. For instance, as most slums were not served by the city systems, 'captive' sewer systems had to be designed for entire slums, by which the slums sewage was collected by interconnecting pipes and then dumped into the nearest city sewer. Cheap latrines were bulk manufactured.

Unfortunately, the problems remain; some 70 per cent of the assets created are out of order at any given time. Thousands of latrines are choked with excreta and garbage. Taps stay dry. Drains overflow after slight rain. Uncovered manholes are safety hazards. Street lights are without power, the blocked latrines and leaking sewer pipes have sometimes led to a spate of cholera and diarrhoea in recent years.

"It would cost only Rs. 3 per household per year to maintain these services," says Moitra. "The existing system of entrusting contractors with responsibility for maintaining services is inefficient and costly. "The CNDA has spent Rs. 25 crore on a Rs. 5 crore job. The rest of the money has simply been siphoned off on contracts". A World Bank mission even took note of this recently. Proposals to involve the community in maintenace have not been welcomed, in fear that the politicians would lose their clout,

Madras

In 1971, the Tamil Nadu Slum Clearance Board (TNSCB) decided to resettle all hutment slum dwellers of Madras city—about 0.74 million people at the time—in pucca tenements to be constructed with public funds by 1977. By 1978, instead of the 1,64,000 tenements planned at an estimated cost of Rs. 164 crore, barely 23,000 were actually completed at a cost of Rs. 25 crore.

The authorities shifted their emphasis to upgradation from 1977 onwards, encouraged by a World Bank loan. The two major schemes are the Environmental Improvement Scheme (EIS) and the World Bank Improvement Scheme (WBIS).

Under the EIS, 293 slums were covered to benefit about 1.05 lakh households at an estimated cost of Rs. 5.12 crore. The local authorities soon realised that the funds would not provide adequate services for all the beneficiary households. The amount spent varied considerably, less than Rs. 100 per household in some slums to Rs. 1,300 per household in others. The average expediture was about Rs. 500 per household.

A survey of the actual improvements in 23 slums was carried out in 1980 by the Operations Research Group (ORG), an independent research unit, which found that less than Rs. 625 per household had actually been spent in 90 per cent of the slums. The report contrasts the case of VOC Nagar where expenditure was the highest, Rs. 1,300 per household, with Karunanidhi Nagar, where it was Rs. 500 per household. Whie VOC Nagar's 490 households had been provided with baths and lavatories, stormwater drains, roads and wells, their less favoured counterparts in Karunanidni Nagar were provided with only stormwater drains and roads.

Maintenance has been neglected. The 1980 survey of VOC Nagar found "the baths, latrines and wells under disuse, the roads broken and full of ditches and stormwater drains in no condition to clear stormwater". "Of the households contacted nobody today finds any use for any of the above facilities", says the report. The water from the wells, say the slum dwellers, is filthy and dirty and without water being available, the baths and latrines cannot be used either. Their old lifestyle for relieving themselves and bathing in open spaces is no worse.

Karunanidhi Nagar was just as bad if not a little worse. "In fact, stormwater drains and some parts have even ceased to exist and at other parts mud sand have totally filled them up. The hard surface of the road is barely visible at places and at some places ditches have formed which contained water and dirt" the ORG survey found.

Slum Population Growth in Some Major Indian Cities

Metropolitan City	*Year*	*Population of City (millions)*	*Slum Populations (millions)*	*% of Slum Population*
Calcutta	1974	3.19	1.44	45.00
	1981	9.16	3.24	35.35
Greater Bombay	1976	7.94	3.24	41.00
	1981	8.2	4.20	51.03
Delhi	1977	5.29	1.32	25.00
	1981	5.71	3.32	53.00
Madras	1977	2.47	0.99	40.09
	1981	4.27	1.36	31.86
Hyderabad	1977	1.99	0.4	20.00
	1981	2.52	0.54	21.28
Ahmedabad	1976	1.89	0.42	22.00
	1981	1.51	0.66	25.16
Kanpur	1976	1.33	0.56	40.34
	1981	1.68	0.68	40.34

The maintenance of TNSCB facilities was to be the responsibility of the Madras Municipal Corporation. But such a clear division did not materialise. The result, concludes the 1980 survey, "which was observed at the slum level was that though occasionally MMC staff would come far cleaning roads in the areas under their jurisdiction, they would not clean or look to the upkeep of the facilities created by the TNSCB (in the same area)." Within four years of provision of services, the entire investment had come to nought.

The WBIS seems to be relatively better planned and more comprehensive. Part of the multisectoral Madras Urban Development Project-1 taken up in 1977-78 under the coordination of the Madras Metropolitan Development Authority (MMDA), the scheme, whose implementation is the responsibility of the TNSCB, is intended to benefit 30,000 households in 36 slums at an average cost of Rs. 1,300 per family. Unlike the EIS the WBIS has standard guidelines, the standard being uniform for all slums. In addition to routine provisions, there is provision for primary and secondary schools, plots for children's creches and sheds providing work space for suitable cottage industries. Preliminary reports indicate that this upgrading project has been significantly more successful than the previous one.

Slum City

Dharavi. Bombay's and Asia's largest slum, sprawls over 4.5 sq. km. of prime land in Central Bombay. This prime real estate, where huts today fetch a price of Rs. 95,000, was originally creek land, where families of potters and tanners – the Bandra Slaughter House, then existing, was close by – had pitched their shanties. They were soon followed by garbage, rag pickers and scrap dealers and others who brought in their wake common goondas, thugs and slumlords.

The encroachers originally occupied private land, but by and by took over municipal as well as government-owned land. Today 2.5 sq. km. of Dharavi land belongs to the Corporation, and one kilometre each to private landowners and the government. Unofficial estimates put the number of shanties, pucca and semi-pucca constructions at over 65,000. The area has a stable population of over 3 lakhs and a floating population of over one lakh, which means that a substantial 4 per cent of Bombay lives here.

The slum is a melting pot : 67 per cent of the population is Hindu, 26 per cent Muslim and the remaining 7 per cent divided among the Christians, Sikhs and other minority groups. Though some of Dharavi's 44 colonies are divided community-wise, most are grouped together by states.

Hut sizes vary from 150 sq. ft. huts to over 500 sq. ft. Income groups range from Rs. 400 · about 60 per cent make this much to those whose earnings could be pegged at anywhere between Rs. 500 and Rs. 1,500 per month and even more. Besides shopowners, a large chunk of Dharavi's population is in government service.

It is loosely estimated that over five persons live in a hut, without counting the 'guests' in 'permanent residence'. In addition, the slum dwellers tend to hide their kids, whenever a survey is undertaken, because they think that they will be forcibly sterilised. The average per hut might be as much as eight persons.

Hyderabad

Hyderabad's population in 1981 was estimated at 2.2 million of which approximately 500,000 lived in slums. The city's slum dwellers have participated in an urban community development (UCD) project which covered barely 4,000 people in nine slums in 1967 but embraced 400,000 people in 459 slums by 1981. It was adopted in 1974 by the Hyderabad Municipal Corporation as its main policy for slums, and over the period 1974-84 has proved to be a highly successful slum upgrading programme which now covers over 81 per cent of the city's slum population.

Hyderabad was one of the 20 cities along with Ahmedabad, Baroda, Calcutta and others, selected in 967 for a Central Government-sponsored series of pilot projects in urban community development. But it is only in Hyderabad that progress was made in converting a successful pilot project into a relatively successful city-wide programme. The main components have been emphasis on self-help in bringing about and maintaining improved civic amenities; granting of land tenure to the slum residents by the municipal corporation; and a flexible responsive and dedicated approach on the part of the project organisers.

The slum dwellers simply determine the location of taps and latrines, the width of the paved passages and soon. They have contributed in greater measure, with labour and money, to improve housing and build community centres - each household pays upto 75 per cent of rebuilding costs of their home through bank loans. The slum dwellers also contribute to community centre activities; balwadis (creches), adult literacy, classes and cottage industry cooperatives.

What made this high degree of involvement and self-help possible was the granting of ownership rights (pattas) to each household for their plot. This not only assured the slum dwellers of the benefits of whatever they invested in rebuilding their homes, it also made it feasible for the city's banks to grant them loans by holding their pattas as mortgage. Between 1977 and 1983 the Andhra Pradesh government gave pattas to 13,000 additional households, which enabled 13,000 new houses to be built in various low income areas.

The resident families are far more concerned with maintenance than they would be if the Government had built the houses and handed them over. Moreover out of a total cost of Rs. 9.6 crore, the government has spent only Rs. 1.1 crore. Nationalised banks have advanced loans of upto Rs. 8,000 per household to individuals recommended by the project staff at very low rates of interest, repayable over 20 years. The actual expenditure per house incurred by the people has varied between Rs. 10,000 and Rs. 12,000. The difference between the loaned amount and actual costs has been made up by the people tapping their own and their relatives' resources, by subsidies from the UNICEF at the rate of Rs. 200 per house for private lavatories and by subsidies from local organisations like Lions, Rotary and Jaycees at the rate of Rs. 500 to Rs. 1,000 to cover the costs of reinforced concrete slabs.

Housing Problem in Delhi

Delhi is one of the fastest growing cities of India. The population of the Union Territory has increased from 2.65 million in 1961 to 6.22 million in 1981-an increase of 3 56 million in just 20 years. It is expected to go up to 9.26 million in 1991 and 13.26 million by 2001. The proportion of population living in urban areas has been increasing every decade-from 93 per cent in 1981, it is likely to go up to 95 per cent in 1991 and 97 per cent by 2001. There was a net in-migration of 1.1 million persons in Delhi during 1971-81.

Net migration into Delhi accounted for 53 per cent of Delhi's growth during 1971-81 and it is likely to account for 54 5 per cent of its growth during 1981-91 and 59.0 per cent during 1991-2001. The 1981 census results have shown that roughly two-third of males and three-fourth of feamels who had migrated into Delhi for employment were either illiterate or were below matric. Obviously, such migrants could have been absorbed only in unskilled jobs. Prima-facie, this class of migrants poses the biggest problem for housing and shelter.

The 1981 census, which covered housing conditions too, revealed that only 41 per cent of the households in Delhi had roof of concrete RBC/RCC, while 38 per cent had roof of stope, bricks, or lime. Twenty-one per cent of the households had roof of completely substandard type such as grass, reeds, wood, tiles or state, corrugated iron sheets or asbestos cement sheets. The census does not categorise the houses into pucca and kutcha houses. However, for our purposes we have taken those houses as pucca houses which had got floors of cement, bricks, stone and lime and mosaic tiles, combined with walls burnt bricks or stone or cement concrete and roofs made of concrete RBC/RCC.

Using the definition, we and that 73 per cent of the households in Delhi city lived in pucca houses. Of these, only 40 per cent had got roof of concrete, RBC/RCC, 17 per cent had got roof of brick, stone and lime and 16 per cent had root of stone only. That is, even using a very liberal definition of pucca houses, roughly one-fourth of the households in Delhi city lived in sub-standard housing conditions while only 40 per cent of the households lived under RBC/RCC roofs in 1981.

About 4.6 per cent of households lived in completely kutcha houses whose floors are made of mud, wood, bamboo etc, walls are made of grass, reeds, mud, wood, or G.I. sheets and rools are made of grass reeds, thatch, tiles, state/shingles corrugated sheets or asbestos cement sheets. Apart from this, 1.36 per cent of the households were found to be houseless.

We can look at the ons from another angle *i.e.*, by the number of rooms occupied by a household. Fifty-four per cent of the households lived in one-room accommodation, while 26 per cent had two-room houses. Only 18 per cent of the households had more than two-room accommodation with them. It was also found that 44 per cent of the households living in one-room houses had five or more members in the household while 61 per cent of the households living in one-room houses had five or more members in the household while 61 per cent had four or more members in the household. Population-wise, 48 per cent of the population of Delhi lived in one-room houses and 29 per cent lived in two-room houses. As for the tenure status of the houses occupied by the household, 51 per cent lived in their own houses and 49 per cent lived in rented houses.

During 1981 census, data had also been collected on the availability of drinking water, electricity facilities to the households. Twentyfive per cent of the househols in Delhi city did not have electricity in their houses. while 32 per cent of the households did not have toilet facilities. This is rather appalling. Among the three municipalities, the position was the worst in Delhi Cantt areas and the most satisfactory in N.D.M.C. areas. In N.D.M.C., areas, only 7 per cent of the population did not have electricity facilities while in M.C.D. hi Cantt areas this per centage was 19 and 26, respectively. In respect of toilet facilities, 11 per cent of the population in N.D.M.C. areas, 27 per cent in M.C.D. and 35 per cent in Delhi Cantt area did not have them within the premises of their houses.

Only 51 per cent of the households in the urban areas of Delhi had access to tap water within their premises, while another 17 per cent had access to top water outside the premises Some 27 per cent of the households relied on hand pumps or tubewells for drinking water supply. While in N.D.M.C. areas, 84 per cent of households had access to tap water within the premises, in Delhi Cantt areas and M.C.D. areas, this percentage was 69 and 54, respectively.

Since about 55 per cent of Delhi's growth is attributable to migration largely consisting of illiterate and semi-literate persons from the rural areas of neighbouring States, the position of housing in Delhi would have become much worse since 1981. To work out the dimensions of the housing problem during 1981-91, we have to add to this the estimated net migrants (16.55 lakh) to Delhi during 1981-91 and also that component of the natural increase of Delhi's population in 1981-91 which occurs among the slum-dwellers. Out of this expected natural increase of 14 lakh population, around five lakh can be assumed to occur among those 27 per cent of the households of Delhi in 1981 who lived in substandard housing conditions. In case of electricity and toilet facilities, the data the percentage of both the households and the population with or without these facilities are available and they are approximately the same, as noted earlier. But in case of housing data and data on water supply, only the percentage of households are available. It is assumed here that these percentages closely approximate the percentages of population also.

Thus, the dimensions of the housing problems of Delhi during the decade 1981-91 would be :

(a) Provision of suitable houses to 38 lakh persons;

(b) Provision of tap drinking water to 40 lakh persons;

(c) Provision of electricity to 35 lakh persons;

(d) Provision of toilet facilities to 39 lakh persons.

We have also to take into account that part of the expected natural increase of population (around 9 lakh) which would occur among the population living in good housing conditions in 1981. Provision of standard houses, electricity, tap water and toilet facilities to such large populations is a stupendous task. The Administration would have to gear itself to this task falling which Delhi would have large slums all around the city with its concomitant social, economic and political consequences.

Projecting this scenario still further into the next decade, we would be faced with a very staggering picture. Delhi's population is expected to increase by 40 lakh during 1991-2001, out of which, 23.64 lakh are likely to be due to net in-migration and 16.455 lakh due to natural increase. Adding these 40 lakh persons, the housing targets for 1981-2000 decades would be :

1. Provision of housing to 78 lakh persons,
2. Provision of tap drinking water to 80 lakh persons,
3. Provision of electricity to 75 lakh persons and
4. Provision of toilet facilities to 79 lakh persons.

Urbanisation is a concomitant of economic development and in the economically advanced countries at least, it ensures better life for the inhabitants of the urban areas. In the developing countries, however, the situation is not rosy. The rapid increase in population of the bigger cities in the developing countries largely due to a combination of the push and pull factors. Migration of illiterate and semi-literate economically depriled population from the rural areas to the urban areas in search of employment tends to create a class of people in the urban areas who are both socially as well as economically deprived of the benefits of urbanisation and they tend to live in very poor and shabby conditions.

We have already noticed that the housing conditions are pretty bad for at least one-fourth of the households of Delhi.

This is true of most of the bigger cities in India as well as in other developing countries. Special attention needs to be paid for provision of cheap housing and other civic amenities to these people, otherwise the modern cities of the third world would become centres of poverty, squalor and disease.

Hill Horror

The British planned hill stations for their own enjoyment with utmost care. With independence theywere open to more and more people and in recent years the traffic has increased enormously. Ironically, the authorities' ambitious residential schemes and the 'development' of tourist facilities have destroyed neighbouring forests, critically reducing the towns water supply. As a result the hill stations, a have from the heat and bustle of the plains, are dying.

Acute water shortage is common to most hill resorts. Ootacamund (Ooty) near Bangalore, situated in a district which produced 50% of the State's hydel power, in 1982 faced one of the worst droughts of the century. Darjeeling, a summer time retreat suffers an annual shortage of drinking water.

In Panchgani, on Mahabaleshwar's eastern ridge, the forest department is left with a ridiculous 25 hectares of forests and Mahabaleshwar slopes are now more or less totally denuded. As such Vennalake, only water source for the town is silting up.

Matters are made worse by a tremendous increase in both the permanent population and the tourist traffic. The population of Shillong in 1971 was 1.22 lakh, today it is double. In Shimla, the summer capital of the British, civic facilities have all but collapsed Every tourist season, one lakh people are added to their swell. Today, it is difficult for natural resources to keep pace with growth. The natural charm of a hill resort has been battered. For instance, in Shillong, instead of towering pines there are rows of houses.

A Tale of Three Cities

Bangalore, Pune and Dehra Dun were for long praised as idyllic towns, cool, green and quiet. Today all three are boom towns, noisy, dusty and hot.

Bangalore is one of the 10 fastest growing cities in the world. It was a small village in the 16th century. It grew to an important textile production centre under Tipu Sultan. Under the British the city saw some limited industrial development. Since Independence, massive investment in public sector industries has completely changed its character.

Between 1971 and 1981, the city's population exploded by 76 per cent from 1.64 million to 2.91 million. Inevitably, the total number of slums grew from 159 in 1972 to 287 in 1982; the slum population grew from 1.3 lakh to about 3 lakh, packed into 3.451 acres or 11 per cent of the city corporation areas. By metropolitan standards, Bangalore can still be described almost a slumless city, but the trends are frightening.

The former Punyapur of the 8th century Rashtrakuta kings, where Shivaji once rode his horse, has not fared much better. Pune is today the base of some of the country's biggest industries, and has in the process turned into a giant sprawling along the ribbon tying it to Bombay, with a population of 1.46 million. The number of slums grew from 127 in 1950 to .. in 1976.

Dehra Dun has suffered similar changes. The lush green hill resort used to be the preserve of government institutions like milliiary academies, the Forest Research Institute and the Survey of India. But in 1960s the area was declared industrially backward and concessions and incentives announced. Since then a large number of quarries and industrial units based on mined produce have come up. Of the 26 major industrial units in Dehra Dun, 50 per cent are limestone based, and 210 small scale crushers are causing serious air pollution problems. The population has grown from 30,000 to 40,000 in the early 20th century to 0.42 million and is expected to reach 0.6 million by 2000.

The problems that growth has brought in its wake are manifold. Till 1895, Bangalore City got its water from a number of tanks such as Darmambudhi, Sampangi, Ulsoor and Sankey, supplemented by local wells and ponds. Most have been swallowed up and the disturbance in the watershed has accelerated the drying up of tanks in 1925, the Hesarghata lake became almost dry and today the Sankey tank is one of the few that has not been swallowed up. These miniature lakes also contributed to local cooling.

Cities as a Collection of Shacks

Many major cities have moved from slum demolition and mass eviction to slum assistance. Self-help is an important ingredient making government intervention in the housing sector more effective. But projects have fallen short of expectations; production of serviced sites is low and they are not accessible to the very poor; slum upgrading has been superficial, of no lasting significance.

Land is of central importance. Governments have gone in for vacant, peripheral plots, avoiding the issue of finding lands by means which can be replicated on a large scale. In slum upgrading projects, lanc issue, except in Hyderabad, has been avoided altogether. This has had serious consequences: the upgraded house, joint and housing market uncontrolled by the authorities, and the poor can be squeezee out if they can't afford to pay the higher rents, or can be tempted into selling.

If the urban poor are to have significantly improved access to upgraded slums and new serviced sites, the government must flood the city's low-income housing market with self-help housing stock. Otherwise even the few sections of the urban poor who have benefited from protected pilot projects will soon be displaced by market forces. That this danger can be avoided and real progress made in self-help housing for the urban poor is demonstrated by the case of Hyderabad's upgrading programme which recognised the importance of granting land tenure from the very beginning.

At present, government policy wavers between slum improvement and slum bulldozing. There is no clear national plan to give every city dweller a proper roof. A fundamental choice has to be made. As M.N. Buch argues, either Indians can plan for "a built environment in which every citizen has reasonable access to water, lives in developed lots where drainage is satisfactory and where roads are so designed as to provide efficient and cheap public transport" or for "one in which there are areas of excellence and other areas of total degradation". If it is accepted th t the former choice alone will lead to a better quality of life for the maximum number of people then Indians will have to reconcile themselves to the fact that for years to come our cities will be just a collection of shacks".

This means that the perceptions of planners and architects will have to change dramatically. The power of the upper classes, who want to use the cheap labour of slum dwellers but want to deny them a home, will have to be curbed. Ultimately, urban land must become a national asset used in the interests of the majority, not of a small minority.

12

People

Introduction

Is India an overpopulated country, its large and growing population a major cause of environmental destruction ? It has been repeatedly argued that the population pressure has outstripped the country's capacity to cope. But while there is a plethora of loose statements, there is no rigorous scientific study on the production potential of the country's lands. The UN study reported in this chapter-the most rigorous undertaken as yet-shows that India can feed three to four times its current population, but this will be possible only if the Indian government and people can learn to manage their soil and water resources properly. If there is poverty and hunger in the country today, or even apparent overpopulation, it is only because the country's scientists, politicians and planners have not learnt how to use the country's environment at a high level of sustainable productivity.

The environmental destruction that has followed is making the lives of Indian people humanly impossible, particularly of women who have to fetch daily household needs from the immediate environment. To survive, the poor often have no alternative but to turn upon the very environment with which they lived in full harmony in the immediate past. Millions of people today see no alternative for survival other than selling firewood from forests – a new profession that has emerged with amazing rapidity. A people based development process-especially a women-based process-would not only demand the restoration and enhancement of nature but would also bring about family planning and stabilise population.

Population and Environment

Consider the bare figures. Between 1901 and 1951, India's population grew from 238 million to 361 million, an increase of about 52 per cent over 50 years. Between 1951 and 1981, it expanded from 361 million to 685 million, a whopping increase of 90 per cent in just 30 years. India literally added a second India in the 35-odd years of its independence.

In 1967, William and Paul Paddock wrote a book called "Famine 1975-America's Decision : Who Will Survive ?" They classified developing countries, based on their population growth rates and food production potential, into three categories : 'can't be saved', 'walking wounded' and 'should receive food'. India suffered the ignominy of being classified as 'can't be saved' with the comment that it "is the bellwether that shows the path which the others, like sheep going to slaughter, are following'. India was a country, according to the authors, to which the US should not have bothered either to provide aid or food.

Fortunately, India's food production managed to keep ahead of population growth, which helped it through some of the most critical drought years like 1979. But the apparent stagnation in agricultural production-though in 1983-84 a new record foodgrain production of over 150 million tonnes was recorded-once again created despondency.

The 1981 census results turned out to be even more of a shocker than usual. The Planning Commission had just put the finishing touches to its Sixth Five-Year Plan when the Registrar General of Census announced that the country had some 13 million more people than the number estimated by the planners and that the birth rate was not 33 per 1,000 people but around

36. The Planning Commission had to hastily restructure the Sixth Plan to bring it into line with the new figures.

The key question for expert and lay person is : can India's lands support such a large and growing population ? This is where the 'carrying capacity' comes in. It is the number of people or animals that an area of land can support on a sustainable basis. Unfortunatel , not one expert in the country has attempted to quantify the carrying capacity of the area under a single development block leave alone the whole country.

Indeed, it is an extremely complicated and uncertain business trying to set a figure to the number of beings a piece of land can support. Take the case of Fiji where, in 1973, UNESCO and the United Nations Fund for Population Activities (UNFPA) launched a pilot project to study the population-environment relationships in the eastern islands. The number of ecological variables were found to be enormous, putting paid to all efforts to put a figure on the small islands' carrying capacity. Only specific inferences were possible. For instance, eastern Fiji is caught in the stranglehold of declining copra industry and is suffering from poverty and emigration. If the colluvial soils which have been planted with increasingly unproductive coconuts for copra export were converted to farming of taro, a root crop, it was estimated that up to five times more people could be supported with this economy than if they continued with copra. Any talk of carrying capacity must clearly be marked with great caution.

In a modern interdependent world, where trade plays a, major role the population supporting capacity of a particular area can, in fact, be increased enormously by importing a few critical items like plant nutrients from outside. Densely populated Europe todoy is heavily supported by the production of a range of items outside European countries, especially in the developing countries. The UN has never dared to estimate how much land in the developing world is today producing food and other items like timber for the developed world and how the quantity of this land has been changing over time. If such a study were undertaken, it would clearly show that, despite decolonisation in large parts of the Third World over atleast three decades, many times more land is being used in developing countries to meet the needs of the developed countries as compared to 19 0, and today this land use is a major cause of environmental destruction across the Third World. For instance, the first major attack on India's natural forests came with the demand to establish tea plantations to satisfy colonial Britain.

Indian Food Situation over the 73's

	Average 1969-71	*Average 1978-80*
Population (million)	552.2 (1970)	684.5 (1980)
Population Index	100 (1970)	124 (1980)
Food Production Index	100	127 (1979-81)
Food Production per capita index	100	102 (1979-81)
Calories per capita per day	1999	1998
Protein per capita per day (gm)	49.5	48.5
Fat per capita per day (gm)	29.2	29.5

Notes : (1) It is important to note that the quantities of food available indicated in the table relate to the quantities of food reaching the consumer but not necessarily to the amounts of food actually consumed, which may be lower than the quantity shown depending on the extent of losses of edible food and nutrients in the household.

(2) The figures represent only the average supply for the population as a whole and do not indicate differences amongst socio-economic groups and ecological zones and seasonal variations.

Source : 1981 FAO Production Yearbook Vol. 35, Rome, 1982, FAO.

Indian Prospects

But what of India ? It is useful to ask just what our land is physically capable of producing on a sustainable basis, without going into the thickets of social organisation and land reform. In late 1983, the Food and Agriculture Organization (FAO) released an extensive study on the subject, 'Potential Population Supporting Capacities of Lands in the Developing World', the result of almost a decade's work that began with preparation of a world soil map.

The picture of India that emerges is both exhilarating and sobering. Simply put, India has enormous problems but also has an equally enormous natural resource base to solve them with. Assuming a low level of inputs; local low-yielding varieties, no fertiliser use or chemical pest, disease and weed control, long fallow periods, no long-term soil conservation measures, high rates of soil erosion, steadily declining soil productivity, mannual labour with hand tools as the only power source, high labour intensity, subsistence production and fragmented landholdings.

Caste and Environment

The question has troubled sociologists for years : How has India's caste system, so elaborate an estimated 40,000 castes rigid, hereditarily determined, hierarchical and oppressive a social structure, lasted for so many centuries ? Madhav Gadgil, an ecologist, and anthropologist Kailash Malhotra feel that the answer probably lies in the discipline that the caste system brought to the use of natural resources. It was a system which, on the one hand, forced its members to share natural resources and one the other, created the right social millieu in which sustainable patterns of resource use were encouraged to emerge. In other words, it was a social system which both forced and cajoled the social being right from birth to adopt sustainable cultural mores.

Within the caste system, birth determined a human being's occupation. An "ecological space" and its natural resources could only be used by a definite occupational group. This "resource partitioning" helped to reduce competition and, hence, conflict among human beings over scarce natural resources, and to create the right psychological environment; the allottees of an ecological space developed sustainable patterns because they had no worry that their resources would be snatched away from them and probably also because they knew that if they exhausted the resources in their own space, they would not be allowed to use any other.

Malhotra and Gadgil are quick to point out that even though the caste system lasts till today, it has stopped "resource partitioning". The growing monetarisation of the Indian economy, first encouraged by the invading British then by independent India in the name of development, has destroyed the ancient sustainable and shared patterns of resource-use evolved over centuries. Social tensions are increasing as people fight over scarce natural resources, and because the caste system continues even after losing its raison d'etre, it has become a source of social power and oppression.

These conclusions emerge from a 12-year study of rural society in western Maharashtra. Here the Western Ghats rise abruptly, some 50 km inland from the Arabian Sea; to an altitude of 1,000 metres to 1,500 metres. Their 15 km to 20 km wide crestline is marked with heavy rainfall, more than 3,000 mm a year and it supports low but stable agricultural and pastoral productivity. The simple society here, mostly marked by single clan settlements, consists of a few sedentary castes of cultivators, pastoralists and huntergatherers, which include Gavils who are mainly cultivators, and Kunbis who are buffalo herders and shifting cultivators.

Further eastward, the Western Ghats merge with the Deccan Plateau through a series of broken hills, and the rainfall in this tract falls to about 1,500 mm to 1,000 mm near the edge of the plateau. This tract supports rich agriculture and neighbours a complex rural society with a number of artisians and service castes. Still further east, the rainfall decreases further, to less than 600 mm in a semiarid tract of 100 km to 150 km width.

Agriculture here is practised only in the river valleys. Pastoralists and huntergatherers use large tracts of uncultivated land.

Because the terrain favours sedentary existence during the four monsoon months, and uncultivated lands are available for grazing, several pastoral as well as nonpastoral nomadic castes have established their base villages in this semi arid tract. They migrate to the higher rainfall tracts, both to the east and west of the semi-arid tract, for the eight months of the year. Among those who have their base villages in this tract are Hatkars, Tirumal Nandiwallas, Fulmali Nandiwallas, Vaidus and Phase-pardis.

Niche Diversification

The first example of "ecological niche diversification" is exhibited by the Kunbis and Gavli Dhangars living on the crestline of the Western Ghats. The Kunbis cultivate paddy in the river valleys, practise shifting cultivation on the lower hill slopes and hunt extensively, they barter their cereals for butter produced by the Gavlis and other goods. The only cattle they keep are for drought purposes.

The Gavlis, on the other hand, live on the upper hill terraces. Although they practise some shifting cultivation, their main occupation is keeping buffaloes and cattle. They curdle the milk, consume the butter milk locally and barter the butter for cereals from the Kunbis. As they get their protein supply from butter milk, they do not need to hunt. Thus, the cultivation of valleys and lower hill slopes is restricted to the Kunbis and of hill terraces to the Gavlis. Maintenance of domesticated animals and exploitation of all fodder and grazing is restricted to the Gavlis and the hunting of wild animals to the Kunbis.

Another instance of niche diversification is provided by the three nomadic hunting communities of the semi-arid tracts of western Maharashtra : Triumal Nandiwallas, Phasepardhis and Vaidus. The primary occupation of the Nandiwallas is entertainment and fortune telling, that of the Vaidus dispensation of herbal medicines, while the Phasepardhis are specialist hunter-gatherers. Unlike the settled castes the Tirumal Nandiwallas and Vaidus also hunt in addition to their primary occupation. But the three castes use distinctly different hunting techniques and specialise in different prey species.

The Tirumal Nandiwallas use dogs to hunt wild pigs, large carnivorous mammals (leopard cat, hyena, fox) and aquatic animals (fish, carb, turtle), Vaidus use baited traps for smaller carnivores (mongoose, civet, jackal and cat), monitor lizards and acquatic animals. The Phasepardhis specialise in black buck, wild pig and birds such as doves, quails, partridges and peafowl.

Wild meat means a lot to these castes. Gadgil and Malhotra studied the Tirumal Nandiwalias in greater detail and found that the median annual biomass of hunted animal is 2,180 kg per family or Tirumal Nandiwallas, who are thorough in their use of the hunted animals. They even eat portions like the brain and the viscera, normally rejected by other people. Almost two-thirds of the live weight of the prey is consumed. In terms of per capita daily consumption the median works out to 57 gm of protein and 850 calories from hunted wild meat. For the Tirumal Nandiwallas, the hunted meat provides a more than adequate level of protein and over one-third of the calorific requirements of an average person.

What is striking is that while the hunting techniques used by the three groups are significantly different, node of them are sophisticated that they cannot be adopted by another.

Women and Natural Resources

Prabably no other group is more affected by environmental destruction than poor villages women. Every dawn brings with it a long march in search of fuel, fodder and water. It does not matter if the women are old, young or pregnant: crucial household needs have to be met day after weary day. As ecological conditions worsen, the long march becomes ever longer and more tiresome. Caught between poverty and environmental destruction, poor rural women in India could well be reaching the limits of physical endurance.

Biomass plays a crucial role in meeting daily survival needs in the vast majority of the rural households in the country; food, fish, fuel (firewood, crop wastes and cowdung, organic manure; green mulch and forest litter), building materials (timber and thatch) and medicines (herbs), are all different forms of biomass. Water is another crucial element in human survival and its availability is definitely related to biomass in a country where rain falls only four months in the year. When the biomass in the surrounding evironment disappears, water sources like streams and ponds also tend to disappear soon after the mansoon. An important aspect of this biomass-based subsistence economy is that it is mostly non-monetised. Water, fuel, fodder building materials and even food to a certain extent are all gathered freely from the immediate environment. Production and processing of biomass – agriculture, forestry and village crafts based on biomass as raw material are also the biggest sources of employment.

Unfortunately industrialisation, urbanisation and the advance of the cash economy have greatly affected the country's biomass base. The destruction of biomass with deforestation and devegetation – and its transformation away from rural and household needs and towards urban and industrial needs (planting of teak or eucalyptus in place of sal or other natural vegetation, for instance) is having a major impact on the lives of all those who live within the non-monetised biomass-based subsistence economy. Marginal cultures like that of tribals and nomads are the first to face total social destruction. But women, not only within these cultures but in all rural cultures, face the maximum threat.

The special difficulties that women face – their extraordinary work burden or lack of access to health care, for instance – do not arise out of ecological deterioration per se. They are located in the sexual division of labour marked by a double work burden (at home and outside) and by the specific nature of tasks they do, and in the unequal distribution of resources like food within the household, which stems from women inferior status in the household and from a lack of control over cash and productive resources like land. But given this situation, environmental destruction exacerbates women's already acute problems in a way very different from men.

Technology Introduction

With the growing awareness of energy and environment problems, several efforts have been made in recent years by the government to promote new technologies like biogas plants, fuelwood plantations, fuel-conserving smokeless chulhas, latrines and handpumps. These can reduce the rigours of rural living, improve health and meet fundamental household needs. Srilata Batliwala has calculated that if cooking fuel and water were readily available close to the household in Pura villages and the efficiency of stoves was improved to reduce fuelwood consumption, a saving of nearly 500 calories per day per woman could be affected. Yet success stories about the spread of these technologies remain few and far between.

A major reason for this failure is that these dissemination programmes seldom involve women either when the technologies are being developed or when they are being disseminated. Any technology that supplies cooking energy should obviously be directed towards women. For example, a community biogas plant was set up in Fateh Singh ka Purwa, a village in Etawah district of Uttar Pradesh, the first such plant to serve an entire village. Every household, rich and poor, received gas from it.

Technologically this demonstration plant was a success but socially it was a failure. It ran well for a year but soon the community started losing interest. Most of its members stopped providing cowdung and the plant more or less came to a halt. Male community leaders pointed out they were not interested in energy for cooking. They would rather have energy to power irrigation pumps, chaff cutters and milling machines.

A report by Nirlep Malhans and Jyoti Sanghera in 'Manushi' points out that women were extremely critical of the plant. It was decided, without consulting them, that gas supply would be limited to two hours in the morning, from 8 am to 10 am. By then, the women were already in the fields. Clearly, the plant's organisers had completely ignored their work on the farm. The women also complained that the gas did not even provide 25 per cent of the day's cooking, and they now had to look for wood as a substitute for dung cakes. Further, the new technology also increased the women's dependence on men : they had to depend on men even for routine cleaning of the burners.

After the Fateh Singh ka Purwa plant, the government has established several dozen community biogas plants as demonstration projects in various parts of the country. But not one is under the control of women. Because of the failure of community biogas plants to involve the 'community', there are now several proposals to use biogas energy to develop small/cottage industries, in order to improve their economic viability. In an energy-starved environment, if questions of economic viability are raised, women's cooking energy needs will probably be relegated to second place.

Drinking Water

Water carrying is another extremely strenuous activity undertaken by women and it consumes an enormous amount of their time. Available information indicates that women spend long hours and traverse long distances, especially in the hill areas and in the arid and semi-arid parts of the country. In the deltaic plains, the time taken may be less than an hour a day. But this varies : in villages of Karnataka, a study shows this can be from one to 1.4 hours per household a day, whereas in villages of eastern Uttar Pradesh it can be as much as one to 3.9 hours a day on an average.

Contrary to the popular belief that women enjoy fetching water and washing clothes in a communal place because they can gossip, economist Bina Agarwal points out that the few studies, that have bothered to look into this aspect, note that, in fact, women usually prefer to wash clothes alone, and seek to avoid time when washing places or wells are overcrowded.

Vina Mazumdar, Director of the Centre for Women's Development Studies in New Delhi, points out : "Irrigation gets high priority in most plans for rural development, but the supply of water for drinking and washing receives scant attention." M.S. Swaminathan, former Member of the Planning Commission, in his J.P. Naik Memorial Lecture, admitted : If men had to fetch drinking water, then 230,000 villages would not have remained without provision of drinking water after 30 years of Planned development." Very few attempts have been made even to combine irrigation and domestic water supply schemes. They are typically conceived of as separate and independent projects.

Bina Agarwal points out that irrigation schemes can in fact have an impact on drinking water supply and, hence, on women as primary carriers of water, even if they are conceived of as separate projects. Seepage from canals can raise the water table and unless this is accompanied by increased salinity, this would replenish drinking water wells and have a beneficial impact. But in several parts of north and central India, the rise in the water table after the construction of irrigation canals has also led to increased salinity in drinking water wells in the vicinity of canals. These wells have to be abandoned leading to severe drinking water problems.

Deep tubewells, on the other hand, can soak up drinking water wells by lowering the water table and increasing work for women who then have to seek alternative and more inaccessible sources. If the drinking water wells of untouchable households falls into disuse, they may be left with no other alternative but to depend on the mercy of the upper caste wells. They obviously cannot purchase their own tubewells. Bina Agarwal gives the examples of Bulandshahr in Uttar Pradesh, where the drying up of village drinking water wells in untouchable communities associated with the sinking of deep tubewells in the vicinity, has meant that the untouchable women now have to depend entirely on the goodwill of the higher caste to pour out for them water they need for their everyday use.

Lack of clean drinking water most affects the health of children. But among adults, women are more exposed to hazards of polluted water than men. Firstly they are the primary carriers of water, and secondly they wash clothes and utensils. Moreover, since childcare is primarily the women's responsibility, when children get infected women are more likely to catch the infection than men. Even if a woman did not fall sick herself, she would have to spend more time caring for the sick. Poorer households will be clearly affected the most, as they have less access to clean water sources and are less likely to have help in domestic chores.

Malnourished Workers

How does the extraordinarily tough work burden of poor women affect their health ? Firstly, studies show that they eat much less than they should, which leaves them extremely malnourished and susceptible to disease. Secondly, repeated child bearing children are earning members of the family - has a major effect on their health Thirdly, the work burden is so high that most women do not seek medical care even when they are ill or pregnant.

Batliwala has calculated the amount of energy spent by men and women in their daily chores in Pura village and compared this to their dialy food intake. The amount of energy expended by women is about 2,505 calories daily and by men 2,473 calories. But when food intake is compared; men get nearly 800 calories more than their expenditure while womed get 100 calories less. This difference in energy intake between men and women has also been found in a study by Leela Gulati of the Centre of Development Studies Trivandrum, in a typical agricultural labour household in Kerala. On employed days the woman's caloric intake fell short of the recommendations made by the Indian Council of Medical Research by 20 per cent, while the man's fell short only by 11 per cent; on unemployed days the respective shortfalls were 50 per cent for the women and 26 per cent for the man.

But Batliwala argues that this apparently small deficit of 100 calories a day amongst village women in Pura must be seen in the light of other realities of village life and of women in particular. Environmental sanitation is so bad and access to clean drinking water so limited, that the vast majority of Indian villagers have worm infestations. These parasites can 'steal' as much as one-fourth of the total food intake.

During pregnancy and lactation, women need an extra 500 calories to 600 calories a day. At any point of time, one-third of the women in the country are in this condition, but most of them continue to work without any additional nutrition. The woman is the last to get food in an Indian household.

Child Power

In this constant grind of survival, several studies have pointed out that children begin to acquire an extremely important economic role. They do many crucial tasks like caring for younger children, fetching fuel, fodder and water and grazing animals, so that the adults can undertake waged labour. In Pura, for instance, children provide 29 per cent of the total time spent on fuelwood gathering, 20 per cent on fetching water and 34 per cent on livestock grazing.

With time, the importance of children cannot but grow. As Amulya Reddy and Srillata Batliwala, who have studied energy consumption patterns in Pura in detail, point out : "Children have become the unwitting victims of the continuing energy hunger in Third World villages . a family below the poverty line, which does not have the means either to hire labour or purchase labour-saving devices, is compelled to meet its energy needs only by producing several children."

But this will not only have important implications for the education, health and nutrition of children but also for the country's massive family planning programmes and the health of women. If underfed and overworked women are also expected to bear many children, the impact on their health will obviously be drastic.

Women and Male Migration

Male migration, a common phenomenon in large parts of rural India, is an important factor affecting the work burden of women. A number of factors have led increased migration in search of job opportunities; population growth, increased pressure on land, and a technological transformation in agriculture itself that has promoted landlessness.

Introduction of irrigation and high yielding varieties in specific parts of the country have created regional imbalances in agricultural development. Migration of agricultural labour from Bihar and eastern Uttar Pradesh to Punjab now takes place on an extensive scale. Similarly, in fishing villages in Kanyakumari district men have begun to migrate more and more. Kanyakumari has become a labour catchment area for mechanised fishing in Kerala and eastern Tamil Nadu. Fishermen from Kanyakumari now spend anything between two to seven months of the year on mechanised boats in Cochin, Trivandrum and Tirunelveli districts, sometimes going as far as Orissa and Karnataka.

The destruction of forests and grazing lands has put extreme pressure on tribal and nomadic people, forcing them into landlessness and migration in search of jobs. For instance, tribals are migrating in large numbers from Dhenkanal district of Orissa to cities like Cuttack, Bhubaneshwar and places like the Assam tea gardens.

Very little research has been done on the effects of this migration on the women left behind. Most studies of rural-urban migration have emphasised the positive economic benefits which accrue to the village because of remittances by the emigrants. But Andrea Menetee Singh of ILO points out that life of women left behind in the village can be hard. "If the oral tradition is any indication (folk songs, for instance), the emotional price the women pay for their husbands' emigration is very high," says Singh.

Divorce rates are high among emigrants, which means that the proportion of women deserted after emigration may be fairly high. Surveys in Bombay slums have also shown that men who have migrated without their families have a high incidence of venereal disease, which probably means that they are transmitting these diseases to their wives on home visits.

Household Fuel

Collecting fuel is an important daily chore for most rural women. In Pura, to gather 1.74 tonnes of firewood every year, each firewood gathering household spends an average of 2.51 hours daily, makes 172 trips in a year and each round trip is about 8.54 kilometres.

In several north Indian households, there is greater reliance on cowdung than firewood. But this does not necessarily reduce the time spent by women. A study carried out on the health of pregnant women in two villages in western Uttar Pradesh show that women spend long hours in animal husbandry and making dung cakes. For instance, one pregnant woman spent a 14-hour working day, in which three hours went on animal husbandry and half an hour on making cowdung cakes In addition, this woman spent three hours on fetching water, 2.5 hours on cooking and five hours on other household chores, including milling grain.

Women's work in the hills and deserts is even harder. Agricultural production in these areas is extremely limited and alternatives to fuelwood like cowdung and crop wastes are also limited. Available studies clearly show that women put in an inhuman amount of time in the hills and deserts just searching for fuel and fodder. Madhura Swaminathan, in a study of an Himalayan village Dwing in Chamoli district, Uttar Pradesh, points out that one or more members of each household daily walk 5 km uphill and spend six to 10 hours on the job. The average daily time spent by each household is 7.2 hours. An old widow often takes up to 10 hours. Wild animals are a constant danger and two people generally go together. In 75 per cent of the households, only women go out to called wood. Generally three trips are made every four days though many women go daily.

Pakhi is a fairly prosperous Himalayan village with a major road passing through it. Here, some households purchase wood, but in over 86 per cent of the households women go out to collect firewood. Even though the forest is nearer the village than it is to Dwing, the average distance to be travelled to collect firewood is 3.1 km, and means a four hour walk. More than half the households make a daily trip to collect wood.

Swaminathan comments on the life of women in these villages : "Leaving their homes at dawn, after nearly an eight-hour day, they return to the village with their load of wood. What is the normal working day all over the world is just the first of the day's tasks for these women. On returning home, the women go about their household work which includes rearing of cattle. And finally they put in a few hours of work on the land." Dwing also has acute water scarcity. In both the villages, three meals are cooked every day. Because of heavy male migration from the area, women are also being forced to put more and time into agriculture.

Headloaders

It is a usual sight these days to see hundreds of weary women with heavy bundles of wood by their side at the Ranchi railway station early in the morning. For these women, mostly tribals, this is just a part of a gruelling two-day cycle of work.

Increasing numbers of tribals and other poor are taking to unauthorised cutting and selling of firewood. Although the work is hard and the money earned a pittance, it is an important source of income for these poor people.

The Xavier Institute of Social Studies has published a striking report on headloaders in Ranchi by T. Bhaduri and V. Surin,

A study of 170 households spread across nine villages, if shows that headloading has emerged as an important profession only over the last 15 to 18 years.

More than a fifth of the households list headloading as their major occupation; in some villages as high as 40 per cent. For nearly 20 per cent of all the employed people in the villages, unauthorised cutting and selling of wood has become the primary occupation. Working as day labourers in farms and forests was secondary, and cultivating their own land came third. For another 50 per cent of the employed headloading is the second occupation and farming the first.

A Two Day Cycle

The Ranchi firewood sellers, mainly tribal women, usually begin their day at 2 am because they must complete household chores before setting out on the 8 km to 10 km walk to the surrounding forests to collect wood. Seven or eight years ago, the forests were just a kilometre or two away.

The wood is sold in the nearest town : Ranch. To reach the market early, the women must leve their villages the previous evening by train or bus and spend the night at the railway station or some other public place. Each women can carry at best a 20 kg headload, which sells for Rs. 5.50 to Rs. 6.50. About a third of this money is lost in bribes to the forest guards and the train conductors. With the remaining money they buy essentials like rice, edible oil, kerosene (for lighting) and salt.

Work begins again as soon as they return because they must cook dinner probably the only freshly cooked meal of the day. The next morning they usually take leftovers from the night before for lunch when they set off once again to collect wood. The back-breaking work of cutting and selling wood apart, poor nutrition undermines their health further. Women fainting at the Ranchi station is not an uncommon sight.

Extensive deforestation, and restrictions on shifting cultivation and sale of minor forest produce has hit the tribals hard.

In Ranchi, most headloading families have little land; what they have is either upland or wasteland which is uneconomical when cultivated. Many of these people took to headloading in 1979-80 when the monsoon failed.

It is ironic that the tribals, who for centuries lived in harmony with forests, are today forced to eke out a living by further destroying their forests. But the Ranchi women headloaders say that even if they used the forest properly, the forest contractor in collusion with forest officials would cut it down indiscriminately. So they would rather grab what they can before this happens.

The headloaders find a ready market for their wood in towns and cities because of the inadequate supplies of commercial fuels like kerosene and LPG. A report from Palamau in Bihar says that tribal women often collect two or three headloaders a day from the forests and sell them at Rs. 1.50 to 2.00 to households and tea shops. Many also work for wood contractors. The wood finally makes its way to markets in Varanasi.

An exact estimate of the total number of headloaders in the country does not exist. The 1981 census covered them under the group "harvesters and gatherers of forest produce" and "log fellers and wood cutters" but the data has yet to be tabulated. The 1971 census estimated all forestry workers (which included headloaders) at about 200,000 but experts say that this is nowhere near the actual number For instance, the Madhya Pradesh government has estimated the number of headloaders in Jagdalpur, the headquarter of Bastar district, to be 100 but when a researcher recently asked the villagers about this, they laughed and said that just two or three villages would have that many headloaders.

It is possible that at least 2 million to 3 million people are headloading today, making headloading India's biggest source of employment in the energy sector. Even if 50 per cent of the estimated 14 million tonnes of firewood purchased annually in urban India is supplied through the organised firewood trade supervised by the forest department, that leaves almost 7 million tonnes which is sold as headloads.

If each person sold three headloads of 25 kg each, collecting and selling firewood would provide an occupation for 2 million to 3 million people. The actual number is probably higher because besides those who do this work through the year, there are many who headload only when no other work is available.

In conditions of acute poverty, headloading is often the only occupation available. In Palamau district of Bihar, for instance, where the incidence of bonded labour is very high, bonded labourers often take to headloading because this is the only activity that can earn them some cash : the wages they get are only a handful of grain. Rehabilitated bonded labourers have also taken to headloading because they find it the only occupation available.

A major attraction of headloading, even though it is one of the lowest paid occupations, is that unlike other jobs like construction work or harvesting, it provides employment throughout the year, except during the monsoon In their paper on employment opportunities for women in forestry, Malini Chand and Rekha Bezboruah of the Institute of Social Studies, New Delhi, show that firewood collecting paid only 20 paise per hour. They had studied 65 firewood collecting households in Betul district, Madhya Pradesh.

Mainly Women

Studies show that large numbers of had loaders are women. In Ranchi district, for instance, the majority of headloaders are women in the 25-40 age group. Tribal tradition credits women with better marketing sense and bargaining capacity compared to men as they are prepared to wait hours for a good price while men prefer quick deals. Also an independent profession like headloading, thus, plays an important part in the nutrition and basic needs of these families. Besides, women are paid less than men in other jobs as construction work.

Collecting firewood is hazardous for headloaders and often means negotiating rough terrain. Besides, these poor people are provided with no medical facilities and any illness is a liability for them. In most headloading areas the women are malnourished.

Headloader's children are often neglected. They are often uncared for even during illness. Although men do help women collect firewood it is usually only when they are incapable of doing any other work. The few men who pick firewood in the villages near Ranchi are mostly above 40 or below 15.

Studies and reports from various southern states like Andhra Pradesh, Kerala and Tamil Nadu tell a similar story. A report on social forestry in Tamil Nadu prepared by the Tamil Nadu Agricultural University, Coimbatore, reported that landless agricultural labour resorted to firewood trade only when farm employment was not available. Of the 18 villages, spread over nine districts, selected for the study, fuel collection and sale either as a regular business and/or intermittently was practised by over 50 households in seven villages.

Like the villagers in Bihar study, these fuelwood collectors operate in groups of four and five but prefer to collect and sell individually.

Most headloaders operate under the pretence that they are collecting wood for their own use. According to the forest authorities this regular group activity is responsible for much deforestation. There has also been a sharp increase in green felling; women cut down some green branches, and wait for a few days for them to dry so that they can take it away as only dead and dry twigs can be removed.

To deter headloaders, a toll system has been introduced in some states. The Girnar forests are surrounded by several toll stations; headloaders have to pay a toll of 10 paise for every headload. In places where forests are guarded and felling is illegal, headloaders have to bribe forest guards and range official.

In many areas headloaders are often punished severaly if caught pilfering wood from protected forests. In the Girnar forests, people caught cutting any of the reserved species of trees have to forfeit their axes to toll station authorities.

The Real Scourge

Headloaders are now labelled "the scourge of our forests" K. P. Sagreiya, a former chief conservator of forests, complains that in Madhya Pradesh, where he worked for 30 years, permission was given for people to remove as much wood as they could by headloaders as long as it was "dead and lying" in the forests. "Now that such wood has long been picked up, live trees are being girdled, killed are removed in headloads with impunity. Such removals alone far exceed the total growth of the forests which are getting progressively depleted. It is time these malpractices are stopped", he says.

Madhya Pradesh forest officials are considering Sagreiya's suggestion that forest departments cut wood themselves and auction it to headloaders. Forest officials argue that they cut trees in a scientific manner in accordance with silvicultural rules, but experience shows that corruption often prevents this from happening. Moreover, if forest departments get involved in supplying headloaders, the firewood trade will become an organized racket benefiting only large contractors. Headloaders will then be completely at the mercy of forest officials.

Some Solutions

The worst thing about headloading is that there is no real policy. Various solutions have been suggested which are listed below :

1. Cooperatives of tribals and forest department would be set up to lessen the mistrust.
2. Another solution is to legalise headloading by issuing permits against a taken payment for specific forest areas specially developed for fuelwood.

Headloaders could be helped and encouraged to grow fuelwood trees on denuded lands, with the assurance that the fuelwood in these areas would belong to them.

3. Forest departments could engage headloaders on a daily wage basis to collect fuelwood and deposit it in a forest godown at a nominal price. Forests inhabited by tribals could be delinked from the forest department and alternative structures could be set up to help tribals to safeguard, develop and exploit their forests. Rather than headloaders buying wood from deposits controlled by forest department. Tribals could afforest deforested lands and control the produce of forests.

4. Headloaders could be provided with alternative employment Women could make handicrafts and their products could be bought by Forest Department and in turn be given wood at subsidized rates.

5. Headloaders could undertake tree plantations themselves. Women could improve their economic status by afforestation programmes.

Headloading is not an easy job. It is a measure of poverty in the countryside that millions of poor women have taken up as a major means of living. The efforts by a few voluntary organizations to ease the burden are commendable but much more has to be done.

13

Ecology and the Future of Man

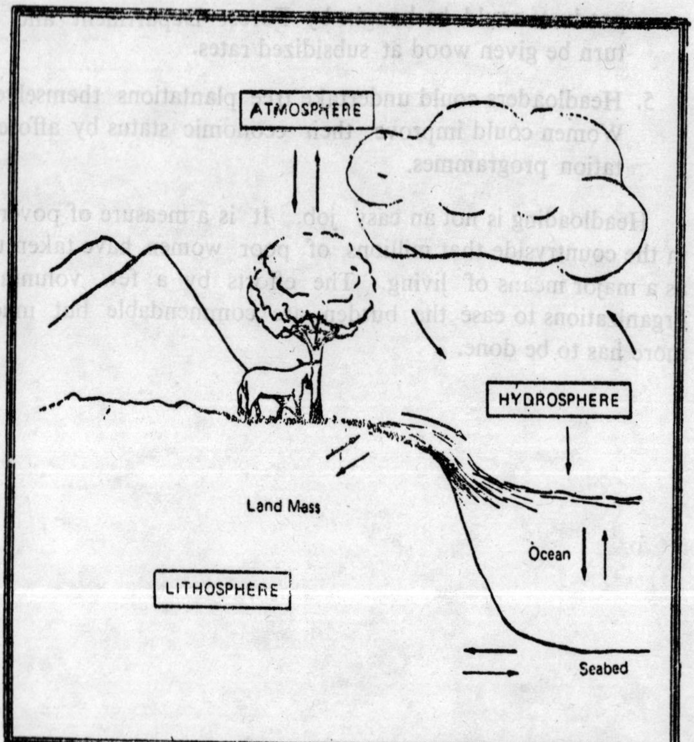

Introduction

Much of the current controversy about ecology concerns the future. Will man continue to progress in knowledge, health and standards of living, or is he headed toward oblivion in the near future ? Is the good life for everyone gradually spreading around the world, and is man coming into better balance with his environment, or is he already on a path of self-destruction.

Other scientists consider these attitudes as overly emotional, fear-producing, and scientifically unjustified They feel that ecologists cannot support many of their prophesies of doom and gloom, and that they are, in fact, raising unreasonable bogey-men. They point out the great strides in health welfare, education and social advance which have already occured, and they feel that there is a good deal of inaccurate sentimentalism about "the good old days." They note that the good old days had no electricity, no easy transportation, long working hours, very little educational and recreational opportunities, and plenty of diphtheria, smallpox, cholera, malaria and tuberculosis. They feel that the alarmist-type ecologists cannot recognize progress when they see it. To this the ecologist replies, "Does progress require polluted rivers and lakes, choking air, birds dying from pesticides, intolerable crowding, ulcers in children, violence on the streets, and heart attacks in young men ?" Thus the contro-versy and arguments can continue almost indefinitely.

In general, ecologists have tended toward pessimistic views of the future for more than 30 years. The first scientific books on the ecologic imbalance of man in the world appeared in the 1930's and 1940's (Sears, 1935; Vogt, 1948; Osborn, 1948). Since these early warnings about the ecologic problem of man, these difficul-ties have generally increased. Ecologists have found ample evid-ence for their fears. They feel that current events show that world population growth will not be amenable to significant slowing before the world is severely overcrowded. They have found that man has not learned the lessons of history and has continued to destroy his environment. They believe that the future survival of man will be, at best, a close and hectic race between these destructive forces and an enlightened society which can find the knowledge and necessary for solutions. They now see little evidence that man will really achieve these solutions beforc it is too late.

Essential Questions for Man's Future

The most difficult and essential problems for man's future lie in broad ecologic and social issues. No two questions are more fundamental or more troublesome to the future of man than the ecologists' inquiry.

"What is the carrying capacity of the earth for man ? In other words, what are the limits of the life support capabilities of the earth in terms of human population ?" are the behaviorists' concern.

"What are the tolerance levels of man to the pressures generated by his own societies ? In other words, if man is his own worst enemy, does this lead to his what stimulus, does this lead to his self destruction ?"

The carrying capacity, or life support capability, of the earth for man may be dependent upon a great variety of factors. We can readily identify oxygen water, food space and shelter as vital requirements for human life. There might be many others as well, such as mineral resources, energy availability, or the concentration of toxic metabolites, which could reach states above human tolerance levels. Most of these, however, would operate through the media of air, water or food. Hence, it is worthwhile to examine each of these briefly.

Oxygen Supply and Air Quality

The atmosphere of the earth consists of about 78 per cent nitrogen, 21 per cent oxygen, 0.03 per cent carbon dioxide, and the remainder, other gases. This composition is remarkably stable despite the fact that oxygen is a highly reactive element which readily combines with a great many other elements in chemical processes of oxidation.

The ecologist Lamont C Cole (1970) has pointed out that oxygen has remained stable in the earth's atmosphere because green plants recycle it in molecular form through photosynthetic process. Approximately 70 per cent of the free oxygen produced each year is through to come from planktonic plants in the ocean, and the remaining 30 per cent from terrestrial vegetation of forests and grasslands. Cole emphasized that if we should seriously pollute the sea, as is already happening in coastal areas, and if we should excessively denude the forests and grasslands of the earth, we shall be in danger of upsetting the oxygen balance in the atmosphere. In other words, we must look upon the oceans, forests and grasslands, as the world's oxygen tank, and we must insure their continued healthy ecologic function.

This point of view is contested by W.S. Broecker (1970), a geochemist at Columbia University, Dr. Broecker points out that each square meter of the earth's surface is covered by 60,000 moles of oxygen gas. Plants in the ocean and land produce only about 8 moles per square meter per year, and this is almost entirely consumed by their own respiration and that of animals and bacteria. Thus other organisms use all but 0.0004 per cent of the oxygen produced by plants. Broecker maintains, therefore, that man does not depend upon the oxygen generated by plants, and that there is no current danger at all to his oxygen supply. "In conclusion it can be stated with some confidence that the molecular oxygen supply in the atmosphere and in the broad expanse of open ocean are not threatened by man's activities in the foreseeable future. Molecular oxygen is one resource that is virtually unlimited". Broecker feels that man's existence will be threatened by other types of environmental pollution long before his oxygen supply becomes endangered, and that we need not worry about total oxygen supply.

This is typical of some of the different opinions which are unfolding between physical and biological scientists. Such differences often indicate the need for more research and better knowledge before we can fully accept one position or another.

Another atmospheric problem of long-term concern is the increase in carbon dioxide in the atmosphere. Between 1860 and 1960 the combustion of fuel increased the CO_2 content of the air by 14 per cent of its formerly stable level (Commoner, 1966). Carbon dioxide tends to reflect infrared rays from the earth, and this provides a "greenhouse effect" for the whole earth; that is, it accumulates heat. This will gradully increase the temperature of the earth - an effect which may have long-term implications in climate modification. A rise of just 2 or 3 degrees in annual mean temperature would have a major melting effect on glaciers and ice caps, and this could raise the level of the seas several feet. Such a rise in the sea level would inundate many of the world's major cities. Some experts predict it may require 4,000 years, and others feel it will not happen at all. The latter believe that any occurrence of the greenhouse effect due to increasing CO_2 in the atmosphere will be offset by increaing particulate pollution in the air and increasing capacity of the earth's atmosphere. Thus less light energy will enter, and we could actually have a cooling of the earth's surface. Once again, we see fundamental controversies on major issues - largely because we do not really have the complete answer.

The atmospheric problem may also become acute in terms of various toxic elements released into surface, air and water; carbon monoxide, sulfur dioxide, lead from automobile exhaust, hydrocarbons from industrial operations and numerous other products of modern technological societies. For example, 500 million pounds of lead are discharged into the atmosphere of the United States each year from automobile exhausts alone, and significant amounts of this become incorporated into living organisms. A recent study in Sweden showed concentrations of lead in mosses to be as high as 300 to 500 ppm along roadsides. Some roadside plants in the United States may have lead levels up to 3,000 ppm. This lead enters the plants *via* both air and water and can then enter man and animals. A hundred years ago, before the automobile, concentrations in roadside plants were less than 20 ppm. Lead levels over 0. ppm have detectable physiological effects in man and animals, and 1.0 ppm levels are often toxic.

In conclusion, there seems to be little or no immediate threat that we will run out of air for breathing, but several warnings have appeared regarding the quality of our air. We apparently have an adequate supply of total oxygen for many centuries to come, though even this is a debatable point. In some localities the situation may be critical at certain times. If stationary inversions of air trap the polluted atmosphere over some of our largest cities such as Tokyo, London, New York or Los Angeles, toxic air pollutants increase to dangerous levels. In August of 1969, a massive still-air inversion occurred in eastern United States, causing serious air pollution in urban areas from New Orleans and St. Louis, north to Chicago and Milwaukee, and east to Washington and Baltimore (Brodine, 1971). At no time was there an absolute shortage of oxygen, but dangerous level of carbon monoxide, sulfur dioxide, hydrocarbons and other air pollutants accumulated in several cities. Twenty million people were affected by varying degrees of eye irritation and respiratory diseases. Fortunately, this great blanket of smog was dispersed by upper air currents in the first week of September, and no large-scale health tragedies occurred. More serious events of this type are likely to occur in urban area unless strong measures against air pollution are undertaken.

Water Supplies

Worldwide water supplies represent a concern for the future of man in terms of both the quality and absolute quantity of fresh water. Many cities have reached or are coming close to their total available fresh water supply and are reaching further and further to obtain new supplies. Los Angeles, for example, is planning tremendous canals and conduits which will bring water from the Pacific northwest over 1,000 miles to the growing megalopolis of southern California.

Two Russian geographers and hydrologists, Kalinin and Bykov, have pointed out (1969) that "The world's water resources form a single entity. The basic hydrologic cycle of the world depends upon solar energy, atmospheric circulation, heat reflectivity and surface retention. The latter is strongly influenced by the biotic community, Kalinin and Bykov point out, on the assumption that the moisture content of the earth remains constant that the total gains and losses from the land and the sea are in balance. Thus, the total precipitation equals the total evaporative loss from land and sea. However, the relative retention time and thus the total amount of water in terrestrial and fresh-water environments can change substantially in relation to the biota and terrestrial storage capacity.

A large proportion of the world's fresh water falls and resides in relatively unpopulated portions of the globe, particularly Siberia and northern Canada. Here, man's activities are limited by other factors, and he cannot readily obtain these vast stores of water. Secondly, the matter of water quality is of immediate concern. The principle of water reuse is now becoming accepted, so that people recognize much of their water has been used previously by other municipal systems. Each reuse cycle, however, is increasingly costly in terms of treatment, and each reuse cycle requires disinfectants and other industrial chemicals, combining these problems of distribution and quality, it seems that the water problem of the future is not going to be one of a global water shortage on a planetary basis but one of local crises due to the concentrations of people, industry and water use. Certainly, the immediate concern is finding local water resources of adequate quality to meet the burgeoning needs of urban populations. To the city dweller in New York or Hong Kong whose water tap runs dry, or to the refugee in the Middle East who is dying of thirst, it will be little comfort to realise that 29,000,000 cubic kilometers of fresh water are locked in glaciers and polar ice caps.

World Food Suplies

Nowhere is scientific controversy more evident than on the issue of world food supplies. Some scientists predict famines and mass starvation within the next 10 years. They feel the stage is irrevocably set for millions of people to die of starvation in the 1980's. In fact, we already have famine in many parts of the world, if one defines famine as people dying of lack of food. Even at the height of the Green Revolution in Asia in which new strains of rice and wheat were producing phenomenally larger yields, famine conditions existed in parts of Rajasthan in India; in sections of Burma and in southern Thailand. Starvation is now a grim reality in much of the world and it may well get worse despite the Green Revolution.

Other scientists refute the pessimism and find evidence for a hopeful outlook in the world food situation. They feel that the Green Revolution which was based on new strains of rice and wheat in Asia and Africa, and which produced a doubling and tripling of yields in one or two years, shows what can be done with modern agriculture. For the first time in many decades, agricultural output increased more rapidly than population growth. India began to approach self-sufficiency in food production, and the Philippines began to export rice again for the first time since 1900.

A word of caution about the Green Revolution is also necessary. It is quite true that new strains of agricultural crops can sometimes dramatically increase yields, but often these new strains require more careful management, more intensive care, more water, more fertilizer and more insecticides. They may be more vunerable to plant viruses or fungi. Although the plant geneticists try to breed in every possible disease resistance, these new strains do not have the centuries of natural selection and regional adaptation behind them as do the local strains which the villagers have always used. In West Bengal and Orissa, there were reports that some of the newly introduced high-yield strains of rice were succumbing to a disease that was not affecting the old traditional strains. If this happens to a major extent, the Green Revolution could collapse and a severe blow would be dealt to the villager's faith in the agricultural scientist.

Another factor in the world food picture is one of pure climatic good fortune.

Green Revolution in India was accompanied by the best monsoon seasons in over 10 years. A long series of droughts had broken in southern Asia and adequate rains fell in the right pattern. The possibility exists, therefore, that a good part of the success of the Green Revolution was due to good weather, combined with new strains and more fertilizer. This is a question that will be more adequately answered in future years to see if increasing agricultural production can be maintained in years of poor rainfall.

In India too much rain came and it fell to early. The wheat crops of northern india were magnificient in April and May just before harvesting and threshing, but heavy premature rains in many areas caused the wheat to rot before it could be threshed. Bihar had serious crop losses from flooding. Once again, this was a tragic and dramatic demonstration that even the Green Revolution remains dependent on climatic factors outside the control of man.

Throughout all of these arguments on how many people the world can feed, one must ask the realistic question: "Is it necessary that there always be a race between population and food production ? Is it desirable to be teetering on the brink of starvation of all times ? Is it the best representation of the human condition to be struggling in this manner to feed more and more people at a subsistence lever ?"

One also has the impression in reviewing the literature on world food supplies that too much emphasis is placed on future predictions, and not enough on current conditions. We already know that starvation and malnutrition exist throughout the world, even in the most advanced countries. We furthermore know that malnutrition of any kind has profound effects upon growth, development and behaviour. Children growing up with malnutrition from early infancy definitely show reduced growth rates and possible retardation of physical and mental development. They furthermore show increased susceptibility to infectious disease. Thus, we return to the issue of the quality of life, and the realization that merely preventing death by starvation does not by any means insure a quality nutritional base for adequate growth and human development. It is apparent that the problem of world food supplies will continue for some time as a problem of both quantity, and quality and especially one of distribution.

Space on Earth

In 1958 Paul Sears wrote an important essay entiled, "The Inexorable Problem of Space," in which he concluded that "Our future security may depend less upon priority in exploring outer space than upon our wisdom in managing the space in which we live". With a shortage of space crowding in upon us in the 1970's, and 1980's, his words gain validity every year.

The earth's total surface is 196,940,400 square miles of which about 30 per cent or 57,230,000 square miles is land surface. Of this total land surface, approximately 5,000,000 square miles are in Antarctica and are too cold to be easily inhabited except by scientists and explorers 9,000,000 square miles are deserts, and 11,000,000 square miles are in mountainous regions. Of deserts and imountains, at least one-fourth of their jano area cannot be linhabited, so this eliminates another 5,000,000 square miles of land surface for human habitation. We are left with approximately 47,000,000 square miles of land surface, at the most, for the world's population.

The world now has about 3.5 billion people, and will probably reach 7 billion within the next 35 years. This means we now have about 75 people for every square mile of inhabitable land surface on the face of the earth, and this will approach 150 people per square mile by the year 2000 AD.

As with water, air and food, it is important to distinguish between total or potential amount of any given resource on earth and that amount which is actually utilized or practically available to man, it was pointed out in the sections on water and air that the earth's total amount is apparently adequate for many years to come, but the supply in any given city may not be of sufficient quantity or quality to support the standard of living we would like. So it is with space. Although we theroetically have one square mile of land for levery 75 people (about 9 acres per person), the human population is not evenly distributed. In the great urban concentrations of the world densities often exced 50,000 people per square mile. The amount of living space per person in some cities is less than 20 square feet per person. Theoretically, we could redistribute these crowded people to provide more even spacing around the world. Obviously this cannot be done, and if it were attempted it might wreak hovoc with agricultural production.

The economic and behavioural needs of man are such that he must cluster together, but in so doing, he creates super-densities such as Calcutta, London, New York, Paris, Tokyo-Shanghai and Buenos Aires. The slums of which are definitely deleterious to human health and welfare. Why, we might ask, does man insist on this degree of crowding and then complain about it so bitterly ? Why do people continue to pour into the cities of the world when they are already known as centers of population, crime and illness ?

Perhaps we need to know more about the real spatial and social needs of man. What are his requirements for space and privacy, and how do these interact with his needs for social stimulation and group activity ? Does man become conditioned to crowding in early childhood and then seek further crowding throughout his life ? Does he require the social pressure of a crowd because he feels a compulsion to be in touch with the group and thus be able to predict what the group is going to do next ? Have were produced, in our urban centers, self-destructive, systems which so modify the behaviour of their inhabitants that these inhabitants exaggerate their own problems and can no longer achieve rational solutions ?

These are all tough questions cannot be answered in simple or easy terms. The answers will require the best intelligence we can muster from many disciplines. Some sorts of answers will certainly be necessary before we can even begin to solve the ecologic and behavioral problems associated with our own use of terrestrial space.

Distributional Inequality

With all of the last four life-support requirements of man – air, water, food and space – we have seen that the major problem of immediate importance lie in distribution more than in total amount. Technologist therefore look upon these problems as soluble by technological means. They feel, they are essentially matters of economics and transportation Behavioural and environmental scientists see them as more basic, however, relating to man's fundamental ecologic behaviour. They feel that certain basic social and ecologic tendencies of man prevent equitable distributions of the earth's resources. Obviously this type of argument can go on forever, and it is perhaps necessary to take a pragmatic approach and look at the current picture for the best glimpse of the future.

Many of the industrial nations of the world, especiallly the United State, the national northen Europe and Japan, have enjoyed rising prosperity for the last 25 years. In ecologic timé, of course, this is but a flash in the pan, but let us assume that it can be sustained for another 25 or 50 yeass. It is worth examining where the material and energetic support for this prosperity arises.

This distributional inequality is another major symptom of ecologic instability in the world. One would hope for some narrowing in the gap between the haves and the have-nots but unfortunately the opposite trend is still occurring. The wealthy nations of the world are far outstripping the poor nations in their rates of economic gain and human betterment. There are, however, self-corrective mechanisms within this system as, for example, when some segment of the labour market in a wealthy country can no longer compete with its foreign counterpart which does the same work for much less. The world economic enterprise continually strives for the cheapest labour it can find which will do a satisfactory job Thus, there has been a great expansion of manufacturing in Japan, Hong Kong, Taiwan and Singapore in industries as diverse as steel, electronics, textiles, toys and optics. The

Western factory worker can thereby suffer a decline in his economic security and standard of living while the Asian factory worker experiences for the first time an improving situation. The western industrial worker continues to refuse his own working prospects by striking for higher and higher wages. This will certainly be one mechanism by which wealth is more adequately balanced between nations.

Stil the mass of poverty throughout much of Asia, Africa and Latin America is so great that it is difficult to imagine a painless equilibrium. The differential distribution or wealth remains sharp and represents a type of ecologic situation which cannot last indefinitely without severe consequences and reverberations. To the ecologist it means either a successional stage which is not yet mature, or a system headed for a major breakdown. Economic and ecologic maturity should ideally be in the direction of broader and more equitable distribution of the earth's resources—a narrowing of the gap between the haves and the have-nots, both within and between nations provided the total system does not pass beyond the state of corrective change.

Violence and War

Any ecological system or biotic community can be overloaded to the point of collapse by a variety of factors. Certainly one of the factors for man of most immediate and realistic danger is that of social violence. Either by war or by civil riot, man threatens his own existence.

To maintain a sense of physical security man has built an incredible stockpile of destructive power. The nations of the world spend more than 200 billion dollars per year on armaments—more than on health education, welfare, agriculture, child care and care for the aged combined. Twenty tons of explosives are stored in the world for every man, women and child. Russia and the United States have stockpiled 6,000 nuclear warheads possessing thousands of megatons of explosive force. Each megaton has more destructive power than which existed in any of the previous wars of history. The world spends almost 8,000 dollars per year training each soldier and less than 100 dollars per year educating each child.

The question is, of course, how long can man continue to preside over armament like this without using it? And the answer seems to be, not very long. The fact is we do use part of this power in an attempt to achieve social goals. It is being used daily in various parts of the world to maintain territory and dominance ranks. It is justified as a spare and deliberate use in hopes of avoiding the all out effort. How long this can continue is anyone's guess, and the ecologist can claim no more omniscience than the political scientist. The ecologist points out, however, that the ecologic and social pressures which generate international military tension may rise with increased population pressure. It is not likely that man can solve the problems of war and violence until he approaches and understands their basic origins in terms of the behavioural and ecologic relationship between social groups, and the environment.

Adaptive Capabilities of Man

Many scholars feel that one of the great problems of the present and the future is the rapid rate of change which is occurring in society and technology. There is not, however, agreement on the consequences of this rapid change, changing life patterns are too rapid for man to adapt successfully. Mankind is so overstimulated by the increase in new sight and sounds, so bombarded by worldwide mass media, and so constantly beset with noval situations and new problems that he will break down psychologically and physically. His responses will be maladaptive and pathologic. He will thus collapse under sensory overload.

It is quite possible that each of these points of view has large elements of truth in it. In the present world one can find many examples of both. Our mental institutions are already filled with those who could not adapt to sensory overload and who experience a personal breakdown of physical and mental health. Yet one can find people are living a sub-human existence-where the quality life is so low that the amount of human pain and misery seem unbearable. Yet man has adapted successfully to these in human conditions.

In many ways, the future is with us now. We can find throughout the world an abundance of high quality life in high quality environments, but we can also find an abundance of low quality life impoverished environments. Many of the questions about the future revolve, therefore, around the question of relative balance and proportions. Which way are we heading and which way will the scales tip ?

A major aspect of the future centres around our sense of responsibility to posterity. No representative of posterity is here to speak for that segment of society. How do the present populations of mankind regard their obligations to future generations ? Are we willing to be more conservative in our life requirements, and are we, in fact, willing to reduce our rates of personal economic growth to insure a better environment and a greater abundance of resources for future generations ? This may be one of the central issues of the next twenty years, and one on which the quality of present and future environments depends.

New Approaches to Environment and Development

Human progress has always depended on our technical ingenuity and a capacity for cooperative action. These qualities have often been used constructively to achieve development and environmental progress : in air and water pollution control, for example, and in increasing the efficiency of material and energy use. Many countries have increased food production and reduced population growth rates. Some technological advances, particularly in medicine, have been widely shared.

But this is not enough. Failures to manage the environmen. and to sustain development threaten to overwhelm all countriest Environment and development are not separate challenges; they are inexorably linked. Development cannot subsist upon a deteriorating environmental resource base; the environment cannot be protected when growth leaves out of account the costs of environmental destruction. These problems cannot be treated separately by fragmented institutions and policies. They are linked in a complex system of cause and effect.

First, environmental stresses are linked one to another. For example, deforestation, by increasing run-off, accelerates soil erosion and siltation of rivers and lakes. Air pollution and acidification play their part in killing forests and lakes. Such links mean that several different problems must be tackled simultaneously. And success in one area, such as forest protection, can improve chances of success in another area, such as soil conservation.

Second, environmental stresses and patterns of economic development are linked one to another. Thus agricultural policies may lie at the root of land, water, and forest degradation. Energy policies are associated with the global greenhouse effect, with acidification, and with deforestation for fuelwood in many developing nations. These stresses all threaten economic development. Thus economics and ecology must be completely integrated in decision-making and law-making processes not just to protect the environment, but also to protect and promote development. Economy is not just about the production of wealth, and ecology is not just about the protection of nature; they are both equally relevant for improving the lot of humankind.

Third, environmental and economic problems are linked to many social and political factors. For example, the rapid population growth that has so profound an impact on the environment and on development in many regions is driven partly by such factors as the status of women in society and other cultural values. Also, environmental stress and uneven development can increase social tensions. It could be argued that the distribution of power and influence within society lies at the heart of most environment and development challenges. Hence new approaches must involve programmes of social development, particularly to improve the position of women in society, to protect vulnerable groups, and to promote local participation in decision making.

Finally, the systemic features operate not merely within but also between nations. National boundaries have become so porous that traditional distinctions between matters of local, national, and international significance have become blurred. Ecosystems do not respect national boundaries. Water pollution moves through shared rives, lakes, and seas. The atmosphere carries air pollution over vast distances. Major accidents—particularly those at nuclear reactors or at plants or warehouses containing toxic materials—can have widespread regional effects.

Many environment - economy links also operate globally. For instance, the highly subsidized, incentive-driven agriculture of industrialized market economies generates surpluses that depress prices and erode the viability of the often neglected agriculture of developing countries. Soils and other environmental resources suffer in both systems. Each country may devise national agricultural policies to secure short-term economic and political gains, but no nation alone can devise policies to deal effectively with the financial, economic, and ecological costs of the agricultural and trade policies of other nations.

14

Living Resources

Introduction

India has a uniquely rich diversity of plant and animal life, which contribute in no small measure to the human economy. It is not just the wild living resources that are today threatened because of the destruction of their habitat, domestic species of plants and animals are also facing genetic attrition, partly because of those very activities that are undertaken by human beings in the name of economic development, such as the Green Revolution. Indian farmers probably grew over 30,000 varieties of rice. But with high yielding varieties becoming popular, these varieties are disappearing, and by the end of the century there will be no more than 50 varieties being used, with the top ten alone occupying over 75 per cent of the acreage. This genetic erosion, if carried too far, could threaten the very future of the human society. Protection and conservation of living species including domestic species, must, therefore, become an important issue for public policy and public pressure.

In recent years, the government has begun to take conservation of wildlife seriously. But the manner in which these conservation programmes are being conceived and implemented is creating more tension than peace and harmony. Whereas conservation is an issue of crucial importance, it cannot be undertaken without the involvement of the people, especially those directly affected. Nature conservation programmes cannot be promoted in the same ham-handed way as those programmes which have destroyed naure. The result will inevitably be conflict and failure.

Genetic Resources

Plant and animal life, as it has evolved on this earth, has produced a bewildering diversity. There are millions of species of plants and animals, but even within a species, there exists enormous diversity. For instance, there are an estimated 1,20,000 types of rice plants alone in the world and this is only a guess. There can actually be many more.

This massive diversity, or "gene pool", as scientists like to call it, is a common heritage of all human beings. It is also their insurance for future survival. Increasingly, as new varieties of plants and animals useful to human beings are bred, this masssive gene bank provided by nature becomes important. Breeding is essentially an exercise in finding the desired gene from nature's vast genetic reservoir and introducing it into the genes normally used. The Green Revolution, for instance is essentially based on a strategy of genetic reconstruction aimed at producing new varieties of cereal plants which can produce high yields.

Scientists involved in this work, however, have also realised that the Green Revolution is something akin to riding a tiger; once you are on it, you have to stay on it to survive. To keep the Green Revolution going, breeders must produce new varieties every five to seven years as older high-yielding varieties succumb to pests and diseases, which themselves keep changing in intensity and character. If one variety covers too large an area for too long a time, it becomes an open invitation to disaster. A disease can sweep through a genetically uniform crop like wildfire leading to widespread crop failure and famine. The brown rust infection, for instance, affected large areas in north India planted under the wheat variety Kalyansona during 1973-74. Other than high yields, a high priority for both plant and animal breeders is the development of resistance to diseases, and screening available plant and animal varieties for these traits accounts for substantial amount of time and resources of breeders.

Threat to Genetic Diversity

Ironically using "superior" genes on a massive scale to increase production, as exemplified by the Green Revolution in food crops, blue revolution in fish production, while revolution in milk, and massive afforestation with high yielding monocultures, has posed one of the biggest threats to the world's genetic diversity in livestock, fish, crops and trees. In each of these areas, human beings are promoting a narrow genetic base – a few genes with highly desirable characteristics – to increase production.

In this process, farmers can replace thousands of wild varieties and primitive cultivated varieties overnight with one high yielding variety. If the displaced varieties are not collected and properly stored, they could be lost forever, India has already lost thousands of traditional varieties of wheat since the beginning of Green Revolution.

Simultaneously, environmental destruction caused by shifting cultivation, deforestation and dam building is destroying many wild varieties of even domesticated crop plants, trees and animals. Water pollution is affecting fish stocks and probably even genetic diversity. Overgrazing is affecting grass and tree populations.

However, no gene in this diverse nature can be considered useless and, thus, deserving to be driven to extinction. India possesses an extremely rich diversity of plant and animal life. The Indian region is considered one of the 12 centres of genetic diversity in the world. The total Indian flora consists of about 45,000 species of which about 30,000 are non-flowering plants like algae and fungi which are usually inconspicuous and small. The flowering trees and plants of India consists of about 15,000 species but this number should not be taken as the final figure as several areas in the country remain unexplored and under-explored. From a very small area, Silent Valley in Kerala, covering less than 9000 hectares, the Botanical Survey of India recently discovered nine species new to science and a new genus of plants.

About 7000 endemic species are found in India, that is, they do not occur anywhere else in the world. Of these the Himalayas and the Khasi Hills in Meghalaya account for about 3000 endemic species, and the Deccan Peninsula for about 2000 species. The area of maximum diversity in India is northeastern India which is also the most threatened. The second major area of genetic diversity is the lower region of the Western Ghats around the Nilgiris. Over 1500 of the rare flowering plants are subjected to a lesser or greater degree of threatened extinction.

Valuable Indian Genes

Just as India has benefitted from plant and animal genes from all over the world, India, too, has given to the rest of the world many valuable genes. Probably the most dramatic case is that of a single sample of wild rise, Oryza nivara collected by botanist S.D. Sharma from eastern U.P. in 1963. Grassy stunt virus is one of the four major diseases of rice plants. During the early 1970s, grassy stunt epidemics destroyed more than 116,000 hectares of rice in Indonesia, India, Sri Lanka, Vietnam and the Phillippines. At the international Rice Research Institute in Los Banos (Philippines). Gurdev Khush and K.C. Ling searched 6723 samples of cultivated and wild rice varieties looking for resistance to grassy stunt virus. Only one sample, that of Sharma's displayed the resistance they were looking for. Today, the genes from this sample are used in resistant varieties grown on 30 million hectares in India, Nepal, Bangladesh, China and Southeast Asia. The disease has almost ceased to exist In Indonesia these genes have helped push average production per hectare to six tonnes and in some cases to 10 tonnes. Two or three crops have become common and rice production overall has doubled.

In an interconnected world, Sharma's rice sample has helped India to repay a debt. Indonesian wild genes have helped India to set up a sugarcane industry. The limiting factor to commercial sugarcane production in India used to be the red rot diseases. The resistance was acquired from the wild Indonesian cane Saccharum spontaneum. Indonesian S. spontaneum genes are now present in important commercial varieties of sugarcane used in such top sugar-producing countries as Cuba, Mexico, USA, Argentina, Brazil, Columbia, China, India, Indonesia, Pakistan, Philippines, Thailand and Australia. Today equally widespread are S. spontaneum clones from India, which are sources of resistance to gummosis and root rot.

The list of valuable genes discovered in India is long. Large areas of the world, for instance, are covered by unproductive saline soils. Tolerance to salinity in rice has been found in several traditional varieties from southern India and Sri Lanka. A wild melon from India has provided resistance to a form of mildew and has saved the melon crop in California, USA.

Indian Animal Breeds

Among animals, too, many Indian breeds are known all over the world. The home tract of the famous Sahiwal cattle is in the Sahiwal district of Pakistan. At the time of partition, only a few Sahiwal herds existed in India. A few more animals were brought by migrating Sikh and Hindu families. Today, the Sahiwal population in India consists of about to 1000-1100 breeding females. The breed is being used for crossbreeding in many countries. Sahiwal semen is being produced on a large scale at a farm in Queensland (Australia) for export to New-Zealand, which, in turn, after contract matings by use of Sahiwal semen on Jersey and Friesian cows, exports crossbred heifers to Southeast Asian countries.

The famous Murrah, Nili-Ravi and Surti buffaloes are being used to upgrade buffaloes in many countries of the world. In Vietnam, a consignment of Murrah buffaloes is widely regarded as the most valuable gift from friendly India. The Jagarabadi has been successfully introduced into lands in which buffaloes did not exist until recent times, for example, Brazil and Trinidad. The Jamnapari goat breed of the Chambal ravines in Etah district (UP) has been used for improvement of native breeds in several countries. The famous English Anglo-Nubian breed, which has been extensively used for crossbreeding, is itself based on a cross of the Jamnapari.

Insect Germplasm

Even germplasm of economically important insects needs to be preserved. The two insect livestock groups of major world importance are honey bees and silkworms. There are many strains of honey bee, and there is great variation within the strains for characteristics such as honey production and disease resistance. Though efforts to collect wild bee germplasm seem to be a long way off because of various technical reasons, the need is being increasingly felt.

Work on breeding has already been done with the mulberry silkworm. In India, wild forms of the tasar silkworms (Antheraea species) are being used for improvement. Antheraea roylei from India and A. pernyi from China have been crossed to produce the hybrid A. proylei. Silk from the hybrid is claimed to be the finest ever produced in India and the new species also outyields its parents by almost 170 per cent in weight of silk and 94 per cent in filament length. The hybrid is expected to play a major part in efforts to expand the tasar silk industry and provide rural employment.

Conserving Micro-Organisms

In a recent lecture on genetic conservation, Dr. M.S. Swaminathan, one of India's leading agricultural scientists and Director General of the International Rice Research Institute, drew attention to the need to conserve even micro-organisms. There has been an enormous growth of interest in biological nitrogen fixing organisms because of the ill-effects of chemical fertilisers on the environment. Among the major sources of biofertilisation are blue green algae and azolla. Azolla germplasm from all over the world is already being collected at the international Rice Research Institute. Azolla is an aquatic fern capable of fixing atmospheric nitrogen through a symbiotic blue green algae living in the cavities of its leaves. It has long been used in China and Vietnam as green manure in rice fields and its use is growing in other Southeast Asian countries now.

Blue green algae (BGA) is being extensively promoted in India as a cheap source for fixing nitrogen in paddy fields. Unfortunately, there is no organised effort to collect BGA for genetic conservation purpose. Strains recorded during ecological studies are not kept by the scientists because maintenance of a culture collection is time consuming and expensive. Scientists have achieved considerable success in selecting strains of Rhizobia nitrogen fixing bacteria which live in the roots of leguminous plants, and give them their nitrogen fixing abilities. This success indicates the possibility that superior strains may exist in other nitrogen fixing organisms, too.

Today, the world's main food supply is determined by only 20 crop species, and only eight forest species provide most of the world's wood and timber. Similarly, about 20 animal species provide most of the world's milk, wool, hide and draught power. A much larger number of plants do provide food, such as oil, vegetables, fruits and fodder, and meet other needs, such as species, beverages, medicines, ornamentals and industrial raw materials. However, all these plants and animals constitute only a fraction of all the plant and animal species in the world.

It is impossible to guess what this natural wealth will yield in the future. But the potential can be seen from the vast variety of medicines alone that plants have given to human kind. The Ayurvedic system of India gave to the world reserpine, a tranquiliser and a treatment for hypertension, anxiety and schizophrenia. Reserpine is a therapeutic agent extracted from a herb, Rauwolfia serpentina, which is now almost impossible to find in forests.

Medicinal Herbs

In India, at least 2500 plants are used for medicinal purposes. The roots of the berberry shrub found in the Himalayan foothills are an important source of berberin, a compound used widely in tropical Asia and Japan against diarrhoea. A single species, Strychnos nuxvomica, indicates the commercial value of drug producing plants. A source of the compounds strychnine and brucine, this species earns India over Rs. 50 lākh yearly in exports. India's total exports of plant-based chemical products, mainly medicines and pharmaceuticals, are worth over Rs. 55 crore yearly, a trade which increased five-fold during the 1970s. It is distressing that overexploited medicinal plants face probably the greatest threat of extinction in India today.

From an ethical point of view too, there are strong arguments against human beings driving any life form into extinction. But protecting the wonderful gene bank provided by nature is an issue on which both utilitarian and ethical perspectives coincide.

Crop Genetic Resources

India is known especially for its rich genetic wealth of bamboo, fruit trees (mango, banana and citrus), vegetables like eggplant, okra and cucurbits (melons and gourds), grain legmes (mung beans, urad or black gram and cowpeas), rice, pepper, cardamom, ginger, turmeric, sugercane, jute, and various medicinal and aromatic plants. There is also or rich diversity in wheat, maize, millets, oilseeds and various other fruits, spices, vegetables, fodder and plantation crops.

Plant species from India have not only influenced agricultural development in the region but also in many other parts of the world. Plants were taken from India to the ancient civilizations of the Tigris, Euphrates and Nile river basins. Many Arab traders took plant materials to Mediterranean countries in the 8th 10th centuries AD. Buddhists took several Indian cultivated plants to Southeast Asian Countries.

It is estimated that about 250 Indian species have wild plants of agricultural and horticultural importance and of these about 60 are rare or endemic. Though many wild varieties have been collected and used in breeding programmes, only a fraction of the job has been done. India is also very rich in its natural wealth of forage legumes and grasses. About 2000 such species occur, of which two-thirds are grasses.

The northeastern Himalayas are a veritable mine of valuable genes in many crops and fruit trees. Unfortunately because of shifting cultivation and widespread habitat destruction, this genetic wealth faces the threat of extinction. Dr. Bhag Singh of the National Bureau of Plant Genetic Resources (NBPGR), for instance, discovered Citrus Indica, an endangered species of citrus, in the Tura range of the Garo Hills, Maghalaya. This species is the most primitive of citrus species and perhaps the progenitor of cultivated meterials. Citrus fruits are native to South Asia and the northeastern Himalayes hold a notably important position in the world with regard to citrus wealth. The Khasi and Garo tribes grow citrus trees in and around the vicinity of their houses.

Varieties of Rice

A major crop in India and Asia, rice has attracted the maximum interest amongst Indian scientists interested in plant genetic resources. The areas with maximum diversity in India are the northeastern tracts and the hilly regions of Orissa, Bihar, Madhya Pradesh and Andhra Pradesh, which are mostly inhabited by tribal populations. Various local rice varieties have also been found in the Malabar region and in the cold Himalayan tracts.

A national survey estimated that although about 50,000 local rice varieties are still in cultivation, they are being rapidly lost, particularly in the irrigated tracts, where new high yielding varieties have been adopted. Growing deforestation poses a serious threat to many wild varieties of rice. Tropical rain forest habitats have been important sources of disease and pest resistant varieties of rice. The Silent Valley forest reserve area in Kerala, which was to have been inundated with the construction of a hydroelectric project, has been an important source of resistant genes now incorporated into many Pattambi rice lines. The Assam hills in the Northeast are another invaluble source of rice germplasm in the form of the well-known Assam Rice Collection (ARC). Together with the threatened cultivated varieties, these endangered natural habitats should be a priority area for immediate collection of wild varieties. In a germplasm convervation programme, the wild species are as important as, or more important than, local cultivated varieties.

A total of nearly 15,200 local rice varieties are being maintained at the national rice gene bank at the Central Rice Research Institute (CRRI) at Cuttack, and an estimated 35,000 local varieties are held as working collections at 40 rice research stations. Thus there is a total collection of 50,000 cultivars (cultivated varieties). Assuming 50 per cent of the collection to be duplicates, the national collection is now estimated at 25,000 with scope for further collection of another 25,000 varieties.

The genus Oryza (rice) consist of about 24 species, out of which only two are cultivated : sativa (Asian rice) and O glaberrima (African rice). Collection of wild varieties has lagged behind those of cultivated varieties. The group O. rufipogon, O. nivara and O. meridonalis, is the most closely related to the widely cultivated O. saliva rices and represents the progenitors of the cultivated rices. Although as genetic resources, their importance is paramount, unfortunately, they are fast disappearing.

Endangered Plants of Actual or Potential use in Traditional Medicine

Species	*Common name*	*Use*
Aconitum deinorrhizum	Atis	Medicine
Acorus calamus	Vacha	Sedative
Alpinia galaanga	Khulanjan	Drug
Atropa awminata	Angurshaga	Medicine
Commiphora wightii	Guggal	Drug
Coptis teeta	Mishmee teeta	Medicine
Denarobium nobile		Dendrobine
Dendrobium pauciflorum	Picotee dendrobium	Alkaloids
Dioscorea deltoidea	Kins	Cortisone
Diplomeris hirsuta	Snow orchid	Alkalodis
Gentiana kurroo	Kadu	Drug
Paphiopedilum druryi		Alkaloids
Podophyllum hexandrum	Indian podophyllum	Drug
Rauvolfia serpentina	Sarpagandha	Drug
Santalum album	Sukhad	Drug
Saussurea loppa	Kuth roods	Various uses
Obova Ita	Brahm Karnal	Medicine

Source : World Health Organization

This group occurs in one of the densely populated areas of the world, where the habitat is being reclaimed for cultivation of crops, construction of roads and buildings. The cultivation of O. sativa rices is coming closer to the population of these wild species which are hereby losing their genetic identity by being "swamped" by the genes of O. sativa. A time bound collection programme for the wild rices in this group is thus urgently needed.

The group O. Collina, O. officinalis, O. minuta and O. malampuzhaensis are Asian rice species which occur in forests in partial shade, by the side running streams or swamps. With deforestation, this group is also endangered. The present collection of O. officinail from South Asia is not representative of their natural distribution. It is important to collect these species before they disappear.

In India, the CRRI is aided by 10 base centres of Simla, Almora, Jodhpur, Ranchi, Shillong, Cuttack, Hyderabad, Akola, Trichur and New Delhi. These centres collect not just germplasm for rice but also for a wide range of other crops. Each centre has been allocated a state or region in which it actively collects. In future collections of wild rices cultivated varieties, upland rainted rices, flood-tolerant lowland rices, and varieties displaying saline-alkaline tolerance, cold tolerance, pest and disease resistance, and aromatic and quality characters will be emphasised. Collecting these wild species and peripheral populations under severe pressure may require specialised collectors and the creation of "gene sanctuaries" to protect threatened habitats.

According to Dr. M.S. Swaminathan, the collection of germplasm should not only involve professional scientists with expert knowledge of rice but also farmers, the general public and the defence services. In China, for example, each provincial academy mobilises farmers to send in seeds of traditional varieties. Although some samples will duplicate, this system lessens the work of the scientists and lets them concentrate on classifying the material and identifying duplicates and distinct types. As some of the high priority germplasm collection areas and endangered habitats are either sensitive border zones or inaccessible regions, defence personnel could also play an important role in collecting germplasm.

Vulnerable Plant Species

Amongst the most vulnerable plant species in India are overexploited medicinal plants, ornamental plants, like orchids and botanical curiosities like pitcher plants. Some 1250 species of orchids occur in India, of which 300 are found in Meghalaya alone. About 20 species of orchids are becoming scarce in Meghalaya. Scientists from the Botanical Survey of India (BSI) recently made intensive studies on a group of beautiful orchids called Coelogyne, but failed to see even one plant of three species, namely, Calbolutea, C. assamica and C. treutleri in living condition. They have probably become extinct. Even their preserved specimens are not available in any herbarium. The Government of Meghalaya has recently established a "pitcher plant sanctuary" in the Khasi Hills. The Indian Orchidarium at Shillong has brought approximately 350 indigenous orchid species into cultivation for conservation purposes. But much more needs to be done in this connection. For example, a rhododendron sanctuary is urgently needed in Sikkim.

The loss of valuable plants is detinitely a matter of great concern, though new attempts have been made as yet to collect their germplasm. Amongst the endangered medicinal plants are three orchids : Paphiopedilum druryi a native to Kerala, which is endangered or extinct in the wild due to forest fires and excessive collecting; Dendrobium pauctiflorum found in West Bengal and Sikkim in areas open to tree felling which is also endangered, possibly extinct; and Diplomeris birsuta, restricted to a tiny population in West Bengal and is vulnerable to landslides. A decrease in its number has been noted by botanists. The numbers of another orchid, Dendrobium nobile are seriously declining. This orchid occurs in the Himalayan regions of India and China and is the source of dendrobine which is still exported from China by the ton in a dehydrated from.

Several other medicinal plants are being seriously threatended because of over exploitation. Acorus calamus, known as vacha in India, is used in 51 different drug preparations. It used to be common in Manipur and the Naga Hills, but today it is so scarce that it has to be imported from overseas,

Rauwolfia serpentina, the source of resperpine, known to Ayurveda for centuries, was once common and abundant in forest areas all over peninsular India. But after scientists isolated the alkaloid reserpine in 195?, a ruthless search was started all over the country, which came to a halt only when the plant almost disappeared from forest areas. Today it is extremely rare in nature and is mainly found in cultivated stands. Variations within the species may have been lost already.

Two other species that are extensively used in Ayurveda and are today endangered are Saussurea lappa, the source of kuth roots, and Dioscorea deltoidea, the source of steroidal hormone. Today these plants are listed in the Convention on International Trade in Endangered Species and cannot be exported without an export permit. Dioscorea deltoidea, much sought after by private agencies and pharmaceutical firms, has been continuously collected in India except perhaps in the inaccessible areas of the Himalayas. The roots yield cortisone, a steroidal hormone used in treating rheumatic diseases and ophthalmic disorders.

The need to preserve the medicinal plants of India is clearly vital and urgent. Maintenance of botanical/herbal gardens will not be a solution. Efforts will have to be made to conserve the germplasm of each important species of medicinal plants. The National Bureau of Plant Genetic Resources (NBPGR) has only made a beginning by collecting Opium Poppy, Syllium, Senna, and Rauwolfia Serpentina. The NBPGR has a major role to play in the protection of the country's vital genetic treasure. But the Indian Council of Agricultural Research (ICAR), which supervises NBPGR, will have to be more effective in energising this organisation than it has been so far. Germplasm conservation is a complex task. Clear priorities have to be set in collecting germplasm. There should be an understanding of how existing collections or duplicates should be weeded out. Collected seeds should be stored carefully. They must be properly characterised for their attributes and then made available to breeders for use. Finally, there should be adequate facilities for multiplication of seeds while duplicate samples should be kept as an insurance.

Although the NBPGR was set up in 1976, it has still not built a proper facility for collected germplasm. Funds were allocated for the construction of a storage facility in 1978 but it took three years just to chose the site. It has taken a further three years to finalise architectural details, with Central Public Works Department officia's anxious to visit Europe to look at existing storage systems. There is still no firm date by which the storage facility will be built.

Indian National Priorities for Collection of Plant Germplasm

	Priority 1	*Priority 2*	*Priority 3*
1	2	3	4
Cereals	Rice Wheat	Maize Barley	Buckwheat/ Amarnath
Millets	Sorghum Pearl millet Finger millet	Lesser millets	
Grain Legumes	Pigeon pea Green gram Black gram Cowpea Chickpea	Lentil	Winged bean Rice bean Soybean Moth bean
Oil Seeds	Coconut Brassica Species Sesame Groundnut	Safflower	
Fibre Crops	Cotton	Jute	Kenaf, Mesta

(*Contd.*)

1	2	3	4
Vegetables		Okra Egg Plant beans Cucurbits	Non-tuberiferous Solanums
Tuber Crops			Yams and aroids Pachyrrhizus Sweet potato
Fruit Crops	Banana Citrus	Mango Jackfruit	Papaya Phyllanthus Guava Morus Litchi Zizyphus (ber) Syzygium (Jammu)
Sugar Yielding Plants	Sugarcane		
Beverages	Coffee	Tea	
Spices	Peppe	Turmeric Arecanut	Ginger
Medicinal Aromatic Plants		Diooscorea Rauvolfia Cymbopogon species (Lemon grass, Citronella grass, Opium poppy Senna Palmarosa oil grass)	

Major Gene Banks in India

Name of Plant	*Accessions*	*Term of storage*	*Institution*	*Location*
1	2	3	4	5
Rice (Oryza sativa)	30,000	Medium	IARI	New Dehli
	15,249	Medium	CRRI	Cuttack
Sorghum (Sorghum	24,000	Medium	ICRISAT	Hyderabad
bicolor)	14,340	Medium	ICRISAT	Hyderabad
Pearl (Pennisetum Millettypholdes)	2,247	Shooit	AICMIP	Poona
Foxtail Millet	5,017	Short	AICMIP	India
(Setaria it alica)	1,160	Medium	ICRISAT	Hyderabad
Finger millet)	5,904	Short	AICMIP	India

(*Contd.*)

1	2	3	4	5
(Eleusine coracana)	1,241	Medium*	ICRISAT	Hyderabad
Kodomillet	1,405	Short	AICMIP	Poona
(Paspalum scrobiculatum)				
Cassava (manihot)	1,800	Short	CICRI	Kerala
Mungbean	3,000	Short	PAU	Ludhiana
(Phaseolus aureus)	1,000	Short	IARI	Delhi
Chickpea (Cicer arietinum)	13,000	Medium*	ICRISAT	Hyderabad
Pigeonpea (Cajanus cajan)	8,850	Medium	ICRISAT	Hyderabad
Groundnut (Arachis hypogaea)	8,800	Medium	ICRISAT	Hyderabad
Winged Bean				
(Psophocarpus tetagonolobus)	1,000	Short	NBPGR	Delhi

*Long-term facility under development.

IARI : Indian Agricultural Research Institute
CRRI : Central Rice Research Institute
ICRISAT: International Crops Research Institute for Semi-Arid Tropics
AICMIP : All India Coordinated Millet Improvement Programme
CTCRI : Central Tuber Crops Research Institute
PAU : Punjab Agricultural University
NBPGR : National Bureau of Plant Genetic Resources.

In such circumstances seeds have to be sown every year and recollected. One can only guess how well the specimens have been kept. It is quite possible that many samples have been lost or are mixed up. The state of the rice collection at the Central Rice Research Institute should also be a matter of public concern, CRRI too does not have a long-term storage facility. In the humid climate of Orissa, the viability of rice seeds is generally not more than six months. Consequently, many scientists lack confidence in NBPGR and prefer to keep their collections with themselves.

The most extensive collection of rice from the Chattisgarh region was made by Dr. R.H. Richharla, Beginning in 1971, while working as the agricultural adviser to the government of MP, Richaria collected, with single minded devotion, some 18,000 samples of rice. Though many of these may be duplicates, Richaria's vast collection became the nucleus for the breeding activities of the MP Rice Research Institute (MPRRI) which began in 1976 and was later merged with J.N. Agricultural University, Richharia, in fact, argued that it is the higher-yielding among these indigenous varieties which need few inputs, that should be propagated in India's poor areas. Tribal farmers, for instance, cannot afford expensive inputs that are needed to make exotic rice varieties produce high yields. With law inputs, many indigenous varieties will outyield the so-called high-yielding imported varieties. The work of collecting rice varieties in Chattisgarh has almost come to a halt since MPRRI was closed down in 1978 allegedly under World Bank pressure. Doubts have been expressed about the care with which Richharia's collection has been kept.

Animal Genetic Resources

India has a rich diversity of animal life which is threatened by a number of factors. Even as scientists discover new breeds like the Siri cattle in Sikkim, and although many species are nowhere near extinction, they are undergoing a process of genetic erosion. Urban centres attract cows and buffaloes with the highest milk yield from the rural areas. But from there they go to the slaughterhouses, after completion of lactation. Thus, high producting animals (genotypes) are being regularly removed from the breeding tracts, leaving inferior genotypes for breeding.

P.N. Bhat, one of India's leading animal geneticists, believes that the genetic quality of the famous Murrah breed has deteriorated. The maximum recorded milk yield in a lactation for a Murrah buffalo was obtained in 1932 with one female yielding about 4000 kg. In a lactation of 310 days, it is difficult now to find buffaloes recording better than 2500 kg per lactation). It has been estimated that between 1921 and 1979 the loss in average lactational yield of good producers due to the genetic erosion caused by slaughter-houses and other factors has been of the order of 1000 kg.

Cattle and Buffaloes

India has 26 breeds of cattle and seven of buffalloes, together with a number of nondescripts, which account for about 75 per cent of the total cattle population. The improved and specialised breeds are generally found in the northwestern and western dry areas.

Indiscriminate crossbreeding with exotic breeds poses a serious threat to indigenous breeds. Already crossbreading with temperate breeds has made inroads into good draught breeds. Bhat orgues that the need for crossbreading in their breeding tracts needs to be evaluated and stopped if erosion is large, so that the local breed may not be lost.

Some of the finest milk breeds are facing constant erosion, says Bhat, both because of indiscriminate crossbreeding and neglect. There have been no practical breeding programmes for genetic improvement of the indigenous breeds. These breeds are not in as good a genetic condition as they were a few centuries ago. Mass artificial inseminatio and import of exotic stock for cross breeding has now made it impossible even to locate genetically superior animals of Sahiwal, Red sindh and Tharparker breeds in India, "Pure specimens of rathi bread are fast becoming difficult to find; similarly high producing kankrej and Gir animals are becoming rarer and rarer," laments Bhat. Among cattle, the good milk breeds like Sahiwal and Red sindhi and dual-purpose breeds like Tharparker and Rathi are in acute danger of being lost in their pure form India.

Bhat himself believes nhat livestock breeds which can sustain and produce on low agricultural inputs should be identified and selected in animal research programmes in tropical countries. There is litttle point in introducing animals which need high inputs to produce efficiently in conditions of poverty.

Apart from the humped cattle, there are in the northestern Himalyas a number of wild, semi-wild and domestieated humpless cattle. Economically the most important is the mithun (Bos frontalis), raised primarily for meat, sacifice and bride price. The mithun is believed to have arisen from the cross of gaur, the wild cattle of the indo-Malaysion region, with domestic cows. The 1972 census revealed 121,000 mithuns in Arunachal Pradesh, Nagaland, Tripura and Manipur. The mithun lives mainly in semi-wild conditions at an altitude of 600-1600. metres.

Among the lesser known cattle, but extremely important for Himalayan traders, is the yak, or grunting ox (Bos poephagus graunieus Linn). In the wild it inhabits the coldest and most desolate tracts of the Himalayas, from the east to the west, with elevations ranging from 4000 6000 metres. Without the yak, travel and trade in these parts would be as impossible as without the proverbial camel in the desert. According to the 1972 census, there were 43,000 yaks in India.

The yak interbreeds with many types of cattle; bison banteng, gayal, zebu and European cattle, and has been so interbred for centuries. No interbreeding with buffaloes has been recorded so far in India. Yak bulls are mostly used for crossbreeding hill cows. The hybrids are hardier than the yak and are capable of bearing heavier loads, but are interior in staying power. Hybrids with cattle are now twice the number of purebred yarks. Since there has been no organised effort to preserve this useful animal, its population has declined by almost 25 per cent over the last 10 years.

Breeding Sheep

Though sheep are very important in certain regions of India, little work has been undertaken to identify whether any valuable Indian sheep breeds face the danger of extinction.

It may take as much as 25 years before a clear picture of the available genetic material in sheep emerges. Scientists argue that evaluation of indigenous seep breed for crossbreading and their conservation should go hand in hand. It should be mandatory for all sheep development programmes undertaken by state governments to conserve nucleus flocks of breeds used for crossbreeding in their natural habitat.

Goats

Like sheep, goats are also important to the Indian rural economy, and as one goat expert puts it, despite regular attempts by soil conservationsists to exterminate this animal, its population has been increassing. The origin of Indian goat breeds is not clearly known. They are believed to have been derived from wild goats which inhabited the Asian mountains in antiquity, and to have been domesticated around 7th century BC much earlier than cattle.

There are at present about 17 distinct goat breeds in India, of which several like the Jamnapaari are extremely useful. The dwarf goat breeds, such as Black Bengala, Banbari, Maalabari and Assam Hill are famous for high prolificacy (multiple births) and early sexual maturity. The Pashmine goats of Ladakh produce the finest quality of pashmina fibre (cashmere) In the world, But though Pashmina goats support a major cottage industry in the Kashmir valley, little is known about them. Research on goats is of recent origin and specific strains are now being developed for milk, meat and fibre. Crossing of pashmina goats with angora might make possible a goat with mohair and pashmina. Gaddi, a hill goat, produces hair used for ropes. Crossbreading of these goats with angoras yields mohair.

Pigs

India had only about seven million pigs in 1975 but they are an important source of meat especially for enonomically disadvantaged groups like Harijans. Traditionally indigenous pigs produced bristles and meat. In fact, meat was only considered a by-product, until the advent of nylon which replaced pig bristles.

There has been no serious attempt to catalogue the various breeds or strains of domesticated pigs in India, and previously unknown varieties are some times found. For example, P.N. Bhat recently located a number of strains of dwarf pigs in the north eastern region.

Indigenous pigs can survive under extremely adverse conditions of malnutrition and health. They are reportedly more resistant to diseasethan exotic breeds. Pigs are receiveing greater attention under government programmes and indigenous pigs crossed with exotic breeds are becoming more popular particularly in central India. Their higher growth potential makes them attractive, but given the adverse circumstances in which they are maintained, the survivability of crossbreeds is low, and therefore their advantage is marginal. Because of extensive crossing with the Large White, the indigenous pig in the central region is becoming extinct.

Indigenous breeds need to be properly documented and their performance tested so that proper breeding strategies can be decided. Besides domesticated species, there are several wild ones, such as the common Indian wild boar. The present day pigs have evolved as a result of gradual domestication of wild pigs.

Camel Husbandry

India is one of most important countries as far as camel husbandry is concerned. Its population of 1.1 million camels constitutes about seven per cent of the world total. Camels are found mainly in Rajasthan, Haryana and Punjab, the Arabian one-humped camel being the only species. There are three types of camels found in India; the desert camel, the riverine camel and the hill camel. Bikaneri is the most important desert camel breed in India. Nearly 50 per cent of all Indian camels belong to it. Riverine camels are found in districts of Uttar Pradesh and Punjab, where there is a good water supply. They carry heavy loads and are slow in movement with a speed of 3 km/hour. The desert camel can walk 190 km overnight at a speed of 15 km/hour. The hill camel is more compact than the camels in the plains.

The National Commission on Agriculture (NCA) had expressed concern over the changes taking place in the traditional system of camel breeding practised over centuries in Rajasthan. Under the traditional system, camel breeding was undertaken by professional breeders under the patronage of princely rulers and large landlords, who maintained large herds of camel. Excellent camel herds were owned by Bikaner, which possessed some of the best camels in the world, camel breeders were given extensive areas for grazing of camels with no tax levied on their grazing. Drinking water from all the wells in the state was allowed free for the animals. Good stud camels called "Mehias" were branded and let loose by individuals interested in camel improvement. These stud camel would move in herds in winter and during the off-breeding season they were free to graze anywhere, without fear of impoundment even if they damaged standing crop.

The traditional camel breeders, with their keen observation, used to breed camels with a definite objective. The camel breeders in Rajasthan evolved distinct types of camels like the Bikaneri and Jaisalmeri. Today, however, the conditions favouring breeding of camels are fast disappearing. Grazing lands have diminished at a rapid rate, having gone under the plough. As a result most camel breeders are now leading a pecuniary existence with small herds of camels in poor condition. Under these adverse conditions the fine art of breeding has disappeared, reduced to a mere multiplication of stock. "It may not be long before these camel breeders will go completely out of business", warned the NCA.

The NCA was not even sure whether there is a need to make a special endeavour to retain the distinct types of camels that were evolved in the past. "Breeding of animals for genetic improvement without ensuring adequate nutrition is meaningless", it argued. The construction of roads in desert areas is making the animal itself redundant. In the past camels were prized most for their speed of travel and capacity to traverse long distances at a stretch in the desert. In increasing numbers, camels are today being employed for agricultural operations. They played an extremely important role in the construction of the Rajasthan Canal. The survival of the animal, and its various types, is now dependent on whether adequate grazing lands will ever be found for it.

Horses, Ponies, Donkeys

India also has about 0.94 million horses and ponies, 0.075 million mules and 0.99 million donkeys. Uttar Pradesh has the greatest number of horses and mules, and Rajasthan donkeys. Horse breeding has been practised in India for a long time. There are six important breeds of Indian horses and ponies, which include the Kathiawari, one of India's best breeds which is found in Rajasthan and Gujarat; Marwari, which is closely associated with the great feats of bravery attributed to the Rathor Kings of Rajputana; Bhutia, which is bred along the Tibetan border and sub-Himalayan tract from Punjab to Darjeeling; Manipuri, which has been bred in Manipur and Assam for many centuries; the Spiti pony of the Spito Valley; and the Zanskari horses of Leh in Ladakh.

Donkeys are of two types; the small Grey, which is comonly found in most parts of the country and the large white which is found in the Kutch. Unlike the camel, the donkey has never received much attention. There has been no organised donkey breeding in India even though it is an important work animal.

There are no text books or monograghs devoted to the donkey, no research on its physiology, productivity or work ability, no selection experiments to increase its performance, even though next to the camel, it is the best farm animal adapted to a hot, dry environment.

Poultry Development

Chickens constitute over 90 per cent of the total poultry population of India, followed by ducks, six per cent turkeys and geese, 2 per cent. India and the neighbouring countries have been referred to as the original home of the Red Jungle Fowl (Gallus gallus), from which present day domestic birds are believed to have descended. The Aseel or Malay Fowl is reported to have given rise to all the breeds of poultry. There is evidence to show that these birds, some 2000 years ago, moved through the middle East to Europe and gave rise to the present day European breeds.

Eighteen types of indigenous breeds are on record. But in number, they are much smaller than exotic breeds. The former number about 50 million as compared to about 100 million of exotic breeds (including crosses). Several indigenous breeds are facing extinction. The original Aseel, a native Andhra Pradesh, is a medium-sized, round-bodied and well-built bird. However, large specimens have also been bred and developed by the promoters of cock-fighting. The bird is a poor layer but is known for its delicious meat. With the ban on cock-fighting, the Aseel is losing much of its importance, and as a result pure specimens are becoming rare.

The Ghagus is a large sized bird, which is usually found running with thenomads of Karnataka and Andhra Pradesh. It is a hardy breed and is supposed to have resistance to diseases and parasites. Due to neglect, it is gradually becoming extinct.

Information on ducks and other birds is extremely limited. There appears to have been no evaluation of indigenous ducks. At a recent international meeting a scientist expressed concern over government plans to establish breeding farms for exotic Khaki Campbell ducks, both for intensive rearing of these ducks and for upgrading local ducks. There is no specific information as yet that under normal husbandry conditions of limited feed supply and non-existent disease control, the exotics are better than the indigenous birds.

Animal conservation is still, a new area even for scientists. With the failure of several animal development programmes in which exotics have been introduced on a larger scale or crossed indiscriminately with local breeds, the importance of environmental adaptation using local breeds is being recognised. Hence, the importance of conservation of indigenous breeds to support more appropriate breeding programmes is also acknowledged.

Conservation of animal genes is neither easy nor cheap. It can be done either by maintaining a living herd or flock of animals that need to be conserved, by storage of frozen semen from a representative sample of males or by storage of frozen embryos.

In India the need to conserve animal genetic resources is being recognised. Several institutes concerned with preserving animal genetic resources have been set up. They are the National Bureau of Animal Genetic Resources (NBAGR), a National Centre for Studies on the Mithun in Arunachal Pradesh, a National Institute for Research on Goats in Mathura district, Uttar Pradesh and a breeding farm for camels at Bikaner. An enormous amount of work is needed to document and evaluate indigenous breeds of birds and animals and then based on this information, urgent action taken to conserve breeds.

Fish Genetic Resources

Though India has a rich variety of fish fauna only a few species have ever been cultivated. The most important among these are the carps including Catla catla, Labeo rohlta, Cirrhinus mrigala, labeo çalbasu and a few others. Several exotic species have also been introduced into India (the Chinese carps, tilapia, trouts, gourami, etc.).

The natural habitats of the cultivated species are the major rivers where they breed naturally. Under aquaculture, however, though they mature in ponds, they normally do not breed there. Fish seed, therefore, has to be collected from the rivers for aquaculture. As it is being practised today, this will have a deterimental effect on fish and may leave a genetically depreciated stock in the natural habitat.

Since fish species inhabit different ecological systems in geographically and climatically isolated areas, it is possible some of these species could have evolved into definite strains having various advantageous characteristics. Unfortunately, no comprehensive study has been made so far even no these few cultivated fish.

Inbreeding of Fish

Intensive aquaculture through genetic selection and selective breeding result in the species becoming more inbred and with less genetic diversity. Though aquaculture work in India is still limited, original diversity could be eroded in future. Thus to maintain the genetic diversity for present and future requirements, conservation of genetic resources in fish in essential.

Several dams and canals are being built to harness the country's water resources. They can have serious implications for fish genetic resources by permitting inter mixing and breeding of different strains and species which hitherto enjoyed isolation and preservation in different geographical and ecological boundaries.

Transmittance of undersirable species to areas where they are not liked or which may compete for food space and breeding with local varieties; introduction of fish predators and various other organisms : introduction of fish parasites and dangerous pathogens to areas where they were not known before; and, spreading of undesirable and harmful aquatic weeds are some of the more obvious possibilities.

In the case of crop genetic resources, after exploration, varieties are conserved in a central pool. Thus at any given time the required strain is available from a central place. But maintenance of fish at a central pool for a number of years can be hazardous as there can be complete loss of a precious variety through disease and other unfavourable factors. Furthermore, maintenance of certain strains over decades in ponds and repeated breeding will result in partial or complete domestication of the species. This is contrary to the idea of conserving genetic resource.

The best way is to protect these varieties in their natural habitat. This needs legislation to declare certain natural habitats protected and free from fishing; preventing pollution of the protected zone; preventing construction of dams and weirs in these protected zones or if dams and weirs are constructed, suitable and effective provisions for fish movements will have to be incorporated; preventing interlinking of various major drainage systems; protecting fish breeding areas in natural habitats; and, restricting large-scale exploitation of fish seed from riverine sources.

The Indian Council of Agricultural Research (ICAR) has recently established a working party on conservation of genetic resources of fish to consider all aspects of this problem. But this constitutes little more than a beginning of the understanding of the problem.

Tree Genetic Resources

Large areas of forests are in imminent danger of disappearing. Although there is little risk of entire species of trees becoming extinct, populations at the limit of the species range or in isolated blocks of forests are definitely vulnerable. Often these small populations have a high frequency of valuable genes for drought resistance, for example—which could be used in future tree breeding programmes.

Tree genetic resources have been neglected in the past. With a growing interest in forestry in across the world now, it has become essential to preserve natural wealth in trees. South Korea, which has carried out the most extensive afforestation in Asia, has set up an entire research institute devoted to tree genetic resources and development of suitable trees.

Tree populations are most threatened in arid and semi-arid areas where an intense pressure exists because of the demand for firewood. Developments in commercial forestry have put immense pressure on specific trees which have acquired an economic value. The Dang forests in Gujarat, for instance, are a rich reservior of a number of plant species which were not considered of economic value about 20 years ago. Good examples of these species are kakad (Garuga pinnata), which is used for pulp and packing cases, tanch (Ougeninia oojeinensis) used for making agricultural implements, haldu (Adina cordifolia) and kalam (Mitragyna parvifolia) used for making bobbins and furniture, and modad (Lannea grandis) and kudi (Wrightia tinotoria) used for making toys. It is important that tree species in this area, as in many other forest areas of the country, be preserved. Equally, several village woodlots, with trees representing well-adapted genetic material, are being steadily depleted. FAO wants to give priority to collecting the diversity in tree species suitable for firewood and fodder production in dry areas.

Arunachal Pradesh is another area where the richness in tree resources is being steadily threatened. As two scientists of the Forest Research Institute at Dehra Dun put it : "In spite of the richness of the flora, the area, to this date, remains a terra incognita and without a comprehensive published flora".

Depleting Species

A paper published in "The Indian Forester" in 1980 listed about 20 species facing danger of extinction or depletion. This list includes Phyllostachys assamica, a bamboo used for making huts, Gymnocladus assamicus, a small tree which grows up to 20 metres high and whose fleshy pods are used for washing hair as a substitute for soap, and the spectacular Rhododendron dalhousiae which has cream colour flowers with purple bands. R. dalhousiae is such a beautiful plant that it was awarded a First Class Certificate by the Royal Horticultural Society in the UK. It is now facing extinction owing to massive road building and developmental activities in Arunachal Pradesh.

Tree genetic resources can be conserved either in situ or ex situ. Conserving in situ, that is preserving the entire ecosystem together with its trees is definitely more desirable but often extremely difficult and expensive. When effective in situ conservation is not possible, trees have to be conserved in artificial stands established ex situ in their country of origin or elsewhere.

Conservation of tree genetic resources has been practised in India through the system of "preservation plots". The first preservation plot was set up in the sal forests of Bihar in 1906. Some 188 plots have been established since then to protect mainly commercial species like Shorea robusta (sal), Tectona grandis (teak), and various conifers. Most of these plots cover natural forest types and range from a few hectares to 4000 ha. Preservation plots in plantations are small. The coverage of these plots is also patchy. Many forest types like southern tropical forests, semi-evergreen forests, and tropical thorn forest, need particular attention.

A beginning is being made in the preservation of trees important for social forestry. Leguminous shrubs, which can enrich soil fertility through biological nitrogen fixation and at the same time provide food, fuel and fodder, are particularly worthy of greater attention. Leucaena is already receiving worldwide attention. Several international conferences have recommended the collection and conservation of casuarina and neem germplasm. Two conferences held on neem in West Germany in recent years have pointed out that the potential for improvement of the neem tree in yield and quality characteristics is considerable.

India is participating in an international project to collect tree species used in arid and semi-arid areas. According to the list of priority species that has been prepared for this project, first priority will be given in India to three indigenous species (Acacia nilotica, Acacia senegal and Prospopis cineraria) and several exotic species of acacia, prosopis eucalyptus, and leucaena. Among the second priority indigenous species are Albizza, lebbek, pongamia pinnaka (karanj) and Azadirachta indica (neem). Among exotics given second priority is Prosopis juliflora, which was introduced into India nearly a century ago but which is now widespread and shows considerable variation of from. Indigenous species listed for future consideration are hardwickia binata, Kingiodendron pinnatum and Dendrocalamus strictus (bamboo).

STRANGE CREATURES OF INDIA

Gangetic Dolphin

Born completely blind, the Gangetic dolphin (Platanista gangetica), remains sightless for the rest of its life. Nevertheless it does get by perfectly well, having adapted to a sightless existence in its murky, river habitat. Not to be confused with marine dolphins (Delphinidae), the fresh water dolphins (Platanistidae) have lost all contact with the sea. They search for fish crustaceans and other such organisms by sweeping the river bed with their sensitive muzzles. Freshwater dolphins must come upfor air every one or two minutes. During the day the dolphins prefer to stay in deep water, coming out at nightfall to feed in the shallows. Hydroelectric projects and the large fish nets of fisherfold pose considerable problems for freshwater dolphins but in recent years positive steps have been taken to identify and protect some of their known habitats in the Brahmaputra, Indus and Ganges river.

Civets

Civets are cat-like creatures with elongated bodies that resemble those of weasels. Solitary and nocturnal, these carnivorous creatures spend the dayin hiding, coming out to hunt small birds, mammals or reptiles at night. To accommodate their 32 to 40 teeth civets have evolved a longer snout which further differentiates them from the cat family. In India, Sri Lanka and even further east lives a civet that has come to be known as the 'tody cat'. This creature, the Plam civet (Paradoxurus hermaphroditus) has a tail as long as the body and derives its name from its odd preference for fermented today. It often lives close to human habitation and, as many have discovered to their unpleasant cost, can give off an extremely foul, clinging odour from its anal glands if disturbed. The palm civet is less carnivorus than its cousins and has specially adapted claws suitable for climbing the straight palm trunks.

Ant Lion

Very often one comes across small sand-ringed holes in the soil. These structures are the trap-cum-homes of insects known as ant lions. An ant lion larva digs a steep tunnel in loose earth and lies in waiting in the pit thus created with only its deadly pincers and head showing above the surface. A small insect which wanders too close, slides down the rim and cannot recover its balance as its legs get no purchase from the crumbly slopes. The ant lion larva shoots up a spray of sand, which histens the hapless victim towards its end. Having fed heartily on the digestible insides of its prey the tiny predator spits out the husk and patiently waits for the next victim.

Indian Pangolin

The Indian Pangolin (Manis crassicaudata) has evolved a scaly body covering, fabricated out of modified hair. As protection against predators, it rolls itself into a ball and raises its sharp scales. These anteaters are mainly terrestrial, have no external ear openings and can seal their nostrils to protect themselves from ant bites. Their long, extensile tongues are coated with a sticky secretion which allows them to capture ants, termites or larvae by probing deep into the labyrinths of termite mounds which they rip apart with their strong claws. Pangolins have no teeth as their diet demands no chewing. A young pangolin will perch on its mother's back, just above the base of her tail, when it needs to be transported from one place to anther.

Great Indian Hornbill

The great Indian hornbill (Buceros bicornis) is a large, ungainly bird with strange nesting habits. The female seals herself into a hollow in a tree whereupon her feathers soon moult. The male dutifully brings her food for the entire period during which her eggs are incubated; feeding her through a narrow slit in the 'prison' nest.

When her young grow and their appetite increases, the female breaks down her prison doors to help the male obtain food. The young now rebuild the damaged portion of their home to seal themselves in. The hornbill's nesting behaviour serves to protect defenceless young efficiently.

Indian Giant Squirrel

This large nocturnal rodent, the Indian giant squirrel (Ratufa indica) lives in the upper canopy of trees and seldom ventures down to the forest floor. Flaps between its body and its extremities act like parachutes and enable the large mammal to sail or glide from tree to tre covering as much as 20 feet in one 'lead'. The flying squirrel evolved the art of gliding as an evolutionary adaptation for predator escape and also as a means to reach food sources (other trees) by expending the least amount of energy.

Mouse Deer

This is a deer which stands just 12 inches off the forest floor. It has no antlers but the maleh as downward pointing canine that look like tusks. Its small size helps it to evade predators by hiding beneath even very low bushes and low vegetation. The chevrotain (Tragulus meminna), also called the mouse deer, has actually been observed bolting up the hollow of a tree when pursued by dogs. Timid and gentle, these dwarfs of the Indian jungle lend enchantment to the forest and are the subject of several fanciful tales in Indian lore.

Flying Fox

The flying fox (Pteropus gigantus), found commonly all over India, is one of the world's largest bates. Its wingspan approximates five feet and the furry mammal has been recorded flying distances of over 150 miles. Contrary to popular belief, flying foxes do not come out in search of human or animal prey. They rely instead on a diet of overripe fruit. The also perform the useful function of scattering fruit seeds, promoting the growth of fruit-bearing plants for the entire jungle community.

15

Natural Resources

Introduction

Whether we run out of a particular resource depends in part on how we define resource. In broad terms, a resource, or natural resource, is anything needed by an organism or group of organisms. In other words, a resource is something useful but for humanity what is useful or useless can change because of technology, economics, and the environmental effects of getting and using a resources.

Technology cannot bring back an extinct animal resource or a paved over wilderness area, but it can extend the supply of some resources by improving them, using them more efficiently, or recycling them. For example, a unit of today's steel provides 43% more structural support that it did a decade ago. Today we get 7 times more electrical power from 1 ton of coal than we did in 1900. Similarly the energy needed to produce 1 ton of pig iron has fallen eight fold since 1800. In the United States in 1900, only 10% of the copper was recycled. Today about 40% is recycled. However, while many matter resources, such as copper, lead and silver, can be recycled, we can never recycle energy resources. Once a fossil fuel resource, such as coal, oil or natural gas, is burned, it is gone forever as a useful energy source. The concentrated energy in the fossil fuel is released as heat, which is eventually dispersed into the earth's atmosphere. From there it flows back into space.

Sometimes technology can solve the problem of a scarce resource by finding a substitute or replacement. For instance, as a structural material bronze replaced stone, iron replaced bronze, steel replaced iron, and now aluminium and reinforced plastics are replacing steel for some structural uses. As an energy source, animals replaced human muscle power, coal (to produce steam) replaced animals, and oil replaced coal for many uses. There is a vigorous debate over whether a combination of coal derived energy and nuclear power (obtained from Uranium) or a combination of solar, wind and, and plant (bio mass) energy may soon replace petroleum and natural gas.

In addition to technology, resource use is tied to economics. Something is useful as a resource only if it can be made available at a reasonable cost. For example, once we deplete the easily available supplies of a resource, we have to look harder and dig deeper to find remaining supplies. If the costs of findings and making a scarce resource available rise, the resource will eventually become too expensive for most people. Higher costs may stimulate a new search for new supplies or make mining and processing lower grade deposits economically feasible. But regardless of what we are willing and able to pay, we can't get a resource out of earth if it is not there.

There can also be an economic limit to recycling. Typically, recycling is cheaper than mining virgin materials, but only if the material to be recycled is not too widely dispersed. For example, if products made from iron and steel, such as cars and toys, are thrown away, buried, and widely scattered through use, labour and energy costs for finding and collecting the objects may be too expensive to make recycling feasible.

The continued use of a resource may also depend on the impact its mining, processing, and use has on the environment. Even if affordable supplies of a resource are available, its use (at least for certain purposes) may have to be abandoned if this use seriously threatens human and other forms of life. Sometimes the environmental effects of a resource use can be minimized and cleaned up. But this costs money, and in some cases the costs may be so high that we can no longer afford to use the resource.

Natural Resources : Definition

The sum of all physical, chemical, biological and social factors which compose the surroundings of man is referred to as environment and each element of these surroundings constitutes a resource on which man draws in order to develop a better life. Thus, any part of our natural environment - such as land, water, air, minerals, forest, rangeland, wildlife, fish, or even human population - that man can utilize to promote his welfare may be regarded as a natural resource.

Ramade (1984) defined a resource as a form of energy and/or matter which is essential for the functioning of organisms, populations, and ecosystems. In the particular case of humans, a resource, in his words, is any form of energy or matter essential for the fulfilment of physiological, socio-economic, and cultural needs, both at the individual level and that of the community.

The functioning of ecosystem or of technological civilization involves a continuous flow and consumption of energy sun being the source of it as regards ecosystem and fuels as regards technological civilization. In the light of this fact, a resource may be defined as one of the various forms of primary energy present in the nature.

The five basic ecological variables—energy, matter, space, time, and diversity are sometimes combinedly called natural resources As such, laws which govern changes in these quantities have great applicability in the resource use.

Classification of Natural Resources

The classic subdivisions of resources include renewable and non-renewable resources. Renewable resources are those which are reproducible and are obtained from the biomass of living organisms, *e.g.*, forests and fish. Resources which are not reproducible and are obt ined from the finite non-living reserves are called non-renewable resources, *e.g.*, coal and metals. Solar energy although having a finite life, as a special case, is considered as a renewable resource in an much as solar 'stocks' are inexhaustible on the human scale.

Some authors prefer to classify resources into biotic (or living) and abiotic (or non-living) resources, forest, agriculture, fish, and wildlife falling in the first category and land, water minerals etc., in the second.

A broader classification of resources, as proposed by Owen (1971), is represented in Fig. 1. Quantity, mutability, and reusability are the attributes on which this classification is based. Inexhaustible and exhaustible are the main categories of resources based on their stocks in nature. Inexhaustible resources are further divided into two groups · Immutable and mutable resources, depending upon the possibility of their qualitative degradation as a result of man's activity. On the other hand, exhaustible resources, being limited in occurrence, are vulnerable to both quantitative and qualitative degradation. Thus, their availability depends upon the method of use, so these are classed into maintainable and non-maintainable resources. Supply of maintainable resources could be made to last for long through wise use, however, some of them have capacity to reproduce – called renewable, while other cannot reproduce – called non-renewable resources. Non-maintainable resources have static supply and when destroyed or consumptively used are irreplaceable, yet some of them have potentiality for reuse called reusable resources, and rest whose permanent exhaustion is certain are called non-reusable resources.

Conservation and Management of Natural Resources

The term 'conservation' to many means 'hoarding' or to control supply of goods in such a way that some part is left for the nature. Most people interpret conservation as total protection or restriction on consumptive use of resources, therefore, wildlife conservation to them means a total ban on killing of any animal and the forest conservation means a complete moratorium on all fellings the term, in fact, is derived from two Latin words—con (=together) and servare (=to keep or guard), thus meaning 'to keep together'. The dictionary meaning of the word is 'an act of preservation'. The term was proposed by Gifford Pinchot in 1908 (see Smith, 1958) who conceived it from the title of 'conservares' given to British officials in India appointed to look after the protection of natural resources.

Prof. H.M. Rose defined conservation as "the optimum allocation of natural, human, and cultural resources in the scheme of national development, whereby maximum economic and social security will be assured" (see Smith, 1958). In 1967, the American President John F. Kennedy looked upon conservation as "the wise use of our natural environment ... the prevention of waste and despoilment while preserving, improving and renewing the quality and usefulness of all our resources". World Conservation Strategy defines conservations as "the management of human use of the biosphere so that it may yield the greatest sustainable benefits to present generation, while maintaining its potential to meet the needs and aspirations of future generations". Thus, the conservation is a practice embracing preservation, maintenance, sustainable utilization, restoration, and enhancement of the natural environment. Summarily, the true aim of conservation is two fold : (i) preservation of quality of environment, and (ii) to ensure continuous yield of useful material-living or non-living by establishing a balanced cycle of harvest and renewal.

The concept of resource conservation has now come under a still wider field of environmental management which is, in essence, the process of allocating natural and man made resources so as to make optimum use of the environment in satisfying basic human needs at the minimum and more, as far as possible, for an indefinite future, and at the same time preventing depletion and degradation of resources.

Environmental management being a too vague, general, or trivial subject has not developed as a full-fledged discipline so far, further, in many developing countries it is viewed as being irrelevant and counter-productive, while in advanced industrialized countries it is often interpreted as a means of coping with the 'burdenl of environmental legislation and problems put forth by 'over-zealous' environmentalists; what to talk of under-developed world ? The global population, however, cannot and should not overlook and deny the fast deterioration of environment world over resulting in dwindling natural resources, pollution of air, water and land, extermination of a number of life forms, down going health status, danger to cultural assets, and many socio-economic backlashes. The root of the problem lies in too many people using too many resources wastefully.

Since the total natural resources available are limited, the wayout to the problem seems only through control of population and also the per capital consumption of resources. Because it is difficult to bring down per capital consumption, the recycling and stringent conservation can bring hope for sustained development without reducing per capital use of resources.

Land

Soil consists of the weathered layer of the earth's crust with living organisms and products of their decay intermingled. Lying on the rocky core of the earth, this complex mixture is one of the four prime requisites for life. Along with sunlight, air, and water, soil supports all living beings on land, plants as well as animals. All higher terrestrial plants have their roots in soil from which they absorb life sustaining water and nutrients. Man and all land animals, in turn, depend on these plants for food, directly or indirectly. Foods, vegetable oils and fats, forage for livestock, fibre, wood and various other indispensable agricultural products come from s il. Thus, a nation's economic

NATURAL RESOURCES

- **Inexhaustible**
  Unlimited or unending. Quality may be degraded, not quantity.
  - **Immutable**
    Quality not degraded by men's activity, *e g.*, atomic energy, wind power, precipitation, tidal power.
  - **Mutable**
    (Misusable)
    Quality may be degraded by man's activity, *e.g.*, solar radiation, water, hydropower of running waters.
- **Exhaustible**
  Limited. Both qualitative and quantitative degradation possible.
  - **Maintainable**
    Availability depends upon method of use by man.
  - **Nonmaintainable**
    Total quantity static, when consumptively used, not replaced.

(Contd.)

Renewable

Perpetual harvest possible, *e.g.*, fertility of soil; bio-products of land, fresh water and sea; human power.

Nonrenewable

Once gone, no replacement, *e.g.*, species of wildlife

Reusable

Consumptive use small, *e.g.*, germs, non-consumptively used metals like gold, platinum, silver, iron, copper, zine.

Nonreusable

Consumptive use high, *e.g.*, fossil fuel; non-metallic minerals like gypsum, sand, salt; consumptively used metals like lead in gasoline and paints, zine in galvanized iron, tin in tooth-paste containers.

Fig. 1. Schematic representation of classification of natural resources.

well being is inextricably linked with the fertility and abundance of her soil resources. Productive land is since the source of human sustenance and security everywhere at all times, this resource because of mounting demands of swelling population and long mismanagement would put in jeopardy the very survival of man.

Land Use and Abuse

The land although appears to be resource available unlimited but, in fact, its injudicious use would limit the availability of this indispensable life-support system. Thus, a land use may be poor or good. If in a region of law rainfall grasslands are replaced by wheat crops, a temporary desert will eventually result; or if the human habitation and factories are developed on flood plains, the damage is almost inevitable. These are the instances of poor land use. Conversely, if the grasslands are used for moderate grazing by livestock, and if the flood plains are used for forestry or agriculture, these will imply good land use. Ecological principles have great applicability in land use planning. Classification of use should, therefore be based on natural ecological features of the land in question, *viz.*, soil type, slope, biotic communities etc.

Rural Land-Use Planning

To a farmer, soil is the essence of survival and his economic well being is firmly linked with the fertility of his land. In rural land-use planning, concentration is chiefly devoted to creating and developing more farmland by removing forests and natural grasslands, draining marshes and swamps, channeling streams for irrigation, and so on. Unfortunately, no effort is made to save existing prime farmland from degradation by ill-planned development. Rather than exclusive farm, range, or forestry oriented planning, the land use should be made under broad consideration of social, economic, and ecological intricacies.

Urban Land-Use Planning

Deteriorating quality of urban and suburban environment is to a great extent the result of injudicious land use and is a threat to the whole socio-economic system. Thus, planned cities are as necessary as planned farms. Although, the implementation of land-use plans is infinitely more difficult in urban areas than In rural ones because of the human problems involved, yet if not heeded it would lead to chronic pollution, epidemics, and social disorder. In any urban land-use planning it should be seen that : (i) at least a third of all land remains as protective open space in the form of parks, green belts, or refuges etc., including adequate space for efficient treatment of domestic and industrial wastes; (ii) residential housing around villages or cities should be developed in clusters with each unit separated by broad green belts; (iii) typical topographic situations like stream valleys, steep slopes, lakes, and marshes etc., should be retained free from houses and other high-density uses; and (iv) waste disposal areas and industries should be kept apart from human habitations.

Management of Land

From the above account it is clear that major problems emerging out of abuse of land in cities include unsafe and unhygienic dwellings and those in rural areas are of erosion of soil and loss of fertility. While urban problems can be prevented and cured by proper town planning and municipal facilities, the problems relating to rural areas need long term soil conservation programme.

Prevention and control of soil erosion. The primary purpose of soil conservation is to prevent soil erosion and heal the damage where it has not advanced too far to respond to curative methods. This is possible largely through two means :

1. The land should wear a vegetational cover throughout the year. Plan roots bind the topsoil and prevent it from being drifted so easily with water or air. The vegetative cover also slows up runoff water and causes much of it to percolate into ground that promotes more vegetative growth. The vegetation - whether grasses, trees, or cultivated crops-should fit to the capabilities of the soil and water available.

2. Engineering and agronomic practices should be applied conjointly to counteract the erosive force of both water and wind on the soil. These include contour cultivation, cover crops and crop rotation, and contour stripcropping, used alone or in combination with terracing where it is required. When terraces are properly constructed and adequately supported by approved cropping and tillage practices, they prove to be one of the most effective erosion-control measures applicable to cultivated sloping lands.

Reclamation of eroded lands. Conservationists are divided into two camps over the issue of reclamation of land already arrested by ravines. One camp of conservationists and naturalists believes in the policy of 'forgive and forget' and advocates that ravines should be allowed to exist, while the other feels that efforts should be made to regain the lost land through bulldozing or some other means. If the ravines are left in there present state, these will spread further and engulf more of productive land, while the violent means to regain the ravines in their near original condition may provoke numerous agencies of nature to take revange on man. Thus, following considerations are important in negotiating the eroded land :

(i) Ravines should be provided with sufficient and suitable vegetative cover.

(ii) Instead of agriculture, these lands should be reclaimed for forestry, pasture, or horticulture.

(iii) Their deficiency in nutrients and moisture for plant growth should be improved.

(iv) Further misuse of such land shouid be prevented. Over-trempling by man and cattle shohld be avoided.

(v) Post damages should be repaired and recompensed under strict ecological consideration. For instance depressions may be filled up, slopes easened out, cattle trails fenced etc.

(vi) Vegetative cover provided should be protected against reckless destruction by local population.

Water

The origin of water on the earth is not clear so far. However, the current presumption is that the primordial earth had no oceans, and perhaps very little atmosphere. It is believed that the volatile constituents bound in the earth's crust, oozing the surface through volcanos, rock movements, and hot spring, condensed to form the ocean and the atmosphere. This way, perhaps this remarkable combination of hydrogen and oxygen – called water – came into being and eventually became an indispensable component of the earth's environment.

Water moves from the ocean to air to land to ocean in a set cyclic pattern called hydrologic cycle. In other words, this cycle represents a continuous process of exchange of water from the earth's surface to the atmosphere and *vice versa* under the influence of solar head.

It is estimated that of the total available water 97.3 per cent is contained in the oceans and out of remaining 2.7 per cent most is in solid from. The amount of water actually available in the atmospher is a very minute fraction and estimated to be only 1 × 10 – 5 of the world's total water.

The factors which control the process of evaporation and evapotranspiration include temperature, humidity, and wind. It is a continuous process going on day and night from the ocean surface, ground surface in land waters (like lakes and streams), plant and animal surfaces, Rising in the atmosphere this water forms clouds that float around in gaseous or droplet form. The atmosphere above the earth up to varying heights of 10 to 17 km (depending upon latitude and season) is called the troposhere up to which the clouds float and in which all weather changes take place. Water returns to the earth in the form of precipitation, most of which takes place over the ocean and a little on the land surface. Rainfall on the land surface may have following fates : it may percolate through the soil and become a part of ground water regime; accumulate in a pond, lake or reservoir as inland surface water; flow down to the ocean in the form of stream; taken up by plants and animals; or evaporate.

Water Requirements and Uses

Goethe rightly said that, "Everything originated in the water, and everything is sustained by water". Water is needed to fulfill diverse requlrements in so many diverse ways. It is vital to life since for all physiological activities of plants and animals it is essential. On an average it constitutes 80 per cent of protoplasm ranging from 8 per cent in dry dormant seeds to 95 per cent in jelly fish. It serves not only to quench the thirst but also to meet the food requirements because it is an essential raw material in the process of phtosynthesis through which green plants make food that is used by all trophic levels directly or indirectly. Ambient humidity which is indeed the invisible form of water that surrounds us is necessary to prevent dessication of terrestrial life forms. Thus, it can be said that water is a biological essential and it serves as milleu internal as well as milieu internal.

Flood plants of rivers have been the cradle of civilization and centres of population since ancient times. The reason being their high fertility and easy accessibility to water for various uses like drinking, washing, and transport. Egypt was developed on the flood plains of the Nile; Babylon on the flood plains of the Tigris; and North India on the flood plains of the Ganges. Even today the Mississippi plain is about the most densely populated area in the USA as flood plains of the Rhine in Europe, and the Yellow river in China.

Besides temperature, water is the other such key factor that influences the global ecology. Covering about 70 per cent of land surface and circulating in hydrologic cycle water influences weather and climate of any region and thus its flora and fauna.

Agriculture is dependent on water, may it be rainfall, or surface or ground water irrigation. It eventually is a more important commodity in a country like India because her largely agrarian economy depends critically on water. Huge amount of protein rich food in the form of fish, shell fish, and prown is obtained from the inland as well as sea waters worldover. Aquaculture has come up as a very potent industry.

The kinetic energy of gravitational flow of water is tapped and transformed into electricity, what is called hydropower. This concept of electricity generation is one of the oldest and cheapest tricks largely used worldover. Morever, it leaves no pollution burden on nature as in case of thermal or nuclear power generaton.

Hardly any industry can do without water. It has multifarious application in industrial processes *viz.*, a raw material, solvent, chemical reactant, coolent, and cleaning agent.

'Water, water everywhere not any drop to drink' is the situation not only found amidst sea but also at places where industrial and domestic refuse make the benevolent water sources so severely contaminated with pathogens and or toxins that they are left unpotable. Being the best dilution medium,it is frequently used as the cheapest and convenient way for liquid and solid waste disposal. Further it was practically impossible to maintain sanitary conditions without water for it is needed in all common cleaning and washing practices.

Water at many places serves as a very convenient means of navigation. Besides, it adds to the aesthetic value of the landscape, and provides opportunities for recreation and sport.

Water being unevenly distributed over the land, influences interstate and international relations and thus is the cause of what is known as hydro-politics.

It can, therefore, be said that while water is an essential for life it is significantly linked with social, economic, political, and ecological intricacies.

Problems and Management of Water

Management of water implies making the best use available water resources for human benefit while not only preventing and controlling its depletion and degradation but also developing it in view of the present and future needs. Water, like forest, is a multipurpose resource and it is important to see that its various uses should not conflict with each other and it can be enjoyed in its totality by man and others. Thus, its right allocation, and quantitative and qualitative conservation are the primary tasks before water managers.

Floods, droughts, improper use, pollution, and disease transmission are the important problems related to water. Because usable inland water resources constitute a very minute fraction of the world's total water and because man largely depends upon this minute fraction of water for his various activities, it is important to make efficient use of available water and at the same time prevent abuse of it. This has to be considered not only in giant irrigation projects or industrial and municipal processes but also in everyday life. Taps running unpurposefully, leaking water supply pipes, using filtered water for gardens and building construction are a few instances of inefficient and misuse of water in daily life.

Water being the best and convenient dilution medium is the worst affected resource by pollution. To a common man pollution of water means only the drainage of industrial effluents or domestic sewage in natural waters. It may astonish a lay reader that even pure distilled water when released in sea may act as a pollutant since it alters the normal salinity of sea water which adversely affects the marine life. Besides chemical alterations, changes in physical condition of water also have grave effects on the biota of system as during thermal pollution caused by the discharge of heated effluents in a body of water. Pathogens, chiefly entering natural waters through municipal wastes rich in excreta and other organic filth, also render the water unsuitable for human use.

Infact, it is not only the man who has to use the water resources but these are equally important to land animals and more to those fish and other aquatic animals who live in it. Thus, pollution of a body of water does prove detrimental not only to the human society but also to regional ecology and the ecology of very aquatic ecosystem. It is, therefore, essential that the water should not be treated as a simple respository of waste disposal, and if inevitable, the wastes only after proper treatment should be released in it under strict ecological consideration.

The qualitative degradation of water is also in a way quantitative depletion usable resource. Therefore, recycling of waste water after due treatment would relieve the water scarcity to a great extent, especially in regions marred by water shortage.

Future Needs and Alternate Sources of Water

The demand for usable water, like any other resource, is increasing fast with the continuous increase in world population and continuous increase in per capital demand linked with more of sewage disposal and transport, prolific use of water-using gadgets (like air cooling and conditioning devices, washing machines etc.), and increased recreational activities. According to an estimate, the daily per capital demand for water of an American was 600 gallons in 1900 which increased to 1,500 gallons in 1960 and is expected to be increased upto 2,700 gallons by 2000. The fact is no different in developing world. Thus, the question is 'where will the water needed for fulfilling increasing demands come from ?' the answer is not so simple and can only be found through long term planning. Possible means of meeting out the impending water deficit problem include :

(i) Reclamation of sewage and waste water;

(ii) Development of ground water sources and surface storages;

(iii) Long range forecasting of rain;

(iv) Rain making;

(v) Transfer of surplus water; and

(vi) Desalination of sea water.

Agriculture

The practice through which specific plant species are cared and managed so as to obtain maximum yield of consumable parts of the plants is called agriculture. An agricultural field is, in fact, a cropland ecosystem in which ecological principles involving limiting factors and productivity provide great applicability.

The demand for food production, linked with tremendous population growth, has been increasing fast world over. This has been possible to feed increasing number of mouths through increased agricultural production, what is called 'green revolution'. It has resulted chiefly from the industrialization of agriculture, which involves highly domestic plant species, sophisticated chemicals, and large fuel energy subsidies. According to an estimate, to double the crop yield through industrialized agriculture, a ten fold increase in fertilizers, pesticides, and hoesepower is required.

The green revolution through industrialized agriculture (fuel powered) as compared to conventional agriculture (man-and-domesticated-animal powered) has following backlashes :

(i) Such an agriculture involving huge amounts of fuel, and chemicals in the form of fertilizers, pesticides and weedicides, is the chief cause of environmental pollution.

(ii) It may lead to social backlashes when a quick shift from small farms to large industrialized farms takes place. In this process smalls farmers and agriculture workers are driven off the land to cities where they may not get work, food, and housing – a situation of despair and social breakdown.

In retrospect, strong measures should be taken to control the population growth, and to improve the conventional agricultural practices with application of current agricultural research under due social and environmental consideration. For a greater and sustained agricultural production following considerations are important :

(i) Available land acreage should be properly and judiciously utilized.

(ii) Soil fertility should be increased through wise use of fertilizers and organic manures.

(iii) High yielding and disease resistant plant varieties should be introduced.

(iv) Mixed cropping should be practicised wherever possible.

(v) Integrated and balanced use of available water sources, surface and ground water, should be made.

(vi) Weeds and pests should be efficiently controlled. Integrated pest control practices should be preferred over total reliance on chemical pesticides.

(vii) Soil erosion and loss of nutrients should be prevented through maintaining vegetational cover throughout the year.

(viii) Exotic varieties should only after due consideration of long term impacts be introduced.

Forestry

A forest is a natural ecosystem having multispecies and multiaged trees as dominant community. Forests cover about one-third of the earth's land surface of which about 50 per cent is occupied by tropical forest. Of the total geographical area of India, 22.74 per cent is forest as against a minimum of 33 per cent forest cover prescribed under National Forest Policy.

Importance of Forests

Functions of a forest may broadly be classified into following three categories :

(i) Protective functions : These include the protective role of forests against soil erosion, droughts, floods, intense radiation etc.

(ii) Productive functions : Forests are the source of wood and many other products like gums, resins, fibres, medicines, katha, honey, bidiwrappers, pulp, paper etc.

(iii) Accessory functions : These include the role of forests in recreation, aesthetics, and as habitat of diverge wildlife etc.

Thus, it can be said that forests are important in two ways – ecologically and economically.

Ecological significance : Environmentally, the forests are very important. They help in balancing oxygen and carbon dioxide level in atmosphere, regulate earth's temperature regime and hydrologic cycle. Forests increase local precipitation and water holding capacity of soil, thus preventing drought situation. Vegetational cover provided by forest impedes the velocity of runoff on soil surface, checks soil erosion, silting, and landslides, thus reducing the danger of flood. The litter derived from fallen leaves maintains fertility of soil by returning the nutrients. Forests also act as refuge of wild animals and provide protection to them against strong, cold or hot and dry winds, solar radiation, rain and enemies.

Economic significance : Woods, which has various applications in domestic and industrial processes, is the chief product of forest. Wood, when used as fuel, has certain advantages over coal as its sulphur and ash contents are very low but, at the same time, excessive use of fuel wood means pressure on forests which have many other functions to serve. Wood may also be converted into solid (coke), liquid, and gaseous fuels. Timber is an important material in building construction and day-to-day uses. Forest provides raw material for various wood-based industries, *viz*., pulp and paper, composite wood, rayon and other manmade fibres, sport goods, furniture, boat building, matches etc. Miscellaneous products, like bamboos, resins, gums, some oils, fibres, medicines, katha, lack, shellac etc., are also obtained from forests.

Depletion of Forests

Increased urbanization, industrialization, and mining has entailed indiscriminate felling of trees and denudation of forests. Nobody will, however, ever learn that who set the first assault on which tree in the history, but this practice of logging down trees for human use has been going on for centuries unabated. Colonists used logs for firewood and making huts. They also made handles for their tools, furniture, and fence posts etc., from it. For them forest had only a present and a past – no future; thus they resorted to ruthless and accelerated exploitation of available forest – a story that still continues with more serious backlashes earth has ever witnessed. India alone is losing more than 1.5 million hectare of forest cover each year and according to Tiwari (1982) 22 million hectare of forests have been destroyed during the last three decades due to over exploitation and misuse and conversion to agricultural fields. The depleted forest wealth would simply deprive the man of economic and environmental values offered by forest, as discussed earlier.

Management of Forests

In forestry, the yield is chiefly obtained from the accumulated growth of the past, so this industry has to depend upon the annual growth. Forests are managed to obtain a sustained yield which is obviously possible through sustained growth. Following considerations are important in this direction.

(i) A forest should have in correct proportion the trees of different ages and classes so that continuous removal of exploitable produce would not affect the future yield.

(ii) Each tree crop should put on normal growth increments.

(iii) No obstacle should exist to the continuous harvest of exploitable stems.

(vi) Unrestrained felling should be prevented.

For all such planning and decision making, sound ecological knowledge is desirable that include knowledge about climatic conditions, composition of biotic community, successional trend, life cycle of important tree species, methods of arresting growth of undesirable trees and promoting growth of highyield tree species, and biotic interrelationships (such as weeds, competitions, pests, parasites and diseases etc.).

Of various tactics employed in forest management, some deserving special considerations are as follows :

(i) Restrained felling,

(ii) Block and selective cutting,

(iii) Reforestation,

(vi) Control of disasters, and

(v) Recycling of forest products.

Rangeland

A range is the grassland that provides natural pasture for grazing animals. Grasslands occur where rainfall is too low to support the forest community but is higher than that which results in desert community. Generally, in grassland precipitation ranges between 25 and 75 cm. However, depending upon edaphic factors, such as a high water table, grasslands may also occur in regions of forest climates. Soil of a grassland is characterized by having high humus content.

Over 6,000 species of grasses, adapted for a great variety of climatic regimes, cover about 20 per cent of the earth's surface. In size, they vary from as small plants as about 1 cm to over 35 cm tall bamboos.

Importance of Rangeland

Man has been cultivating various grass species as his food and as fodder for his domesticated livestock since time immemorial. History of human civilization is the witness of dependence of man on the seeds of various grasses as his own food, for instance, Indo-Chinese, American and Mediterranean cultures were based on rice, maize, and wheat respectively; while forage was used as fodder for domesticated livestock.

A range, if not serving as a direct source of food for man, still serves the mankind in a number of indirect ways. Solar energy transformed into food energy by the green cover of the range is available to primary consumers (grazers – domesticated or wild) as food, wherein a part of it is converted into meat. Of considerable importance to man are these herbivores and their products, such as meat, milk, skins, hides, horns, and wool etc. It can thus be said that at the root of all biological resources that man utilizes is the forage.

Depletion and Degradation of Rangeland

Man would not destroy an artifical pasture development by input of his hard labour and money than a natural pasture. Thus, like all natural resources, the rangeland, inflicted with human stress, is depleted and degraded. Cause of such stress, may be ignorance, greed or incompetence of man. Phenomenal growth of cattle industry, on one hand, is responsible for ruthless over grazing, while on the other hand, wide expansion of urbanization, industrialization, and agriculture has resulted in encroachment of rangelands, compelling ranchers to over-stock the available limited rangeland. Often, out of greed too, a rancher allows a number of head to graze beyond the carrying capacity of the range, being ignorant to the fact that too many sick and weak head would not fetch him as much value as a few head in prime condition of health after grazing on enough good forage.

Community or ecological indicators provide the most practical approach of determining whether or not a range is being utilized properly (Humphrey, 1962). Range managers use the term "decreasers" for palatable species vulnerable to grazing; their disappearance is an indicator of grazing stress. When such indicator are not heeded, unpalatable 'weeds" invade and flourish. Thus, if overgrazing continues, "decreasers" keep on disappearing replaced by "weeds". Once established, if may become hard to eradicate the weeds and not only this, such vegetation will invite a variety of pest animals like rodents and insects. This makes clear the long-term consequences of over-grazing in a range.

There is increasing need to restore the depleted and degraded rangeland not only out of economic (livestock production) but also ecological consideration since rangeland depletion would also cause problems like soil erosion spread of pests etc.

Range Management

Stoddard and Smith (1955) defined the range management as "the science and art of planning and directing range use so as to obtain the maximum livestock production consistent with conversion of range resources".

Efforts to manage a range land include : regulation of grazing pressure; manipulation of stock (herd) distribution; burning of range; reseeding of rangeland; selection of cattle breeds; and control of weeds, pests and predators.

Wildlife

Even on global level there appears to be a great deal of ambiguity attached to the term 'wildlife'. Webster's Dictionary defines wildlife as 'living things that are neither human nor domesticated, specially mammals, birds, and fish hunted by man'. In American usage, term 'wildlife' is used to designate the game and excludes fish, thus they have Fish and Wildlife Services at the federal level. In India too, the common understanding of the people regarding wildlife is of wild animals and shikar, more particularly mammals, birds, and reptiles. However, a report entitled 'Wildlife Conservation in India' (1970) of the Indian Board for Wildlife states wildlife as the entire native uncultivated flora and fauna of the country. The Wildlife (Protection) Act, 1972 also defines wildlife as any animal, bees, butterflies, crustacea, fish and moth, and aquatic and land vegetation which form part of any habitat.

Thus, a comprehensive definition of wildlife may be given as any or all non-cultivated and non-domesticated life including both, plant and animals.

Importance of Wildlife

The wildlife resources provide aesthetic, recreational, and economic benefits and at the same time are important ecologically. Nature's millions of forms of life on this earth perform some or the other definite biological function which keeps the delicate balance of our ecosphere. In nature flora and fauna go hand-in-hand and one cannot exist without the other. Birds and insects are necessary for the cross pollination of flowers, and these with other animals also serve for dispersal and propagation of vegetation. Not only this, the health of vegetation also depends on natural browsing and grazing. Needless to mention that animals, directly or indirectly, depend on green plants for their food. Besides food, plants also provide shelter to a variety of animal species. Equally important are the relations among animal populations. If too many herbivores are killed, the predator, deprived of their natural food, will leave forest, become vermin, and prey upon domestic lovestock or even man. On the other hand, if too many carnivores are killed, the herbivores, because of release of predation pressure, will become too numerous and acquire the status of pest for agricultural crops. Thus, most important is the balance of nature of which wildlife is an integral and important part.

Abuse and Depletion of Wildlife

One often comes across the headlines in newspapers like 'Last of the blue whales', 'Save the golden langur', 'Tragedy strikes sea turtles', 'Major downfall in rhino population', 'Great Indian bustard on the road of extinction' ...and so on, which narrate the sad story of abuse and depletion of our wildlife heritage. Accord ng to an estimate, in last about 2,000 years the world has lost, by way of extinction, about 160 species of mammals and 88 species of birds.

India is endowed with a rich biological heritage and her various scriptures and epics are full of vivid and beautiful account on wildlife. Over 500 mammalian and 1,200 avian species contribute to the richness of this wealth. Unfortunately today, the Indian wildlife, as that in many other parts of the world, is threatened because of man's unprecedented intervention with nature. Nearly 134 plant species have been declared threatened as against 71,47 and 15 species of mammals, birds and reptiles respectively (Khoshoo, 1984). Great Indian bustard Choriotis nigriceps confined to the semi-arid regions of Rajasthan and Gujarat, and the white-winged wood duck Cairina scutalata inhabiting the Brahmputra swamps and the flood plains of Assam are both on the brink of extinction,

Wildlife Management

The conservation, is the practice embracing preservation, maintenance, sustainable utilization, restoration, and enhancement of the natural environment. Wildlife conservation in this sense is the management of nature's living resources if we consider wildlife as consisting both of wild flora and fauna. Since one cannot manage living resources while being indifferent to the non-living environment, the concept of wildlife management becomes the concept of management of natural ecosystems. It also overlaps 'forestry' insofar as it is based on multiple landuse concept and implies management of wild plants as well as wild animals. However, under the present dispensation wildlife management implies management of wild animals as distinct from forestry, and it implies maintenance of habitats suitable to the different species.

In America and Europe wild animals, particularly mamals, are treated as game crops, as do timber in forestry and food crops in agriculture, and hunting is developed as a viable economic activity. In America, wildlife and wildlife related recreation forms multimillion dollar industry. However, in India it is difficult to develop game as an industry because : (i) our forest areas are already overburdened with grazing pressure of domesticated livestock and because of already scarce pasture lands there is little hope of reducing grazing pressure in the forest areas to develop game; and (ii) hunting as a source of recreation is beyond the means of the majority of population,

During past decades the wildlife management implied the understanding, increasing, and controlling of some species. Sanctuaries and National Parks were mostly meant to protect a particular species or a group of species and, to a limited extent for tourism and recreation. Today, however, the emphasis is shifting from the management of species to the management of ecosystem. The objectives of wildlife management, as outlined in the World Conservation Strategy, are thus as given below :

(i) To maintain essential ecological processes and life support systems in natural ecosystems, on which human survival and development depends. These essential ecological processes and life support systems can be defined to include maintenance of both the biotic and abiotic components of the ecosystem.

(ii) To preserve genetic diversity (*i.e.*, the range and variety of genetic material found in the world's wild plants and animals) on which depend the breeding programmes necessary for the protection and improvement of cultivated plants and domesticated animals, as well as much as scientific advance, technical innovation, and the security of the many industries that use living resources.

(iii) To ensure the sustainable utilization of species and ecosystems. This will include the proper utilization of wild animals for game, forest and grazing, and to achieve proper integration for economy both of rural communities as well as trade and industry.

Fortunately, people have been alerted now and the strong measures being taken in favour of restoring, maintaining, and increasing wildlife population given hope that our rich legacy of wildlife will not only be saved for future but developed further. Ways and means employed in wildlife conservation include : (i) protection of breeding stock by enacting game laws and developing wildlife refuges; (ii) artificial stocking including introduction of exotics; (iii) control of predators; (iv) habitat improvement; and (v) game farming.

Minerals and Their Characteristics

Minerals are the nonliving substances obtained from the earth through the process of mining. Formation of minerals took place predominantly in weak zones like mountains, folded and faulted regions, lakes, troughs, and continental shelves through geological processes going on since billions of years. They hold some peculiar characteristics, as discussed below, which distinguish them from other natural resources.

(1) Exhaustibility

Minerals are non-renewable or depleting assets and once mined out the deposits get depleted and lost for ever without any chance of replenishment Refuse or waste, however, can be consumed afterwards through improved benefication technology.

Simply this exhaustible resource cannot be cropped year after year like agricultural products. Therefore, three distinct stages can be identified in the life cycle of a mine : youth—developmental stage; maturity—established stage; and old age—declining ore reserves. If the rate of mineral consumption is increasing fast, the knowledge about more reserves and potential substitutability of many minerals is also being improved. As such these predictions may prove outlived.

(2) Localised Occurrence

Minerals are very localised in their occurrence and the total area occupied by them is far limited. Most of the mineral deposits are very sporadic in distribution, for instance, Kolar gold fields of India occupy an area of 8 km. by 1.6 km. Formations containing coal and petrol um are, in general, bigger as compared to metalliferous deposits. The coal-bearing formations of India, for example, occupy about 4 per cent of the country's total area.

(3) Unpredictable Occurrence

The quantity and quality of a mineral in any region is not easy to predict. Uncertainities prevail with respect to their occurrence, extent of depth, cost of mining (often linked with depth), and the grade.

(4) Risk Involvement

The process of mining involves large number of problems for miners. Abnormal temperature and pressure variations, rock burst, fire and gas hazards, lack of ventilation and drainage, collapse of roof and side walls are the peculiar problems encountered only in mining industry.

(5) Dependence on Other Factors

Mineral production is largely affected by factors like the changing needs of consumer industries, availability of stock, fluctuations in prices, and reuse of metallic scrap in production process. A technologically poor country possessing richer grades of ores may have to face keen competition in international market with technologically richer countries may, however, pogsessing poorer grades or ores since in later case the cost of production is considerably low due to improved mining methods.

(6) International Character

Minerals know no political boundaries. Virtually, no country within its own boundries has adequate production and reserves of all the minerals needed for its industrial development. The unequal distribution of minerals makes the international trade in its inevitable. Some countries are bestowed with monopolistic rights in respect of certain mineral deposits which result in what is known as geopolitics.

Development and Conservation of Minerals

Geologists are of general apprehension that the known reserves of minerals may not last for long and most of them will be exhausted within 100 to 200 years. New possibilities of mineral reserves in land and beyond the land (as in the seabed are being explored which may reduce, if not totally remove, the futuristic fears of exhaustion of the resource.

Presently, ilmenite, rutile, tin ore, diamond, sulphur, coal petroleum, and calcareous and ferruginous sea-sands are largely being mined from the sea-coasts and sea-beds. In India, ilmenite, rutile, and associated zircon, monazite, and sillimanite are obtained from the beach sands of Madras and Kerala coasts. Coral limestone for industrial use like manufacture of calcium carbide, and calcareous sand for cement are being mined from the sea-bed at Tamil Nadu and Dwarka (Gujarat) respectively for quite long time. Indonesia, Thailand, and Burma recover considerable tin ore from the sea-bed, while England recovers coal and Philippines iron from the sea-bed. A number of petroliferous structure occurring off the coast are being mined for petroleum and its products at several places. The 'ferro-manganese nodules' or what commonly known as 'manganese nodules' found over the oceanic floors, generally below 4,000 m depth, are perhaps the most spectacular discovery in this field. The most prolific concentrations of such nodules with high metal values have been found in North Pacific between 139° W to 141° W and 4° N to 16° N. Although precise information about the reserves of manganese nodules is not available but it is estimated that they may run several billion tones and may prove to be potent source of metals like manganese, nickel, copper, and cobalt. It is estimated that 1 million tons of manganese nodules from sea-bed provide about 275,000 tons of manganese, 13,000 tons of nickel, 2,500 tons of cobalt, and 1,000 tons of copper besides other metals like iron, vanadium, molybdenum, silver, and lead. Further, it has now been roughly estimated that the metal resources available in manganese nodules are several times greater than those available in land, such as cobalt 5,000 times, manganese 4,000 times, nickel 1,500 times, and copper 150 times.

The conservation of mineral resources does not mean preservation of minerals under ground to be used by future generations but it implies efforts to utilize minerals in best possible way, depending upon the industrial needs and changing technological requirements, so that the life of the deposits is increased without sacrificing out own needs. From the utilization point of view, minerals are placed under two classes : expendable and nonexpendable. Expendable minerals have practically no salvage value and once consumed are lost for ever, *e.g.*, coal, petroleum, gas, and most of the non-metallic minerals. Non-expendable minerals are those which have good salvage value and can be used again and again through recycling, except those quantities that are lost due to corrosion and abrasion, *e.g.*, metallic minerals.

Efforts to conserve mineral resources include :

(i) Application of efficient methods of mining to take out every possible tonnage lying under ground.

(ii) Utilization of unmarketable ores through the innovative methods or ore dressing and metallurgical practices.

(iii) Use and reuse of scraps.

(iv) Recovering all associated elements as co-product or by-product.

(v) Economic use of minerals.

(vi) Substituting the use of rare and costly minerals with those which are abundant and cheaper.

A facility (house, hospital, port), or even in providing a service to society (medical care, public transport). Firewood, fossil fuels (coal, petroleum and its products, natural gas), solar energy, biogass, wind and hydropower, and atomic power are chief energy resources. Of these, firewood and fossil fuel are being used for supply of energy since time immemorial and are known as conventional energy sources. Wind and hydro-power, solar energy, biogas, and atomic energy are the new and non-conventional sources of energy. Fossil fuels being limited in supply may be termed as exhaustible sources while others, like solar, hydel, wind and atomic energy, that would be available to man for infinite period are regarded as inexhaustible sources. Firewood and other biofuels (such as, biogas derived from biological wastes, domestic garbage etc.) are indeed renewable energy sources but exploitation of fire-wood has great ecological implications since we are already short of forests and whatever forests available should not be squeezed or thinned. However, use of other biofuels is known to be the best option in the light of two important facts : their steady or even increasing supply at low cost; their use in energy production reduces pollution of environment.

Because of rapid depletion and poor regeneration of the exhaustible sources of energy the society should concentrate on tapping the inexhaustible and renewable sources under strict economic and ecological consideration.

Conventional and Exhaustible Energy Sources

Firewood : No one can trace back in the history that whose are made the first assault on which tree. However, it is evident that man has been logging down the trees for various purposes, including to get firewood as an important one, since long. Because, in olden days man's knowledge and access for various natural resources was limited, he ruthlessly exploited his own surroundings to meet out his day-to-day requirements. The story still continues in various parts of the world at slower or faster pace, but often not because of the dearth of knowledge but out of vested interests or because of having no other option. Felling of trees for firewood is still a common practice in interior where it is the cheapest and, at times, the only available source of energy. This had led to thinning or denudation of woodland that need serious consideration from both ecological and economic viewpoints. Obviously, the only wayout to the problem is making available to the villager the cheaper non-conventional energy sources, such as biogas and solar energy in usable form.

Fossile fuel : Fossil fuels (coal, petroleum, natural gas) are the most extensively used sources of energy today. Like for many other natural resources, increase in population and per capital energy demand coupled with the industrialization at a faster pace, are the factors responsible for depletion of fossil fuel world over. According to a report the earth's original stocks of coal and petroleum were about 6,000 billion tons and 1,250 billion barrels respectively, of which by the year 1960, 100 billion tons of coal was mined and 100 billion barrels petroleum extracted. The story of natural gas is no different which has acquired the status of irreplaceable commodity of domestic use in modern times. Petroleum and natural gas today fulfill 60 per cent of the world's total energy requirements. It is evident that if the exploitation and consumption of these resources continues at the present incredible rate, there supply may last only for a few more decades.

Nonconventional and Inexhaustible Energy Sources

The rapid depletion of conventional energy sources has promoted governments and people to concentrate on finding and tapping some non-conventional energy sources that may last for long. In fact, inexhaustible energy sources, like solar, hydel, tidal, wind and atomic power, can only bring hope for the sustained socio-economic development of humanity for years to come when exhaustible sources will become either inaccessible or too costly to afford.

Solar Energy. Research and application in the field of solar energy have opened new vistas in the direction of fulfilling world's future energy requirements. It has especially drawn attention for its practically free steady supply and pollution-free use. This energy may be transformed into heat using reflectors and condensors, or into electric energy using solar photovoltaics. The resource can especially be tapped more effectively in the regions where there are long sun-shine hours.

Solar heaters and cookers are becoming popular these days in industry and household. Passive solar architectures provide comfortable housing in extreme cold and hot climates. This engineering has also been successfully applied in warehouses and cold storages. A multipurpose solar dryer-cum-warehouse is under construction at Ganaur (Haryana, India) where chillies, potato-chips and other foodstuffs could be dried and stored. Solar energy has also been effectively used in desolination of water. College of Engineering, Pune has designed a solar furnance that can generate heat upto 2,000°C which can be used in various metallurgical and other purposes.

Department of Nonconventional Energy Sources of Govt. of India has prepared a perspective plan that envisages generation of energy through non-conventional sources, chiefly solar, to the extent of 250 million tons of coal replacement per annum including 15,000 MW of electricity by the year 2001. Indian's first solar power station, soon to be operated near New Delhi, would be a great breakthrough in this direction. Rural electrification through solar photovoltaics is already in operation at some places in the country. Moreover, photovoltaic irrigation pump sets have also been installed at some places.

Wind power : The power of wind since olden days is being used in running mills, irrigation of fields, and carrying out other farm activities, but its real potential is not so far being realized. According to inventories of Department of Nonconventional Energy Source of Government of India 20,000 MW electricity can be generated from the wind alone in the country. Wind power plants are set up in various parts of the country but they still account for the total capacity of 3.3 MW only. Obviously, this potent resource is still to be tapped optimally.

Hydropower : Use of gravitational flow of water in generation or electricity, what commonly known as hydropower or hydel power, is quite common in various parts of the world. In India, first hydropower station, having a potential of 3000 KW, was construted in 1897 at Darjeeling. Later on, many stations with far high potentiality were set up. The country's economically exploitable hydropower resources have been estimated at 41 million KW (or 41,000 MW) of which only 18 per cent of the potential was utilized by 1979. The Brahmaputra basin has the highest hydropower potential and nearly 30 per cent of the country's total; next to it rank the Indus, Godavari, and Ganga basins respectively.

In world, a total of 4.910×10^9 KW electric power is generated of which hydropower forms a fourth. In highly industrialized countries like USA, USSR, Canada, Japan, France, Sweden, and Norway hydropower is nearly full developed whereas in the rest of the world it is yet to be developed fully.

Hydel power is considered to be the cheapest source of electricity with the cost of generation ranging from 2 to 3 paise per unit, compared to a minimum of 5 paise per unit of electricity generated by coalbased thermal power plants. Although the initial cost of hydel project is generally 25 per cent higher but owing to lower recurring costs, the cost of generation of hydropower is the lowest. Apart from such economic considerations, development of such projects is beneficial for irrigation and other purposes too.

Tidal power of water in coastal areas of India is, however, not so far being realized. According to an estimate a tidal power potential of 9,000 MW has been identified, of which 7,000 to 8,000 MW is in the Gulf of Cambay, 800 to 1,000 MW in the Gulf of Kutch, and rest in Sunderbans.

Atomic power (Nuclear power) : As already delineated, the fossil fuels being exhaustible, such sources of energy are not to last for long with the current trend of consumption. Atomic power appears to be the only hope for large scale energy requirements when fossil fuels are exhausted. The reserves of atomic fuels are said to be over 10 times greater than fossil fuels and the advantage of former is that their very small amounts would produce enormous amount of energy. A ton of Uranium^{235}, for instance, would provide as much energy as 3 million tons of coal of 12 million barrels of oil. Atomic energy has its application not only in generation of electricity but has successfully been used in marine vessels, spaceships, and even in chemical and food-processing industry.

Important constraints in atomic energy generation are :

(i) Cost of construction and maintenance of plants; and

(ii) Disposal of radioactive wastes. Cost of construction of an atomic power plant and its maintenance is very high as compared to any other electric power plant. For running one atomic power unit of 235 MW capacity, heavy water of a worth Rs. 10 crore is required per year. Further, as a consequence of frequent failures, the running cost of such plants come to be very high. Moreover, after the working capacity of an atomic power plant is over (in about 25 years), these have to be destroyed so as to avoid radiation spillage. Surprisingly, the cost and time for destruction of these plants is almost equal to that for construction. Disposal of radioactive wastes is such a problem that still has no satisfactory solution for long term.

Bioenergy (Energy from biological matter): Organic wastes provide an important renewable source of energy. Biogas generated from the decomposition of organic wastes, like manure and domestic wastes, is becoming increasingly popular as a source of heat and light in many parts. It is considered to be advantageous in view of its relatively cheaper supply, and use of organic wastes in its generation reduces the othewise impending threat of pollution due to their release in environment. As such, it serves two purposes; fuel production and waste disposal. It has more of practical applicability in villages where organic waste, in the form of cattle dung, is available in plenty. Biogas so generated is thus economic and convenient to use as compared to conventional practice of burning of dry dungcakes. Further, the dung-cakes serve only as a source of heat for cooking and allied purposes while biogas has multifarious applications. The slurry so produced in a biogas plant can well be used as manure for agricultural fields.

Under the National Biogas Development Programme, 8.2 lac biogas plants were set up by the year 1985-86 in different parts of the country. A large-scale urban waste recycling plant at Okhla (New Delhi) generates about 0.6 million cubic feet gas per day having heat value of 700-800 BTU per cubic foot. The gas is being supplied to about 800 households.

Conservation of Energy

The need for conservation of energy becomes essential in

view of the following simple facts :

(i) Most of the activities of present day large and growing population require massive energy inputs, thus per capital and total demands for energy are fast increasing.

(ii) The fossil fuels, we are not endowed with which much, are not to last for long at the present rate of consumption. At the same time, much talked about possible alternate sources of energy are, in fact, still in tentative state of development so far their practical applicability is concerned.

(iii) There are economic constraints for many countries even if they want to develop conventional sources of energy, what to talk about sources like nuclear energy.

(iv) Unplanned exploitation and use of energy result in widespread environmental damage, the cost of which is to be borne by humanity.

Conservation of energy means not simply the curtailment of energy consumption but to curtail unnecessary or wasteful consumption of energy and make thrifty use of energy producing resources to ensure their steady supply for generations.

16

Biotic Communities

Introduction

Biotic communities consist of all the plant and animal populations inhabiting a given area. They represent a higher order of biologic organization than populations, yet, since communities refer only to living organisms, they are not as inclusive as ecosystems. In other words the study of communities involves the entire biology of an area, but strictly speaking it does not involve the specific study of the interactions with the nonliving components of the environment as does ecosystem analysis. This of course, is a somewhat arbitrary distinction, and of more value for academic convenience than ecologic reality. As we shall soon see, it is impossible to talk about biotic communities without frequent reference to the nonliving environment or total ecosystem.

As with populations, it is again convenient to think of communities in terms of structure and dynamics. Structure refers to the spatial organization of communities and the way different populations are related in shape and form. Dynamics refers to the interactional processes, energetic relationships, and patterns of change within communities.

Biotic communities assume a variety of forms as diverse as life itself. If one's imagination alternately pictures a tropical forest a coral reef, a grassland, a temperate desert, an arctic slope, or the benthic depths of the ocean, one can quickly grass the tremendous variation which occurs in the tremendous variation which occurs in natural communities. Major terrestrial communities, each characterized by certain types of plants aind distinctive life forms are known as biomes. We shall consider some principles of commtnity structure and function as they relate to a familiar biome, the temperate forest, and then to briefly, describe a few of the common community types found throughout the world.

The Temperate Forest Community

A good starting point in the study of the forest community is to examine its structure in physical and stratigraphic terms. The forest may be thought of as a layered system with each layer possessing characteristic populations and a typical organization.

The structural base of the entire community, of course, is the soil. It supplies the fundamental reservoir in terms of nutrients and water. It is both on aquifer and basic food bank. The soil itself is composed of strata or layers, known as horizons, each with definable properties. The upper layer, or A horizon, commonly called "top soil," consists of the living roots of vascular plants the mycelia of fungi, a great abundance of bacteria, protozoa, algae, and other microorganisms, and a surprising variety of burrowing animals (nematodes, molluscs, annelids, insects, various other arthropods, amphibians, reptiles, moles, shrews, etc.

Of nematodes alone, it has been said that if all plant tissues were instantly dissolved and made to vanish, the outline of trees, shrubs, grasses and herbs and their roots would still be discernible by the presence of nematodes which live within all plant tissuses.

The A horizon of soil also consists of dead and decaying organic material of plant and animal origin. The organic material is undergoing decomposition in the process of human formation of humification. The A horizon is further subdivided into various layers representing different stages of humus formation and these layers are commonly called leaf litter, duff, leaf-mold, humus, and leached zone.

The second major stratum of soil, the B horizon or subsoil, consists of a mixture of the underlying parent material and inorganic compounds which have filtered downward from the process of decomposition in the A horizon. Nitrates, phosphates, acids, bases, salts and other compounds resulting from organic decomposition have moved downward by the leaching action of water into the B horizon. The plant and animal communities of the B horizon are considerably smaller in numbers and species than those of the A horizon.

The third major layer of soil, the C horizon, consists of the basic parent material or substrate. If may be of sedimentary origin resulting from water deposition (alluvial), glacial activity (glacial till), or wind deposition (eolian or loess), or it may also be igneous, metamorphic or volcanic in origin. In general, soils forming on alluvial, glacial or eolian depositions are the richest, and some of the most productive agricultural areas of the world are found in river deltas of glaciated grasslands.

The nature of the soil profile obviously has a major influence on the animal community living within it and *vice versa*. Forest soils tend to have large populations of those organisms adapted for processes of slow organic turnover and decomposition such organisms as fungi, isopods, millipedes, soft larval insects, and so forth. Grassland soils have higher population of those organisms which depends upon high rates or organic turnover - bactetia, nematodes, annelids and xerophilic insects.

Animal communities of soils also have major influences on the process of soil formation. Many burrowing animals play a vital role in soil aeration, physical cultivation, hydration and decomposition. Earth worms not only aerate the soil and contribute to its water holding capacities, but they bring deeper soil materials to the surface and contribute to soil mixing and decomposition. Darwin, in his classic book. The Formation of Vegetable. Mould through the Action of Worms published in 1881, estimated that earthworms brought 20 tons of soil ro the surface per year. More recent studies have shown that the amount of soil brought to the surface by earthworms may very from 2.1 tons/acre/year in hot dry soils to 10.7 toh/acre/year in prairie soil, to 107 tons/acre/year in tropical delta soils (Kendeigh, 1961).

The second major stratum of soil, the B horizon or subsoil, consists of a mixture of the underlying parent material and inorganic compounds which have filtered downward from the process of decomposition in the A horizon. Nitrates, phosphates, acids, bases, salts and other compounds resulting from organic decomposition have moved downward by the leaching action of water into the B horizon. The plant and animal communities of of the B horizon are considerably smaller in numbers and species than those of the A horizon.

Other burrowing animals are also important in cultivating the soil and preventing compaction. In prairie soils, ants may bring over a ton of the surface per acre per year, and prairie dogs may excavate 30 to 40 tons per acre in and around their burrows. Not only do all burrowing animals till the soil physically, but they contribute to its chemical maturation by processing organic material through their digestive systems.

Soil formation is a slow process of ecologic maturation. Forest and grassland soils involve many centuries of organic accumulation, and they represent the stored capital of very long-term ecologic processes. The rate at which these processes occur varies greatly according to climate, topography, substrate, biologic community and all those ecologic factors influencing the processes themselves. It is thus impossible to give a precise figure on the rate of soil formation. Some temperate soils are literally millions of years old. In fact, several studies have shown that temperate soils require 10,0 0 years to develop recognizable soils profiles and some profiles and some profiles are not really well-developed in less than 200,000 years. Tropical soils develop much more rapidly, of course, but they are thin and susceptible to desturction. It is thus important for modern man to realize that soild is not a synthetic commodity, or one that can be manufactured in a short time.

In proceeding upward from the fioor of the forest, the biotic community thing out to a certain extent, animals become more widely spaced in three dimensions, and they become more mobile. The plant community is dominated by herbs and shrubs, and the animal community by insects, birds and mammals. Mammals, including deer, rabbits, mice, squirrels, shrews, racoon, possum, etc. actively forage through the lower layer of this comunity.

Many animal taxa including annelids, some molluscs, myriapods and solid-welling arthropods do not enter this realm and are seldom if ever found above the surface of the ground. There are exceptions, of course such as certain snails (molluscs) which climb trees.

Species Diversity

In 1958 the eminent ecologist of Yale university, G.E. Hutchinson, delivered an address of the annual meeting of the American Society of Naturalists and used the title, "Homage to Santa Rosalia or Why Are There So Many Kinds of Animals ?" He took the title from the sicilian saint, Santa Rosalia, near whose sanctuary on Monte Pellegrino, Sicily, Hutchinson had found a small pond rich in animal life and limestone cavers rich with the bones of pleistocene animals. The setting gave Hutchinson addittional inspiration to reflect on the ecology of evolution. Why has a natural selection produced such a great variety of living organisms – over a million known species of animals alone, three-quarters of which are insects ?

This topic has been a major question in ecologic and evolutionary theory for many years. In discussing the evolution of species diversity, Hutchinson emphasized the mosaic nature of the environment.

When the environment is viewed in very small units of space, it may be considered an infinitely variable mosaic of different conditions. Hutchinson pointed out that the small physical size of many animals, such as insects, has permitted them "to become specialized to the conditions offered by small diversified elements of the environmental mosaic, and this clearly makes possible a degree of diversity quite unknown among groups of larger animals." Furthermore, the addition of each new plant or animal to the community increases its diversity, thus creating new conditions on which natural selection can operate. Diversity, then, becomes a self-stimulating phenomenon.

The limiting factors on diversity are the severity of the physical circumstances to which life forms can adjust. All life having a common biochemical basis, has certain central tendencies in the conditions around which life processes operate with the greatest efficiency. This does not infer thatt all species of plants and animals have the same environmental requirements but it does mean that more plant and animal species are going to survive and thrive of environmental conditions approaching the normal requirements for protoplasmic processes. In other words, the biochemical unity of life insures that more plant and animal species are going to do well at 80. to 90. F (27 .. to 32 C) than at 32 F (0 C). Plant and animals can, of course, adapt to the latter but the selection for this will be substantially more rigorous, and the available niches much fewer.

Seen in this light, it becomes reasonable to theorize that those environments whose physical circumstances are harshest and most inimical to life will have the least diversity. Thus we would expect desert, ocean depths (below the light zone). polar regions and snow-capped mountains to have the least diversity in their biotic communities. This is indeed, exactly true. These are biotic communities characterized by the lowest numbers of different kinds of plants and animals, though there may be very large populations of these few species.

On the other hand, we would expect those environments with conditions most nearly approaching the ideal for the processes, that is, warm temperature, ample light and plenty of moisture, to have the greatest diversity of plant and animal life. This also is true. The tropical forests are far richer in number of species of plants and animals than any other biotic community on earth.

The relationships between species diversity and community stability have been mentioned previously. In general, the greater the diversity, the greater the stability. Communities with many species have a more complex web of food relationships, and each species has greater diversity in its food and cover resources. Thus each species has alternatives in meeting its life requirements. If has various ways of meeting new environmental shortages and contingencies. There is in other words, a cushioning network of checks and balances. It must be remembered that we are speaking in broad terms and there are undoubtedly many exceptions to these generalities. Nonetheless, these generalizations are valid as overall principles of community ecology. As a result, any major change in a single plant or animal population in a polar community is likely to have a far greater impact on the total community than similar change in a tropical community. It thus becomes important to think of communities in terms of their vulnerability to ecologic upset if change occurs in any one of their member populations.

Succession

Biotic communities change with time as their plant and animal communities change. This process is known as a succession, and it involves a sequence of community types from pioneer stages to mature or climax stages. Each community in the series is known as a seral stage. One of the best ways to appreciate succession is to consider the sequential growth and development of biotic communities on a cleared forest.

If a forest is cut, and the land cleared to the soil, a succession of plants will invade and grow on the exposed surface. The first plants which invade are typically those capable of seeding in quickly on disturbed land or capable of generating from viable roots left within the soil. This depends not only upon the former vegetation, but also upon seeds from surrounding plant communities, as well as the characteristics of the soil and climate. Typical invaders in the United States are herbaceous weeds such as ragweed (Ambrosia sp.), lambs quarters (Chenopodium album) T dock (Rumex sp.), horse weed (Erigeron canadensis), plantain (Plantago sp), crab grass (Digitaria sp. et al) and woody vines such as honeysuckle (Lonicera sp.) and poison ivy (Rhus taxico-dendron). As soon as the first plant germinates and becomes established, the community increases vastly in physical complexity. A surface of bare earth now has shaded areas which differ in light, moisture and humidity. This enables other seeds to grow, and new species become established. Woody plants appear, again those capable of withstanding disturbed environments and harsh conditions, such as black locust (Robinia sp.), sumac (Rhus sp.) and howthorn (Crataegus sp.). Pioneer animals also arrive, including ants beetles and flying insects. Birds begin to forage over the new community searching for seeds and insects. Small mammals may venture forth from adjacent forests or grasslands Lack of animal traversing the area adds organic nutrients. Within a year or two a surprisingly complex community has arisen, and is growing rapidly. As plants grow, they create more shade, and thus continuously modify the light and moisture conditions on the surface of the ground. These initial pioneers which thrived on total sun exposure are now less favoured and more shade-tolerant seeds can germinate more successfully. Woody shrubs emerge above the herbaceous layer and compete more successfully for the ambient light.

The shade developing beneath the herbaceous and shrubby layers permits the growth of other tree species such as maples, elms or oaks which could not tolerate the initial exposure of sun and wind. This represents the first beginning of the forest. Further succession continues to involve all the dynamic processes of plant and animal competition. Over a period of years, the initial invaders and pioneers begin to drop out. Trees nature first the fast growing locust and sumac, but as they reach the limits of their growth and their relatively short life-spans the slower growing oaks (Quercus sp.), maples (Acer sp.) and elms (Ulmus sp.). maples (Acer sp.) and elms (Ulmus sp.) begin to take over. The forest may now be 40 to 50 years old, reaching a height of 50 to 60 feet.

The ground cover and intermediate layers have now matured to include those shade-tolerant species capable of living in the forest under study. The insect and bird populations have changed from open field forms to forest species. By 70 to 80 years, the forest may be approaching relative maturity, with tall trees in the 60 to 80 foot range, deep shade and cool, moist conditions on the forest floor capable of supporting the rich array of life described earlier in this chapter. The climax first may be approached in 80 to 100 years, when it becomes a stable and self-perpetuating community. Processes of change still occur. Old forest trees die and crash to the ground, creating a microcosm of renewed successional development, but the main forest continues in equilibrium as a permanent community as long as environmental conditions remain favourable. If it is cut, the process of succession begins again. If it is defoliated, flooded, burned or dehydrated, it will change and attempt to seek a new form of equilibrium with its environment. The tree species that finally emerge as the dominant-members of the climax community vary according to the climate, topography and soil type.

It would be incorrect to assert that the process of succession is always consistent and predictable. It may take a variety of forms, and may become arrested at any of several stages. Early in the shurb stage, the biotic community may become overwhelmed by excessive growth of woody vines such as honeysuckle or poison ivy, so that trees cannot develop adequately, or it may stay in the herb or shrub stage if the land is subjected to continuing disturbance by fire, pol ution, or and scarring. Throughout the condtryside, especially around cities, one can find many examples of disturbed and unattractive vegetation that never gets out of the weed, shrub ar vine stage.

The above discussion has mainl concerned a deciduous forest community, but a similar account could be given of coniferous forests, and, in fact, of all biotic communities. Thus, a prairie grassland, a tropical savannah, a desert arroyo, an estuarine marsh, a cedar bog, a pond, a lake, a delta, a mangrove swamp, or a coral reef all have characteristic patterns of development and succession. In some communities, successional development may involve hundreds or even thousands of years before mature climax conditions are achieved. Primary tropical rain forests, for example, are the culmination of thousands of years of successional development. Once cut, they are not recreated in a few decades or a hundred years.

It should be emphasized that succession is also a naturally occurring process and is not confined to disturbed habitats. Theoretically, each biotic cammunity tends toward maturation and stable climax conditions, but there is sufficient natural change in ecosystems so that succession is a continually recurring process in many situations. The grassland and forest are subjected to natural fire as an ecologic process, the desert arroyo to flash flooding, the marsh, bog, pond, lake, delta and swamp to filling, and the coral reef to the continual processes of construction and destruction. Human activities often increase and accelerate these processes far beyond the natural rates of occurrences. In so doing, man impedes the development of maturity and ecologic stability in biotic communities.

An example of natural succession is the filling of a small pond and its eventual conversion into a terrestrial community.

As the pond ages, it becomes more filled with aquatic vegetation silt. Emergent vegetation surrounds the edge creating a marsh community. Accumulations of organic material and sedimentary deposits fill in the water volume, soil is formed at the edge, and terrestrial grasses and herbs become established. Eventually, shrubs and trees complete the process. Man accelerates this process by increasing the siltation process and enriching the waters with nutrients so that more extensive plant growth occurs. The process can be retarded by dredging the pond or manipulating the water level to reduce aquatic vegetation. Lakes proceed through some of the same stages, but if sufficiently deep, a point is reached in lakes where the scouring action of currents slows or even balances the siltation and sedimentation rate.

It is important for man to understand successional processes so that he can manage biotic communities for maximum productivity and aesthetic concerns. In agricultural lands and harvested forests, man has learned to maximize productivity for his own economic gain. This is done by artificially maintaining communities in early successional stages, where productivity is greatest, and by channelling all possible production into the species to be harvested. That is, man maintains a cornfield or a wheatfield in an early productive stage by excluding all other pioneer species. He maintains a forest in an early successional stage and directs all productivity into the species of tree he wishes to harvest. This is all well and good provided the cost of such management is not too high in terms of toxic chemicals that must be applied to eliminiate other species, or in terms of chemical fertilizers that must be applied to support an artificial single-species community.

Quite another principle prevails in the management of parks and sanctuaries where man wishes to maintain balanced plant and animal communities for esthetic and scientific values. Here it becomes essential to understand and promote natural succession so that mature and stable communities are achieved.

In this day of economic pressures it is increasingly difficult for man to appreciate the value of natural biotic communities. What good is a mature forest if it is not harvested, what good is an estuarine marsh, what good is a relic prairie? These are all questions being asked in greater frequency by citizens focusing on economic progress. The challenge the ecologist to explain the message of ecology in factual terms, and to emphasize the importance of balanced biotic communities to the total life support capabilities of the world. The answers are not easy, and are not always available. Much remains to be learned about the full range of functions of an estuarine marsh, for example, in maintaining the productivity of the oceans. This represents one of the greatest current challenges for ecology; namely, to understand and explain in a meaningful way, without great fanfare and unreasonable emotion, the importance and value of diverse biotic communities. Otherwise economic man will see little reason why he cannot continue to convert the entire world to paved streets, grass lawns, grain fields, and greenhouses.

Major Biotic Communities

It is worthwhile at this stage to take a brief look at several different communities and highlight some of their outstanding features. This has been done in various degrees of detail by a variety of authors, and no attempt will be made here to be comprehensive or thorough. Descriptions are offered as capsule summaries of six interesting and representative communities; the arctic tundra, northern coniferous forest, desert, grasslands, estuaries and tropical rain forests. This limited list unfortunately omits many other important and interesting communities : savannah, chapparal, temporate rain forests tropical deciduous forests, fresh-water swamps and marshes, lakes, rivers, coral reefs and open oceans. Volumes could be written about each of these, and many others.

Arctic Tundra

Arctic tundra is the biotic community occupying the northern latitudes, generally above 60 north latitude. The dominant plants are lichens (the combination of algae and fungi known as "Reindeer moss"), arctic grasses and sedges, and some dwarf shrubs, especially dwarf willow (Salix sp) and cranberry (Vaccinium sp.). During the brief summer the flowering season is abrupt and colourful. The vegetation is underlain by chick spongy mass of undecayed deed vegetation. Bacterial action is slow due to the low temperatures, and the substrate is often permanently frozen, a condition known as permafrost.

The animal community is simplified into relatively few species of birds, mammals and flying insects. The dominant mammals are the caribou (Rangifer arcticus) or reindeer (Rangifer tarandus, a domesticated species of caribou), arctic hare (Lepus arcticus), lemming (Lemmus sp. or Dicrostanyx sp.), and a arctic fox (Alopex lagopus). The dominant birds are ptarmigan (Lagopus lagopus, a gollinaceous grouse-like bird), the longspur (Calcarius sp., a sparrow-like bird), and the snowy owl (Nyctea nyctea). Great numbers of migratory waterfowl (ducks, geese, swans and shorebirds) invade the tundra in the summer for breeding. Insects are dominated by the biting diptera, blackflies and mosquitos.

Northern Coniferous Forest

This is a forest belt at latitudes below tundra, occupying major portions of Alaska, Canada, Scandinavia and Siberia, generally between 50 and 60.. north latitude, but with extensions in mountainous regions further south. The plant community is dominated by spruces, firs, pines and hemlock, with a ground cover of mosses, grasses, sedges and cold adapted herbs. Decomposition is slow, as in the tundra, and the forest floor tends to accumulate a layer of dead organic material with excellent waterholding capacity.

Animal populations are more varied than in the tundra, but are still characterized by great seasonal change and wide population fluctuations. Typical mammals are the snowshoe hare (Lepus americanus), lynx (Lynx canadensis), squirrels (Sciurus sp, Tamiasciuruss sp.), marten (Martes americana), mink (Mustela vison), fisher (Martes pennanti), wolverine (Gulo luscus), woodland caribou (Rangifer caribou), deer (Odocoiteus sp.), moose (Alces americana) and black bear (Ursus americana). Typical birds are ruffed grouse (Bonasa umbellus), crossbills (Loxia sp.), siskins (Spinus pinus) and a multitude of other passerines which migrate in for summer breeding. Winged diptera, such as blackflies and mosquitos, beetles and moths are among the dominant insects.

Grasslands

The grassland community covers vast stretches of land within continental interiors. These are areas characterized by plains and 10 to 30 inches of rainfall, per year, generally less than in most forested regions.

The plant community is dominated by grasses, but it contains a great variety of other herbs and forbs as well, particularly legumes such as lupine (Lupinus sp.), trefoil (Trifolium sp.), clover (Petalostemum sp.), and vetch (Vicia sp.); composites, such as aster (Aster sp.), firewood (Epilobium sp.) and daisy (Chrysanthemum sp. or Bellis sp.), and members of the Ronucutacead, buttercups (Ranunculus sp.), anemone (Anemone sp.), columbine (Aquilegia sp.) and larkspur (Delphini sp.). In all a grassland or praairie floro may contain over 120 species in more than 10 families. The dominant green grasses are predominantly or three general sizes : tall grass, often 6 feet in height, which includes tall blue stem (Andropogon gerardi), switch grass (Panicum sp.) and Indian grass (Sorghastrum sp.); medium grass, 2 to 4 feet in height, which includes little blue stem (Andropogon scoparius), feather grass (Stipa short grass) less than 2 feet in height; which includes buffalo grass (Buchloe sp.), blue grass (Poa sp.) and blue grama or mesquite grass (Bouteloua sp.).

As previously mentioned, the annual production of organic material in grasslands is great, organic accumulation is rapid, and a thick layer of humus is produced. Grasslands soils are among the thickest and richest in the world. Roots of the grasses and forbs generate up to 6 feet in these rich soils.

The animal populations of grasslands are also rich and varied. Among the mammals, the ungulates and rodents have flourished with great populations of bison (Bison bison), antelope (Antilocapra americaana), prairie dogs (Cynomys sp.) ground squirrels (Citellus sp.) and gephers (Geomys bursarius), and the mammalian predators are represented by coyotes (Canis latrans), foxes (Vulpes sp.) and badpgers (Taxideaa taxuss). Typical birds are prairie chickens (Tympanuchus cupido), sharp-tailed grouse (Pedioecetes phasianellus), meadow larks (Sturnella sp.) and a variety of sparrows. Insects are abundant, with grasshoppers and locusts being among the most conspicuous.

Deserts

Deserts are arid biomes with usually less than 10 inches of rainfall per year. Deserts usually occur in areas of high pressure (Sahara, Australian), rain shadows of mountains (Mohave, Iranian) or high altitudes (Tibetan, Gobi and Bolivian).

Some of the direct deserts may not have rain for a period of many years. One such area in Chile has not received precipitation for over 20 years. Such deserts may be virtually devoid of living organisms over vast areas. In the Sahara desert of southern Libya, one can travel for hundreds of miles without seeing of living plant or any form of life. Most deserts, however, do have some water resources from occasional rains or from ground water, and they possess a considerable variety of living organisms dependent upon these water resources.

The predominant plants of deserts are succulent species with waxy surfaces, such as cacti, which can conserve water for long periods of time, or deciduous shrubs, also with thick waxy leaves. The growing season is very short, and the flowering period amazingly abrupt, occurring suddenly after flash rains. Many deserts have little or no plant life over extensive areas of sand and rock.

Estuaries

Estuaries are bodies of water where fresh water from the land mixes with and measurably dilutes sea water. They represent dynamic combinations of fresh water and marine communities and they have a number of unique characteristics. They are particularly important to man by virtue of these unique biological attributes.

Several different types of estuaries exist, representing various topographic formations and geologic histories. Some develop from coastal submergence and drowned river mouth such as Chesapeake Bay and Delaware Bay. Other develop as coastal lagons behind barrier sounds, such as Albemarie, Pamlico, and Currituck sounds inside Cope Hatteras. Others arise as steep glaciated fjords such as those along the coasts of Norway, Chile and British Columbia. Finally, many develop within great delta systems such as those of the Nile, Mississippi, Mekong and Ganges rivers.

Tropical Rain Forests

The tropical rain forest is a biotic community which has always stirred the human imagination, and rightly so. It is one of the most dramatic and awesome biologic systems on earth. Such forests occur within the Tropic of Cancer and the Tropic of Capricorn (23° 27 N and S latitude), in areas with more than 80 inches of rainfall per year. Usually such forests have one or more dry seasons per year in which less than 5 inches of rain fall per month. The monsoon forests of southeast Asia are one example, and although it rains throughout the year, concentrations of rainfall occur twice a year in the monsoon periods.

An important characteristic of the tropical rain forest is its relative climatic stability. The temperature variation between seasons is light and usually less than the temperature variation between day and night, or between different microhabitats within the forest. For example, in Singapore at the tip of the Malay Penninsula the diurnal variation in temperature is often 20°F (from 75 to 95°F), whereas the variation in temperature from January mean to the July mean is only 2.5° (from 87° to 89.5°F). By contrast, the variation from January to July mean in Dawson City, Alaska is 81.7°F (from -15°F to nearly 70°F).

17

Wildlife Conservation

Introduction

Wildlife Conservation includes all human efforts to perserve wild animals from extinction. It involves the protection and wise management of wild species and their environment. Some species have become extinct due to natural causes, but the greatest danger to wildlife results from human activities. Thus, we ourselves have created this need for wildlife conservation.

The progress of man throughout has been beneficial for the human race but it is the wildlife that has suffered through the years. Invention of sophisticated weapons, industrialization, urbanization, ever increasing human population have been some of the major causes for the dwindle of our once rich resource. Hunting, clearing of forests, draining of swamps and damming of rivers for irrigation and industry this is what we appraise of man's progress. These activities have vastly reduced the natural habitats of our wildlife and many species are endangered or nearly extinct.

Man's indifference and ignorance has led him to believe that he is independent of natural laws and of natural environments and forgets that he himself is a biological being and is directly and/or indirectly dependent on the natural resources available to him. He exploits, destroys and reduces the renewable natural resources, eradicates wildlife and leads his own race to levels leading to catastrophe. All this can be put to an end by careful, long-range ecological planning.

Importance of Conservation

Alongwith the topic of conservation arises a question—Why should be care about disappearing animals ? The answer to such a question can be attributed to many reasons. Some of them are as follows ;

Survival Value. Every species plays an important role in maintaining an ecological balance among the living systems of the earth. These systems must continue to function if life is to survive. Loss of any species threatens the survival of several species inclusive of man. Man's own survival depends on his willingness and ability to co-operate with the living environment. Human beings turn to nature and wildlife for spiritual, psychological, physical and cultural benefits.

Scientific Value. The study of wildlife provides valuable knowledge about life processes which has helped scientists understand the functioning and behaviour of the human body. Scientists have also gained medical knowledge and discovered important medical products by studying wildlife. By studying the effects of environmental pollution on wildlife it is possible to learn how pollution affects human life.

Economic Value. Wild species of animals provide meat for food and skin for fur. The financial value of wild species is of major importance in the economics of many nations.

Beauty. Wildlife includes a diversity of species that contributes in a special way to the beauty of nature. Such beauty enriches life and increases the enjoyment of camping and outdoor recreation.

On all continents wildlife is becoming an increasingly important recreational asset and tourist attraction. They maintain the health of the living landscape and productivity of the eco-system.

Wildlife represents the result of many years of evolution and constitute an environmental heritage to the past, a world which once surrounded our ancestors. Moreover, why should man deny that animals as species have the right to live and coexist with ourselves ? We are responsible to the coming generation to leave as an inheritance an unspoiled world with intact living beauty and richness.

Apart from the aesthetical and ethical reasons, there are also educational arguments to preserve wildlife.

Reasons for Dwindle

Man in his zeal for progress fails to realize that his own existence is being jeopardised. "Progress", as he terms it, has made him contaminate air; pollute streams, lakes and oceans, drain swamps and wetlands, lowering the water table; rape prairies and forests; spray pesticides over water and land; kill animals excessively for immediate profit, or on false assumptions that they are noxious or for fashionable "sport". Man alone, though often in combination with his livestock, is destroying the natural environment.

"Civilization", as we apply the term, is not just cities, industrial developments, miles of highways, enormous airfields, and numerous other technological conveniences. By destroying wildlife for civilization and culture, we are only negating the meaning of the words.

If he does have to change habits, then he must be able to manage them properly in order to maintain a rich and varied wildlife. Man has to think of the dangerous trop he is setting up for himself human overpopulation.

It is man's ability to think that formulated the idea of conservation conservation is applied ecology and may also be regarded as bioeconomy. In other words, conservation is defined as sound preservation, management and sensible use of the available renewable natural resources. This decision of conservation is in our hands and man seems to be going in the right direction regarding this decision that holds the key to the future of our most valuable natural resource. Governments of several nations have passed laws and set aside national parks, sanctuaries and other wildlife reserves in an effort to save the dwindling wildlife. Such efforts have saved several species from becoming extinct. However, these efforts have to increase several fold, if not for a complete recovery, at least about 3/4 of the species threatened with extinction can be redeemed.

Classification of Scarce Wildlife

There are three major categories of wildlite that are facing extinction.

Endangered pecies. They face the most severe threat of extinction and require direct human protection for survival.

Threatened Species. They are abundant in certain areas but face serious dangers nevertheless. Dangers being caused due to unfavourable environmental conditions, extensive hunting, fishing. trapping or even collecting by hobbyists.

Rare Species. They have small populations. They live in protected environments and their numbers are not decreasing.

Methods of Wildlife Conservation

This depends on the source of danger to the threatened species. In many cases, wildlife can be helped by ensuring that their environment is favourable with regards to availability of food, water and shelter. This method is called Habitat Management. It involves such action as soil conservation, good forestry practices and wildlife management.

Several species have been threatened with extinction due to destruction of habitat by human activities like deforestation for farming, pollution of air, water, land, plants and animals etc. People must set aside seperate areas like sanctuaries, wildlife reserves, parks etc. In which wildlife can survive. Hunting is another major cause for the dwindle in the wildlife and this can be prevented by enforcement of strict laws that forbid or regulate such killings.

Excessive increase in the number of one species in a given area threatens not only its own endurance but also that of the other species by consuming major portion of the available food. Such numbers must be reduced by controlled hunting or by restoring its natural enemies where they have become scarce.

Sometimes species find it extremely difficult to survive in their own once favourable environment. Such species can be protected by breeding in captivity and releasing in a protected area that satisfies the conditions favourable for survival.

A species threatened with disease can be protected by sanitation measures in a habitat.

The success of wildlife conservation depends on the knowledge of the ECOLOGY of the species which involves on understanding of the way of life of the species and its relationship with its biotic and abiotic environment.

History of Wildlife Conservation

The story of the struggle for wildlife conservation is several decades old. If all started with few foresighted men who understood the role of wild animals in living landscapes and the necessity of preserving the threatened species. Encouraged by the pioneering efforts of these men several national and international organizations were formed for conservation of the most valuable of our natural resources.

The first of such organizations was formed by a group of Omithologists in 1922, the International Council for Bird Preservation (ICBP). The first post-War organization to be formed under the sponsorship of UNESCO was the International

Union For Conservation of Nature and Natural Resources (IUCN) in 1948. In 1961 the IUCN formed the first fund-raising organization, the International Wildfowl Research Bureau (IWRB). There are several other organizations like the Food and Agricultural Organization (FAO) of the United Nations, the Smithosian Institute and the Conservation Foundation both in Washington DC and the Fauna Preservation Society in London.

The Zoologists Gessellschaft in Frankfurt, West Germany, has raised important funds for conservation, particularly for Africa.

The IUCN though its Survival Service Commission (SSC), is dealing specifically with animals threatened with extinction. Their function is to collect data on endangered species throughout the world and to initiate action to prevent the extinction of these species. The SSC investigates the status and ecology of a species and advises governments and organizations. Such advice is often continued with direct action financed by WWF. The SSC has long maintained a list of threatened mammals, birds, reptiles, amphibians, fishes etc. called "The Red Data Book".

This conservation expansion has produced positive results in many parts of the world. Inspite of such efforts, the future remains uncertain for the world's wildlife. This uncertainity exists and intensifies with the increase in human population, destruction of wildlife habitats and spread of environmental pollution.

Wildlife Conservation in India

The Indian Wildlife possesses a rich and diverse fauna and is unique in having immense natural beauty in its different ecosystems. The fauna includes about 123 families of terrestrial vertebrates. An estimate puts the number of 400 mammals, 1200 species of birds, 350 species of reptiles and more than 29,70,000 species of insects.

The animals that make up the Indian fauna are Elephant, Rhinoceros, Musk deer (Kastura), Barking deer (Kotra), Spotted dear (Cheetal), Hog deer (Hoghiran), Mouse deer (Indian chevrotain), Swamp deer (Bara singha), Dancing deer (Sambur) Thamin, Kashmir stag. Four-homed antelope (Chau-singha), Indian Gazelle (Chinkara), Blue bull (Nilgai), Bison, Wild buffalo, Himalayan ibex or wild goat (Capra siberica), Wild boar, Wild ass, Nilgiri tahr, Lion, Tiger, Leopard, Striped hyena, Nilgiri langur (Ceropithecus John), Lion tailed macaque (Macaa silenus), Rhesus money (Maaca mulatha), Hanuman monkey (maaca mulatha). Hanuman monkey (Semnopithecus entellus), Indian giant squirrel (Ratufa indica), Porcupine, Pangolin, Peafowl (Pavo cristatus), Junglefowl, Patridge, Quail Great Indian bustard (Choriotes nigriceps), Duck pigeon, Sand grouse, Stroks and Egrets, Pelicans, Eagle, crane, Owl, Hombill, Crocodiles, Gharials, Lizards (Uromastic) and about 216 species of snakes.

Endangered Species of India

The IUCN maintains a Red List of threatened species. This list is updated annually. A 1986 list contained the following species that were threatened with extinction.

These species face the most serious threat of extinction and require direct human protection for survival.

Order	Family	Zoological Name	Common Name	Status	Habitat
1	2	3	4	5	6
	Class Mammalia				
Primates	Cercopithecidae	Macaca silenus	Lion tailed macaque	E	Southern Western Ghats
Lagomorpha	Leporidae	Caprolagus hispidus	Hispid Hare	E	Assam
Cetacea	Platanistidae	Platanista indi	Indus dolphin	E	India
Carnivora	Viverridae	Viverra Megaspila civettina	Malabar spotted civet	E	Southern Western Ghats
	Felidae	Panthera leo persica	Asiatic lion	E	Gir Forest
		Panthera tigris	Tiger	E	India
		Panthera uncia	Snow leopard	E	Himalayas
Proboscidae	Elephantidae	Elephas maximus	Asiatic elephant	E	South NE India

Perissodactyla	Equidae	Equus henionus khur	Indian wild ass	E	Gujarat
	Rhinoceridae	Rhinoceros unicornis	Great Indian Rhino	E	Assam North Bengal
Artiodactyla	Suidae	Sus sulvanius	Pygmy hog	E	NE India
	Cervidae	Cervus duvaucelli	Hungal	E	Terai Kashmir
		Cervus elaphus	Swamp deer	E	Central India
		hungal Cervus eldi eldi	Manipur deer	E	Manipur
	Bovidae	Bubalis bubalis	Wild asiatic Water buffalo	E	Assam Bastar
	Class Aves				
Galliformes	Phasianidae	Tragopan caboti Tragopan	Cabot's Tragopan	E	NE India
		melano cephalus	Western Tragopan	E	Western Himalayas
Gruiformass	Otididae	Ardeotis nigriceps	Great Indian Bustard	E	Rajasthan MP

(*Contd.*)

1	2	3	4	5	6
	Class Reptilia				
Testudines	Crocodylidae	Crocodylus porosus	Estuarine crocodile	E	East coast
	Garialidae	GArialis gangeticus	Gharial	E	N Indian river
	Class Insecta				
Anoplura	Haematopinidae	Haematopinus oliveri	Pygmy hog sucking louse	E	Assam

E-Endangered species.

In order to protect these species from extinction we have to know about their habits and habitats. Once we are thorough with such facts, preservation of the endangerd species can be carried out successfully. The above table shows the list of endangered species in India and a detailed description of each animal is as follows.

Macaca Silenus

Common Name – Lion Tailed Macque.

Local Name – Hind Siah Bundar, Malayalam – Nellamanthi, Tamil – Karungkorungoo, Arakkan.

Size – Head and Body-20-24 inches. Tail-10-15 inches (25-38 cms.) Females slightly smaller.

Distribution – Western Ghats from North Kamara, southwards to Kerala and Kanyakumari district, Tamilnadu.

Distinctive Characters – Great mane of long dark-grey or brownish grey hair from temple and cheek. Glossy black coat,

Habits – In habitats lonelier evergreen tropical forest belt between 2000-3500 ft. (610-1070 m). They are dark coloured, shy and seclusive in habit. They are seldom seen. Macaques are gregarious. They live in herds of 12-20 animals or more. The call of the male resembles that of the human voice – "COYEH" or the "COO" of a pigeon. Their movement through the treetops are slow and deliberate. Newborns are seen in September in South India.

Destruction of habitat and poaching have made this species an endangered species. Trees are the means of transport and form the habitat of this species. Hence afforestation and prevention of poaching are the methods to be adapted for saving this species.

Caprolagus Hispidus

Common Name—Hispid Hare of Assam Rabbit. Not much is known about this species except that they are the same size of the hare with a shorter, dark-brown tail. The fur is coarse and is made of dark brown. Bristles which are paler on the undersurface of the body. The outer side of the ear is brown. The eyes are small. Hindlegs are short and stout. Their claws are long and teeth are large. They feed on bark and roots. Usually seen in Terai, Dooars and Himalayan foothills. They are very elusive.

Panthera Tigris

Common Name—Tiger

Local Names —Hindi—Bagh, Central India—Nahar, Tamil-Poolee, Malayalam-Koduwa

Size —Average length-9 ft.—9.6 ft. (275-290 cm). Female-8 ft. 6 in (260 cms.) Average weight—400-500 lbs (180-230 kgs). Female—100 labs less (45 kgs).

Distinctive characters —Tigers have a rich-coloured, well striped, short coat. Individual variations are great.

Distribution —They are found throughout India from the Himalayas to cape Comerin except in the deserts of Punjab, Sind, and Cutch. Also found in Burma.

Habits —Tigers are found at 10,000 ft. in the Himalayas, humid evergreen forests, in dry open jungle, grassy swamps of Terai, in Sunderbans leading an amphibious life, on trees, in mud and in water. Food, shade and shelter are essential for their survival. Tigers hunt from sunset to dawn, when it is not too hot. Its prey comprises of elephant (female and young), nilgai, deer, wildpig, bears, porcupine, gaur, buffalo, panthers, other tigers and when extremely hungry feeds on fowl or fish. When game is not available, tigers attack cattle and even man. They climb trees and are excellent swimmers. Though they are heavy, they are athletic.

Mating is usually after the rains Young ones are born between February and May Male-female relationship ends when young are born : Gestation period is about 15-16 weeks, Tigresses give birth to 2 to 3 cubs per litter The cubs begin to hunt when they are 6 months old and they stay wi.h the tigress for 2 years. Sexual maturity is attained at the age of 3 years in the case of the tigress and 4 years in the case of the male.

Life span of a tiger is about 20 years. Hunting had greatly reduced the number in India. In 1970, a project was set in motion to prevent the degradation of the tiger population. This project was called "PROJECT TIGER". Nine sanctuaries were set aside as Tiger reserves for rehabilitation of the species. At the beginning of the project the number of tigers in India was less than 2000.

Panthera leo persica

Common Name	– Asiatic Lion.
Local Names	–Hindi Sher, Gujarat –Untia bagh Malayalam – Simham, Tamil–Singham.
D stinctive Characters	– Mane is lesser than of the African Lion, cost is fuller and longer tassel of hair at the end of its tail, more, pronounced tuft of hair at the elbow joints and fuller fringe of hairs on its belly. The mane is light to dark, rarely, black. Cubs are spotted or striped.
Distribution	– Once found in the whole of Northwest and Central India as far south as Namada,

Now lions are restricted to the Gir forest Kathiawar, Gujarat.

Habits — Gir forest in Junagadh district covers about 500 square miles of rugged country and is composed of mainly stunted teak trees, palas, junbal, ber, patches of bamboo with an undergrowth of thorny shrubs and bushes.

Lions hunt at dusk and rest during the day. Their roars are heard throughout the night. The prey on game and cattle.

Lions do not have a particular breeding season. They are known to mate during October and November. Young ones are born during January and February. Gestation period is about 116 days. The male stays with the family and helps to defend the young and to get food for the young. Young ones are produced at intervals of at least 18 months to 2 years. A lioness gives birth to its first litter when it is 21/2-3 years old. There may be 2, 3 or even upto 5 cubs per litter.

In 1978, the number of lions in the Gir forest was less than 200.

Panthera ucia

Common Name — Snow Leopard

Local Names — Simla—Barhal he, Kunawar – Thurwagh, Kumaon Burhel haye.

Size — Smaller than a panther with a relatively longer tail.

Head and Body-3ft —3 in-3 ft 8 in (100-110 cm). Tail-3 ft (90 in).

Distinctive Characters — The snow leopard has a short muzzle, high forehead with a vertical chin. The colour of the coat is soft grey palling to pure white on the underside. Grey is sometimes tinged with buff. Spots are unbroken and distinct on the head, nape and lower parts of the limbs. On the body they break up into larger, paler rosettes. These rosettes are less pronounced in the luxuriant winter coat. The undersides are pale white except for a few black biotches and fur. Newly born cubs are darker than the adults.

Distribution – In India, the range extends along the whole of the Himalayan chain from Kashmir to Bhutan northwards into Tibet, Central Asia and Atlais.

Habits – Not much is known about the snow leopard. Its home is the higher altitudes of Himalayas at heights of 12,000-13,000 ft. (3660-3965 m) above sea-level, above the treeline where the rock and cliffs form the mountaln sides. Lying by day, snow leopards hunt at night preying on wild sheep and goats, musk deer, hares, marmots and other rodents, or even larger birds. In summer, when the upland pastures are open to grazing, they follow their prey to these higher reaches and finding opportunity feed on domestic animals like goats, sheep or ponies. They establish territory near human settlements and feed on domestic animals. In winter, they move down with the animals to about 600 ft. (1830 m). The gestation period is about 3 months. Snow leopards give from 2-4 young per litter.

Uncias are persistently hunted for their valuable fur. Their soft colour and luxuriant beauty is scarcely rivalled.

Platanista Indi

Common Name – Indus Dolphin.

The indus dolphin was considered a separate species. Now considered conspecific with Platanista gangetic. They are found in the Ganges, Brahmaputra, Indus and other large tributories to the bases of hills. They are seen within tidal limits, but do not enter the sea. Its body oil is used for burning lamps and for various other purposes. Not much is known about this species.

Viverra Megaspila

Common Name – Malabar Large Spotted Civet.

This species was once very common in the coastal districts of Malabar and Travancore. It lied in the wooded plains and in the adjoining hill slopes. At present they are scarcely seen and appears to be nearly extinct. Nothing else is known about this species.

Elephas Maximus

Common Name · Asian Elephant

Local Names – Sanskrit · Hasti, Gaja, Hindi – Hathi (male, Hathni) (Female), Marathi – Hatti Tamil – Anai, Malayalam - Ana, Kannada – Ane

Size – Average height-Adult male-9 ft. (2.75 m). Female – A foot lower (30 cms).

Distinctive Characters — The Asian elephant is smaller than the African elephant. While the ears of the African elephant or enormous, that of the Asian elephant are smaller. The major distinctive feature is that of the hollow back of the African elephant to that of the rounded back in the Asian counterpart. Generally only the males have tusks, while females have a protrusion of a few inches or nothing at all.

Distribution – In India, the Asian elephant is found in the Western Ghats, southwards from Mysore, Orissa, Bihar, Himalayan range, almost all over India.

Habits —The Asian elephant is found in areas covered with forests. They are extremely adaptable and live in steamy, humid jungle or in cool elevated forests. They live in herds of 5-60 animals (usually single families). When fodder is scarce herds break up into smaller parties. Big tuskers feed away from the herd. Bulls lead a solitary life and joins a herd when he feels the urge to mate. Elephants sleep when it is hot and feed in the mornings and evenings and during nights invade open forests and raid crops. They raid and retire after midnight. They feed on various kinds of grasses and leaves. stem and leaves of wild bamboo and plantain, all species of crops, barks of some kinds of trees. A full grown elephant eats about 600-700 lbs (270-320 kgs) of fodder a day. The exact breeding season is uncertain, but is estimated at the beginning of the rains. Young appear in Autumn. The gestation period is about 20 months. Elephants give birth to only one calf at a time. Elephants have poor sight but their sense of hearing and smell are well developed.

Hunting and poaching for ivory has very greatly decreased the number of elephants in India.

Equus Hemionus Khur

Common Name —Indian Wild Ass.

Local Names —Hindi - Gharkar.

Size —Height at shoulder-3 ft. 8 in (110 cms) to 4 ft. (120 cms)

Distinctive Characters —The colour of the coat varies from reddish grey to fawn or pale chestnut. The erect dark brown mane is continued as a dark brown stripe extending along the back to the root of the tail, the lower being white.

Distribution -- The Indian Wild Ass is found in the deserts of the Rann of Cutch and Peninsular India. The present population is less than 2000.

Habits —They roam in groups of 10-20, or in 2 or 3's or as individuals in the scorching sun. They are capable of remaining in the open for hours on end and run for a great distance at a speed of 30-32 miles per hour. The Indian wild ass mate in the months of August, September. The gestation period is about 11 months. Young ones appear in the months of July, August, September. They are called FOALS. They are readily tamed when young. They grow vicious and untameable animals. The present population in India is below 2000.

Rhinoceros Unicornis

Common Name —Great Indian Rhinoceros.

Local Names —Hindi-Gainda, Gargadan; Marathi-Genda.

Size —Average height of male-5 ft. 8 in (170 cms) with a girth of 11 ft. (335 cms) behind the wither.

Distribution —The Great Indian Rhinoceros was formerly extensively distributed throughout the Gangetic plain, but is now restricted to the Dooars and parts of West-Bengal and Assam.

Distinctive Characters -- The skin is divided into shields by heavy folds before and behind the shoulder and in front of the thighs. The fold is not continued across the back which is a distinctive character in the Indian species. Another distinctive character is that the hom is smaller in the Indian Rhinoceros.

Habits —Rhinos prefer swamps and grasslands but are also found in wood jungles and ravines of low hills. They lead a solitary life. Rhinos feed on grass. They breed through out the year. Their Gestation period is 16 months. The young ones are born as heavy as 60 kgs and are about 105 cms in length. The female Rhino acquires sexual maturity at the age of 4 years while the male takes 7 years to attain sexual maturity.

The Indian Rhinoceros is a fast diminishing species. They are hunted for sport and their skin is used for making leather products. Their horn is supposed to have medicinal value.

In India, they are found in majority only in the Kaziranga Sanctuary in Sibsagar/Nowgong district in Assam. The Bengal Rhinoceros Preservation Act of 1932 and the Assam Rhinoceros Preservation Act of 1954 were fruitful in protecting the animal and by the end of 1978 the population of the Rhino in the Kaziranga Santuary was estimated to be around 800.

Sus Sulvanius

Common Name —Pygmy Hog.

The Pygmy hog is scarcely 10 inches high (25 cms). They inhabit the forests of the Himalayas in Sikkim, Nepal, Bhutan and Assam. Their habits are similar to that of the wild boar. They live in herds of 5-20 in number. They are nocturnal by nature and rarely seen. Hence nothing much is known about this species.

Cervus Duvauceli

Common Name —Swamp Deer.

Local Names —Hindi—Barasingha, Maha, Central India—Goinjak (male), Gaoni (female), Sal samar, Nerwari.

Size —Average height 54 in (135 cms) at the shoulder, Average weight 370-400 lbs (170-

180 kgs), Average length of the horn – 30 in. (75 cms) with a girth of 5 in. (13 cms) at mid-beam.

Distinctive Characters — The coat is almost woolly and has colour shades from brown to yellowish brown. Stags are maned and darker in colour. Young ones are spotted. The size and shape of the antlers vary.

Distribution —Swamp deers are found in Terai, Uttar Pradesh, Assam, Sunderbans, Madhya Pradesh.

Habits —They live on marsh land, seldom out of water. They are also found on grassy maidans. They are highly gregarious by nature and less nocturnal. Swamp deers feed till late morning and in the evenings, taking rest during most of the hot day. Their eyesight and hearing are moderate, but their sense of smell is well developed. They are an endangered species and strict protection is necessary to prevent their total extinction. The total number in Central India is less than 200.

Cervus Elephas Manglu

Common Name – Hangul.

Local Names – Kashmir - Hangal, Hanglu (Mate), Minyamar (Female) Hindi – Barasingha.

Size - Average height – 40-50 inches (120-125 cms) at withers. Average weight-400 lbs (180 kgs). Average length of horn-40 inches (100 cms).

Distinctive Characters —The coat is light to dark brown fading to a dinghy white on the lips, chin, underparts and buttocks. The white rump patch is divided by broad median stripe extending to the base of the tail. Fawns are spotted.

Tribution — Northern side of the Valley of Kashmir and some adjacent valleys.

Habits — The Hangul is basically a forest animal and leads a solitary life or live in small parties. They roam around in search of food. They give birth to their young ones in late May. They spend their time in summer at heights of 900-12,000 ft. (2750-3680 m). They shed their antlers before they migrate and grow new ones. This is a very regular feature among this species. In winter they descend to lower levels and feed on fresh grass. Another of the endangered species that requires direct human protection.

Cervus Eldi Eldi

Common Name — Manipur Brownanlered Deer.

Local Names — Manipur—Sanga, Burma—Tamin.

Size — Average height 4 ft. (120 cms) at withers, hinds are smaller. Average length of antler —42 inches (107 cm). Average length of head—over 35 inches (90 cms).

Distinctive Characters — The coat is coarse and sparse. The stags have a dark brown, or black coat. Hinds are light brown or fawn. Young ones are spotted. Antlers are almost erect.

Distribution — Manipur.

Habits - The Manipur Deer avoid hills and heavy forests. They prefer open scrub jungle or flat undulating land between rivers and hill ranges. They live in small herds. Their main food comprises of grass on which they feed in the mornings and evenings. Their speed and their sharp eyesight protect them from their predators. Young ones come in the month of October. Gestation period is about 239-256 days. The present number is only about 100 most of which are restricted to Keibal Lanjao Sanctuary on the shore of Logtak Lake in Manipur. This is the reason that they were given the name Manipur Deer.

Bubalis Bubalis

Common Name. – Wild Asiatic Water Buffalo.

Local Names. – Hindi– Arna (male), Arni (female), Marathi - Jungli mhais, Kannada–Kadu Kano Kartee.

Size –Average height of the Bull-5 ft. 6 in. (170 cms) at shoulder to 6 ft. Average weight –2000 lbs (900 kgs) or more.

Distinctive Characters –The wild buffalo is sleeker, heavier had more robust-looking than their tame counterpart. Their body is slaty black in colour. The newborn are light coloured, almost yellow in colour. Their horns are curved or spread outwards. The horns may be flat, triangular in cross-section, the horns of the cows are less massive and vary less in shape.

Tribution. – They are found in the grass jungles of Terai and the plains of Brahmaputra in Assam, Orissa and Bastar District of Madhya Pradesh.

Habits –Tall grass jungles with food, shelter and water to lie in, form the natural habitat of the wild buffalo. They live in small herds. They feed chiefly on grass. Feeding is done only in the mornings and evenings and sometimes in the night. During the day they lie down in the shade. Wild Buffaloes are bold and most savage by nature. They invade cultivation in search of food. Bulls may attack without warning or provocation.

A cow with a new born calf is very dangerous. The chief enemies of this species are man and tiger. Their sense of smell is well developed and they posses good hearing. They have moderate eyesight. Calves appear in tfe months of March, April or May. Wild Buffaloes sometime mate with domestic ones.

Details about only a few species have been mentioned as information about the rest has been hard pressed to find. The above facts have been mentioned only as an example as to how one can proceed with collecting data about an endangered species for the purpose of conservation.

Modes of Wildlife Conservation in India

In India, several laws have been passed and sanctuaries and national parks have been established for the protection of the dwindling wildlife. India was probably the first country to enact a Wild Life Protection Act. The Wild Birds and Animals Protection Act was passed in 1887 and repealed in 1912. For game protection in the states, in 1927, the Foreet Act XVI was enacted. Indian Board for Wild Life was established in 1952 and this was followed by setting up of Wild Life Boards in different States in India. In 1972, a new Wild Life Protection Act was passed. Under this Act, possession, trapping, shooting of wild animals alive or dead; serving their meat in eating houses : their transport and export are all controlled and watched by special staff (Chief Wild Life Warden and authorized officers). This act prevents hunting of femals and young ones. Under this Act the threatened species are absolutely protected and the rest afforded graded protection according to their state of population size.

In India, nearly 200 sanctuaries and national parks have been established for wildlife management. Sanctuaries are places where the killing and caputuring of any animal is prohibited except under orders of the authorities concerned.

National Parks are set up for preserving flora, fauna, landscapes and historic objects of an area. Some well known wildlife sanctuaries and national parks of India are :

Kaziranga Wild Life Sanctuary

It was established in 1926 in the district of Sibsagar, sub-division Jorhat (Assam) on the southern bank of the Brahmputra river. It covers an area of 430 sq. kms of forest grasslands and swamps and supports a fauna of 700 Rhinoceros, besides elephant, wild buffalo, bison, tiger, leopard, sloth bear, sambhar, swamp deer, hog deer, barking deer, wild boar, gibbon and birds like pelican, stork and ring-tailed fishing eagles.

Manas Wild Life Sanctuary and Tiger Reserve

It is located in the district of Kamrup in Assam, covering an area of 540 sq. kms. and is situated at an altitude of 80 metres. The River Manas runs through this sanctuary. It contains the following wild animals : tiger, panther, wild dog, wild boar, rhinoceros, gaur, wild buffalo, sambhar, swamp deer and golden langur.

Jaldapara Wild Life Sanctuary

It is situated in theJ alpaiguri district of West Bengal and covers an area of 65 sq. kms of grassland. Wild life fauna includes animals like rhinoceros gaur, elephant, tiger, leopard, deer and a variety of birds and reptiles.

Palamau National Park

It is situated in the Daltongunj district of Bihar and has an area of 345 sq kms. The fauna of this include tiger, panther, sloth bear, elephant, chital, gaur, nilgai, tiger, chinkara, chowsingha and mouse deer. The Flora is thick tropical forests.

Hazaribagh National Park

This national park was established in 1954 in Bihar. It has an area of 184 sq km of thick tropical forests. The typical fauna of this park includes wild boar, sambhar, nilgai, tiger, leopard, sloth bear, hyena and gaur.

Simlipal National Park

It is situated in the district of Mayuri Bhanj in Orissa and has an area of 2750 sq. kms. It is covered with dense sal forests and has been chosen for Project Tiger. Its typical fauna includes tiger, elephant, deer, pea fowl, talking mynas, chital, sambhar, panther, gaur, hyena, and sloth bear.

Chilka Lake

This largest inland lake has an area of 1000 sq. kms. and is about 1000 kms from Bhubaneshwar (Orissa). Its fauna includes waterfowl, duck, cranes, ospreys, golden plover, sandpiper, stone curlews, flamingoes, etc.

Kolameru Bird Sanctuary

This is a small bird sanctuary near Tadepallegudam in Andhra state. It is a breeding place for pelicans and many marine birds visit this place.

Vedanthangal Bird Sanctuary

This small-sized but very old sanctuary extends over a lake of about 0.30 sq. kms, 85 kms from Madras. Many migratory birds regularly visits thist emporary lake. The regular visitors of this lake are birds like spoon bills, open billed storks, egrets, ibis, cormorant, darter, grey heron, pelicans, snipes and dab chicks.

Guindy National Park

It is situated near Madras city and has mainly chitals and black bucks. A few albinos of black buck are also found here.

Point Calimer Wild Life Sanctuary

This is situated at the southern tip of the Thanjavur district of Tamil Nadu abutting the Palk Strait. Its back water and lagoon is visited by flamingoes and pelicans. It is close to the Vedaranyam forests which has a fauna of numerous black bucks chitals and wild boars.

Mundanthural Sanctuary

It is situated in the Tirunelveli district of Tamil Nadu and was established in 1964. It covers an area of 520 sq kms. The Tamaraparani river flows through the evergreen forests of the sanctuary. Panther, tiger, sambhar and chital form the major fauna of this sanctuary.

Periyar Wild Life Sanctuary

Situated in the state of Kerala, this sanctuary covers an area of 777 sq. kms. It was established in 1949 around the artificial lake which arose behind the dam built across the Periyar river in 1900. This sanctuary of great scenic beauty supports a fauna of wild elephants, gaurs, leopards, sloth bear, sambhars, barking deer, wild dogs, wild boars, black Nilgiri langur and water birds like grey hornbills, egrets.

Mudumalai Wild Life Sanctuary

This wildlife sanctuary was established in 1940 in North Western part of Nilgiris in Tamil Nadu. It is known for its rich forests and diversity of fauna that includes wild elephants, gaur, sambhar, chital, barking deer, mouse deer, four-horned antelope, tiger, panther, bonnet monkey, langur, giant squirrel, flying squirrel, wild dog, jackal, wild cat, sloth bear, porcupine, pangolin, flying-lizard, monitorlizard, rat sanke, python and various birds.

Ranganthitto Bird Sanctuary

It covers an area of 166 sq kms and includes a series of islands in the Cauvery river 15 kms of the Bangalore-Mysore road, near Shrirangapatnam. Its avial fauna includes open bill stork, white ibis, egret, spoon bill, wild duck, pea fowl, night heron etc.

Bandipur Wild Life Sanctuary

This sanctuary was established in 1941 by the then ruler of Mysore (Karnataka State). It is situated 80 kms south of

Mysore city enroute to Ootacamund. It has an area of 874 sq. kms and is at an altitude of 1454.4 metres. Its forest is very thick and has plenty of rain fall. Its wild life fauna includes plenty of gaurs and animals like elephant, leopard, sloth, bear, wild dog, chital, panther, barking deer, porcupine and langur.

Cotigo Wild Life Sanctuary

It is located in South Goa and has an area of 105 sq. kms. It has wet evergreen forest of Elve and Berla trees and supports rich fauna of the following animals, gaur, sambhar, chital, hog-deer, barking deer wild boar, hyaena, panther, leopard, jackal, otter, porcupine, and birds like parakeet lorokeet, woodpecker, kingfisher, bulbul jungle fowl, egret etc.

Bhagwan Mahadev Wild Life Sanctuary

It is located in North Goa and has an area of 240 sq kms and supports fauna almost similar to Cotigao Wild Life Sanctuary.

Sesan Gir

This famous wild life sanctuary for the Asiatic lion is situated in Gujarat state, 468 kms from Ahmedabad and 43 kms from Veraval. It has an area of 1295 km sq of semiarid country with patches of thorn scrub and deciduous trees. Its fauna includes Asiatic lion, spotted deer, blue bull (nilgai), fur-horned antelope, chinkara, striped hyena, wild boar, porcupine, langur, python, crocodiles and birds like green pigeon, partridge, rock-grouse etc.

Kanha National Park

This national park was established in 1955 in former Banjar Valley Reserve (Madhya Pradesh). This park has an area of 939.94 sq. kms and includes hilly terrain ond streams. It is 175 kms away from Jabalpur and has forests of sal trees. Its typical fauna includes animals like tiger, chital, panther, sambhar, black buck and barasingha.

Tandoba National Park

It is located in Chandrapur district (Maharashtra) and has an area of 116 sq. kms. Its fauna include tiger, sambhar, sloth bear, bison, chital, chinkara, barking deer, blue bull, four-horned deer, langur, peafowl and few crocodiles.

Sariska

This is one of the most beautiful sanctuaries situated in the state of Rajasthan near Alwar. It has an area of 800 sq kms. and has dense Dohokara and Solar forests. Its fauna include tiger, leopard, spotted deer, jungle cat, four-horned antelope, langur, porcupine, hedgehog, peafowl etc.

Bharatpur Bird Sanctuary (Koeideo Ghana)

It is located in Bharatpur district in Rajasthan, has an area of 29 sq. kms and harbours all kinds of indigenous birds like nesting water birds, water side birds and migratory birds. More than 328 varieties of birds including cormorants, spoonbills, white ibis, Indian darters, egrets, painted storks, open billed storks, great black necked storks, etc. Many migratory birds like ducks, geese, siberian cranes, etc. regularly visit this sanctuary. Drier parts of this marshy sanctuary have animals like spotted deer, black buck, sambhar, blue bull, wild boar and python.

Sultanpur Lake Bird Sanctuary

This small sized (2 sq kms) bird sanctuary is located in Gurgaon district (Haryana) about 30 kms from Delhi. Its typical avian fauna includes crane, sarus, spot-bill, rudyshell and drake.

Shikari Devi Wild Life Sanctuary

It is located in Mandi district in Himachal Pradesh. It has an area of 213 sq kms and has the following animals : snow leopard, flying fox, black bear, barking deer, musk deer, chakor, partridge, etc.

Bir Motibagh Sanctuary

It is located near Patiala in Punjab. The fauna includes black buck, blue bull, hog deer, hare, jackal and birds like peafowl, partridge, sparrow, babller, myna, parakeet, pigeon, dove, etc.

Dachigam Wild Life Sanctuary

It was established in 1951 in Kashmir, 26 kms away from Srinagar. It has an area of 89 sq kms and has two levels : upper Dachigam at 3692.3 metres altitudes and Lower Dachigam at 1846.2 metres altitude. It mainly preserves hangul or Kashmir stag, musk deer, leopard, black buck, black bear, brown bear and babboon.

Corbett National Park

It is one of India's most famous wild life sanctuaries and was constituted in 1935 as the first national park of India. It is situated between Nainital and Gharwal districts in Uttar Pradesh. It has an area of 525 sq kms and is located within west-south bend of the river Ramganga. It supports a rich and diverse fauna of the following : tiger, panther, sloth bear, hyaena elephant, blue bull, swamp deer, barking deer, Indian antelope, porcupine, birds like bulbul wood pecker, barbet, babbler, and bee eater, and reptiles like crocodile, python, etc.

Shivpuri Sanctuary

It is located in Madhya Pradesh and is an asylum for tigers.

Annamalai Sanctuary

This sanctuary was established in 1972 in the southern part of Coimbatore district in Tamil Nadu. It has an area of 958 sq. kms and supports rich fauna of animals like elephant, gaur, sambhar, spotted deer, barking deer, nilgai, Nilgiri langur, Nilgiri tahr, lion-tailed macaque, tiger, panther, sloth bear, porcupine and pangolin.

These are some of the major sanctuaries in India. There are several smaller sanctuaries and national parks that house some of the important species. A sanctuary or a national park may be large or small, but planning should be based on a scientific background. To provide the basic needs for the wild life the following steps may be taken as far as possible for conservational purposes.

1. The forest inside the sanctuary/national park should be left unexploited as far as possible. It would be an ideal environment for the wild life to get the fullest measure of protection, and in such preserved conditions of life they can multiply freely.

2. If total unexploitation is not possible, the cutting operations of the forest should be in small units, well-distributed over the forest. Frequent light cuts are generally better for wild life than infrequent heavy cuts.

3. The forest should not be planted with or allowed to grow only a single species of plant. Further, fruit-bearing shrubs and trees should be encouraged wherever possible and the natural openings of the forests should be preserved.

4. Poaching needs to be ruthlessly prohibited in the sanctuaries in particular and generally in other areas outside them, where wildlife still exists. Mobile armed units should be attached to all sanctuaries to patrol regularly the entire sanctuary to stop illegal activities.

5. The water holes and stream sides should be managed and guarded carefully for the benefit of wildlife. Such critical areas may be of small size, but their importance is disproportionately great because they supply important elements in wildlife ecology for larger areas during hot months of the year.

6. Annual burning of the grassland in most of our sanctuaries has become an integral part of management, since not burning these might bring about a lack of suitable fodder during the hotter months. Controlled burning has been found to increase forage and to preserve

organic material in the soil Such burning may be done in blocks with unburned areas in between, so that areas are burnt in rotation. The burning is to be confined to definite blocks by means of five-breaks. Burning at night is favoured unless it is likely to extend the fire beyond the desired limits.

7. Provision for dust bath fot animals should also be made by leaving some exposed soil for their dusting. These spots may be prepared in the form of small mounds elevated a few inches from the forest floor to allow for drainage and to dry quickly in wet months.

8. Some artificial salt-licks are also to be provided in the sanctuaries since the wildlife need them at regular intervals for their normal health. Such salt-licks are necessary for animals especially in areas where natural salted clay is scanty.

9. Concentrated grazing by domestic live-stock is to be prevented as wildlife can never compete with domestic stock under ordinary circumstances and domestic animals may be responsible for transmitting diseases like foot and mouth diseases, rinderpest, surra (sleeping sickness), heamorrhagic sepicaemia, arthrax etc.

10. Cultivation near the sanctuary should be avoided as a rule as it may have an indirect effect on the wildlife. The chemical control of certain pests is often desirable for better yield in agriculture, but these pesticides can present ecological hazards for wildlife.

11. Scientific studies of the sanctuary wildlife particularly of threatened species, by qualified persons, should be encouraged in order to improve the status of the animals even by breeding them in captivity to rehabilitate them in suitable habitats.

Most wildlife could be saved by sufficiently rigid protection of their natural habitat, combined with controlled exploitation of species which are of economic importance.

Some wildlife may be driven away from their natural habitats so that their requirements are not in conflict with human interests. The only possible course for maintaining their viable number is in national parks, sanctuaries, game reserves, where they could thrive unharmed.

In the case of the most endangered species which have reached a poins of extinction in their favourable habitat, their survival may be ensured through the maintenance of breeding stock in zoological parks in suitable parts of the country.

Man, An Endangered Species

"Man is part of the vast web of life and cannot escape the natural consequences of his actions ... The sheer power of the population and technological revolutions may make man himself an endangered species in many parts of the world".

—*Stewart Udall*

The present day presents Man as the most successful species in the world. The ability to think, a sixth sense, is what places the species of Homo sapiens at the top of the ladder. It is this ability that has assisted Man to become the most dominant form of life on earth. In the game of survival, Man has come out in flying colours, this being possible only by his dependence on the environment around him. Initially, this progress of Man was very slow, but with time it has increased to a dizzying pace, this being made possible by the explosion of scientific knowledge and its application to every phase of modern life.

Man's dream for the future is machine oriented civilization. This coupied with his greed is definitely going to have an adverse effect on his natural environment, soul and spirit.

An endangered species is one whose population is decreasing In Man's case it is the ever increasing population that is making him an endangered, species. There is a general belief - more people more technology—which if followed blindly, as is being done, is to flirt with ultimate disaster.

The result of progress and technological advancement is the pollution of water, land, air, removal of forests and other natural resources beyond redemption. All these activities tilt the ecological balance on which all life depends.

Each species is a part of a food chain and man is no exemption. Thus preservation of wildlife forms a major decision in the fate of man Man's attitude towards nature only suggests that he is not ready to believe this fact. Unless this attitude changes, all one can decipher is that man is paving roads leading of extinction.

President Johnson of the United States of America in an address to the Congress in 1968 observed that – "Conservation's concern now is not only for Man's enjoyment but for Man's survival". Modern Man, if he is to survive has to improve his relationship with nature. If he succeeds– he must, he will – all wildlife too will have a better chance.

The Future

The future of our wildlife looks very bleak. If wildlife is not preserved then this world would be a paradise lost, to put it in better terms, a paradise destroyed. As Man's life becomes more and more urbanized, mechanised and machinised, the habitat for wildlife is being devoured and wild animals are losing their place on our globe. Preservation of habitat is the most important criteria in conservation of wildlife.

With the ever increasing human population, the Idea of conservation can be scarcely brought about. It is a habit for nature to eliminate anything that exceeds a limit and Man, the most powerful and highly developed of living creatures, fails to recognize the needs for his own existence in the future.

Nature and wildlife are of definite economic value and components of an environment of which Man is a part and on which he is directly or indirectly dependent. Man enjoys music, poetry and paintings, all he has to realize is that the flora and fauna are the music, poetry and paintings created by nature. Even more, they form a part of the food chain which is a guarantee for the continuity of life on earth.

Very few individuals in this world have time for recreation, leisure or intellectual interests. The rest of them think wildlife conservation is a luxury. Each of them is too busy trying his own problems for him to even consider the idea of conservation. Starving illiterates have time to think only of ways to earn their next meal and not about conservation, which for them is non-functional luxury.

Human overpopulation and illiteracy are two of the major problems that bring about the tragic destruction of environment and animal exterminations in the world. For every extinct or endangered species or subspecies of animals, two of the major factors have been the destruction of habitats and excessive cropping by man. Connected with this depletion is the lack of knowledge about the essential needs of a species for survival, its needs for habitat, and its reproductive capacity.

Several organizations are painstakingly trying to bring about a successful turn-about regarding the concept of conservation. They educate and advise on the need for conservation of our depleting natural resources. All these organizations are run by men who are devoted and work their hearts out for saving habitats from destruction and species from extinction throughout the world. They fight against social, economic, political and religious problems in different parts of the world.

It all looks like a hopeful sign that the future of the world's wildlife is of growing concern to more and more human beings. But this is a race against time. Will people and governments be aware of this before too much is lost ?

The future of world's life is still a big question ?

18

Natural Environment

Natural Environment

The perception that man lives in close relationship with all living things, plants and animals, is quite old. Ancient and medieval philosophere have often emphasized the inter-relatedness of the entire living world. Recent developments in science have not only confirmed this view but have gone beyong to assert the inter-dependence of all life, microbial, plant and animal, between themselves and with the physical environment of land, air, water and solar energy. This inter-dependence is embedded in the evolution of life during the past three billion years or more with the origin of various forms of life, the extinction of others, and the formation of our atmosphere, water bodies and land masses. These idea have been sharpened in the last nearly fifty years by the concept of ecosystem (1935) which defines an area in which all living creatures and the soil-water-air environment of this physical region of the earth have developed various inter-relations to produce a quasistable or almost stable system of mutual inter-dependence. The science of ecology is the study of such systems, their inter-relationships and normal slow secular changes of such systems.

There is an even broader idea proposed by some scientists like Lovelock who would like to think of the entire biosphere as a single ecological system which he called Gaiea. Conceptually, this view is reasonably and exciting and recalls the Jain concept of 'Prakriti'. However, the current scientific methods of study are unable to deal with so large and complex a total system. Ecologists, therefore, usually define much smaller systems which are amenable to study, investigation and analysis. Some of these systems may be relatively self contained and small enough to enable a scientific study to be pursued credibly. A forest, a lake, a vally or a river basin are examples of almost self-contained ecosystems. Normally the migration into almost self-contained systems, where it takes place, is restricted both in the numbers of intruding species and in the absolute number of such species. There are other systems which may be more open. The analysis of such open systems is more difficult. Examples of such open systems are oceans and large open plains such as the Indo-Gangetic plains However, certain biogeochemical processes are almost world-wide and their cycles represent and essential part of natural processes occurring within the more restricted area defined by an ecological system. Examples of such world-wide geobiochemical systems are the carbon cycle, the nitrogen cycle (Fig. 1) and the water or hydrological cycle (Fig. 2). On a minor scale, but essential to life support systems, are phosphorus and sulphur biogeochemical cycles.

A new dimension has been begun to appear in the last few years as the impact of man and his activities on nature came to be noted. The word 'environment' was given an additional meaning since about 1964 to express this broader concept of interaction of human activities on the ecological system and the impact of the ecosystem changes on man himself.

The impact of human activity on our physical and biological environment has a consequential effect on man and his society due to the changes in the environment. Illustrations of such changes occur in the archaeological records, such as the gradual desertification of parts of Sind and Rajasthan which one sustained larger populations and greater agricultural activity as well as more extensive grasslands. A dramatic evidence of such reciprocal interaction in our times is the discovery of mercury poisoning cases in some areas of Japan through eating of fish contaminated by mercury. For a few years the effect, or disease, in man was not even diagnosed as mercury posioning but as a new illness. Later, it was traced to the mercury present in the fishes. Normally, mercury is only slightly toxic. One can ingest a few grammes of mercury without servere effects. However, research revealed that the high toxicity was due to poisonous compoinds of mercury called methyl mercury $(CH_3)_2$ $(Hg_2$ and dimethyi mereury $(CH_3)_2$ Hg_2. These were being produced by certain anaerobic micro organisms in the sea bottom of shallow areas off the east coast of Japan and were

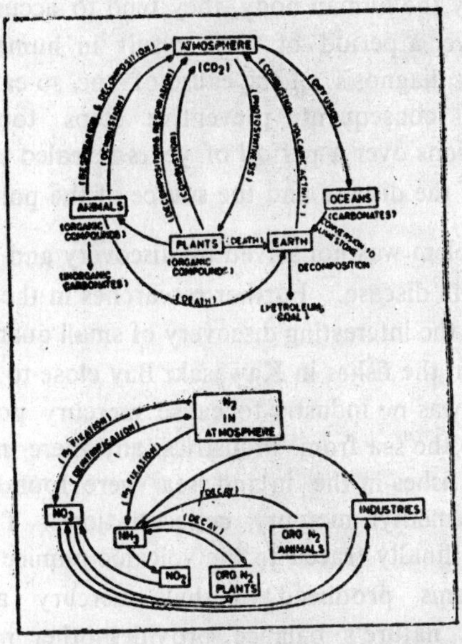

Fig. 1. Carbon and Nitrogen cycles. Combution refers to oxidation of carbon in forms of petroleum, coal, wood etc. and to respiration in animals. Reference to water is chiefly to that in oceans. Nitrogen fixation may be natural from atmospheric discharges and nitrogen fixing organisms, or man-made in chemical factories.

being ingested by fishes along with sea weeds which constitute their normal diet. The Physiology of fish was such that the poisonous mercury compounds tended to accumulate in some of the organs of the fish. The source of mercury was a product of industrial activity of pulp and paper mills which discharged small amounts of mercury in their effluents The entire marine and human food chain is complex in which the microorganisms of the sea bottom play an essential role to convert mercury and to link it with the food chains since the microorgansism enter the marine food chain and work upwards to the edible marine fishes and ultimately to the human population who eat these fishes (Fig. 3). Since methylated mercurys are not easily eliminated by the human body, they tend to accumulate in the body, and over a period of time result in human illness and death. The diagnosis of the cause of the so-called Minamata disease and consequent preventive steps took some time as investigations over a period of years revealed the true nature and cause of the disease and the source of the pollutants.

The problem was not solved by discovery and prevention of the Minamata disease. Further researches in the Inland sea of Japan led to the interesting discovery of small quantities of methyl mercury in the fishes in Kawasaki Bay close to the inland sea where there was no industry to cause mercury pollutants being dumped into the sea from industries anywhere near the area. Further, the fishes in the inland sea were found generally to have lower methyl mercury concentrations. The source of mercury was finally traced to the volcanic pumice in the seabed. Microorganisms produced methyl mercury and dimethyl mercury but nature's balance provides other microorganisms which break down the methyl and dimethy mercury to mercury so that the concentration of the methyl and dimethyl mercury remains law. Recently the presence of mercury has been found in the fishes in the gulf of Cambay and in the coastal fishes in certain parts of the Bombay coast. The source of the mercury is still unidentified.

Less dramatic but, equally if not, more disastrous is the erosion of topsoil in the submontane slopes by floods and landslides preceded by and due to loss of vegetative cover resulting from tree falling (for fuel and logging) and loss in the process is estimated to be 105 mha in India. We are increasingly conscious of problems due to our paper mills effluents in many areas of the country. The acid sulphite process to separate the lignin from cellulose leads to formation of the lignosulphonates. These lignosulphonates are less biodegradable than the lignins and more toxic. Their presentence in water affects not only fish and aquatic fauna but also downstream users of water because of smell, toxicity, colour and decreased utility for human and agricultural use. Pesticides like DDT have been found in human beings in autopsies in Delhi and Bombay and in animals at several places because of their entry into the food chain in a variety of ways.

There are other well-known cases like there lease of a toxic chemical (dioxin) from a chemical factory by an explosion in the factory in Seveso near Milan in Italy and its entry into the surrounding country and subsequently ingested by men and animals in the neighbourhood. More recently in Michigan (polybrominated biphenyl) in bags normally used for packing entry of these compounds into the food chain through cattle

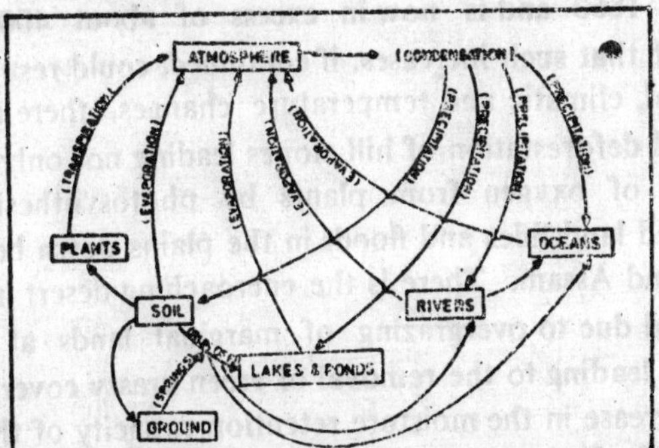

Fig. 2. Water Cycle. The major movement of water is evaporation to the atmosphere and return by precipitation. Apart from the springs, ground water is also brought to the surface by human activity. Rover water also perctolates to the ground directly orthrough irrigatton canals.

feed affecting both cattle and human beings. It was first neurological disorders and skin diseases in man were traced to

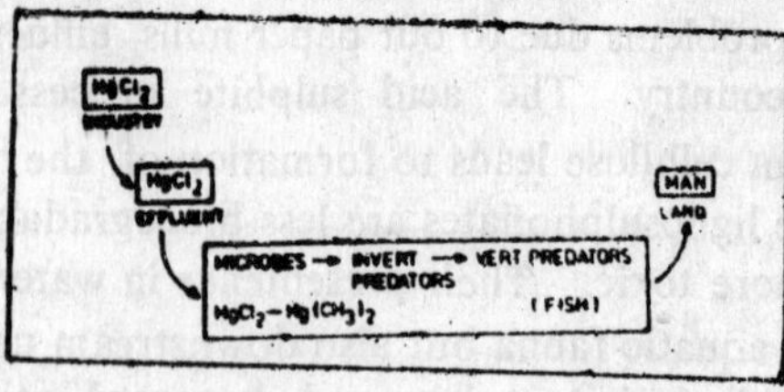

Fig. 3. Mercury in the food chain.

the presence of the chemical. In Delhi and Punjab, DDT has been found in cattle and human beings, children and young people tending to show greater amounts their organs than older people.

There are many important activities of man apart from such accidents, industrial effluents and wastes, which have an impact on man or on the ecological system as a whole. The large scale burning of coal and petroleum as fuel for energy produces vast amounts of carbondioxide (CO_2). There has been a small but significant increase in CO_2 w 600 ppm (parts per million) around 1860 and is now in excess of about 400 ppm. It is thought that such increases, if continued, could result eventually in global, climatic and temperature changes, there are instances of rapid deforestation of hill slopes leading not only to reduced release of oxygen from plants by photosynthesis but also increased landslides and floods in the plains down below in U.P. Bihar and Assam. There is the encroaching desert in Rajasthan and Sind due to overgrazing of marginal lands at the edge of deserts, leading to the removal of green grassy cover of the land and decrease in the moisture retention capacity of the soil. This has resulted in decrease in rainfall. Consequently the land once marginally habitable has become deserified.

Technological activities of man have increased energy consumption Coal and petroleum are the main sources of energy today. In mining coal and extracting other minerals from under the earth, there is often subsidence of land and holes and gashes on the surface of the earth leading to the accumulation of water in these areas as well as changes of water courses and reunoff affecting the ecology of these areas. It also gives rise to many environmental problems by local flooding, abandoned, water logged lands and increase in insect and water-borne diseases. The impact on the health of the community or of ugliness of the countryside is not always quantifiable.

The excessive use of fertilizers and pesticides contributes substantially to the ecological disturbances and environmental pollution. Excessive use of nitrogen fertilizers, for example, can lead to leaching of these excess fertilizers into the water bodies.

These can be transformed by microorganisms into the nitrites and carcinogens such as nitrosoamines and can find their way back to men or animals through the food chain.

Excessive fertilizers can also adversely affect the nitrogen fixing organisms or microorganisms in the soil and reduce the natural nitrogen absorption and production of soils, similarly pesticides, besides killing the unwanted pests. insects or birds and can enter the human system through food or contact and become a health hazard. They can also indirectly come into the human system through animals, which ingest the plants. Other human activities also cause various ecological problems; for instance, over gazing in a grassland can denude it to the extent that it is gradually transformed into a desert. In Rajasthan, it is estimated that the desert in certain areas is advancing at the rate of half a kilometer a year.

Many of the large-scale human activities-industrial, agricultural or transport-tend to alter the natural balance of living organisms in an ecological system. Such a disturbance in the ecology of the system can affect human societies in a number of ways both in the immediate and in the long term. A grass land can be denuded and gradually transformed into a desert. Denudation of forest and grassland vegetation can cause long-term climate changes by reducing rainfall and thus perpetuate and strengthen the desertification process over decades and sometimes over centuries or millenia. The courses of streams and rivers. If can also destroy the ability of trees, but the soil and humus which support underbrush, microorganisms, insects and animals which live in various symbiotic relationships with each other and the larg plants and trees. Once the forest is gone, these animals, plants and other organisms are also destroyed, making regeneration difficult or impossible. Larger trees and animals tend to be more valuable and are often the first to disappear.

Man is largely, if not wholly dependent on his environment for the resources that sustain him and give him various products. In extracting these resources whether by way of agriculture or minerals or obtaining energy resources, the topsoil may be overexploited leading to dustbowl-like conditions or so scarred as to be unfit for use. Forests may disappear if trees are felled faster than replacement. The leads not only to denudation of topsoils but there may well be no fuel for the poor and no paper for the schools. One of the primary concerns of environmental management is the preservation of our natural resource base base and wherever possible enhancing further its productivity in the long term. Unfortunately, the rapidly increasing population of our land has already doubled since independence. It is likely to exceed 100 crores by the end of this century. This large population needs food, energy shelter and other amenities of life and in increasing amounts. The need of land is roughly proportional to the population if other things remain the same. However, the 150 million hectares of current agricultural land is not likely to increase very much. On the contrary, urbanization, developmental projects, roads, etc., are going to encroach more and more on agricultural land and such land will be increasingly precious.

In various ways, therefore, the ecological systems and the human environment are closely connected. These linkages tend to confuse us sometimes, because the words are often used synonymously. Environment in our context will be used specifically from the human point of view. In some ways this leads to a larger concept which also takes into account human interactions with an ecological system and the mutual effects of human activities on nature and nature on man. When primitive man lived as a part of the ecological system, he was one of the many species of animals which interacted with one another and with the physical world. Civilized man has gradually, through the use of technology, transcended these symbiotic relationships to a considerable extent. This his given man a singular ability to interfere with or even damage nature, not always with wisdom or with advantage to his own long-term future. We have, however, become increasingly sensitive through the developments of modern sciences of biology and ecology that ultimately man is extremely dependent on nature for his own survival and vulnerable to ecological changes. In a sense we have rediscovered the ancient wisdom of our land, the relation between Purusha and Prakriti. If the ecological system is destroyed or seriously disturbed, this has the potentiality of jeopardizing human welfare, or if damage is great, of destroying human civilization. This recent awareness has, of course, been he heightened by a persistent strain of philosophical awareness of on intimate relationship between man and naturn in all societies.

Profound alternations of our environment are now being caused by the growth of modern technology, particularly during the last 100 years and at increasingly accelerating rates at present. Such alterations of our environment have been speeded up by the singular growth of human population at an almost exponential rate over the last 250 or 300 years, doubling almost every twenty-five years or thirty years, apart from population growth the other singular factor of our times is the many-pronged development of technology and the expectations that this has created in the process of technoeconomic development of all societies including our country. The expectations of our people call for progressively greater availability of goods and services through processes of development. However, development requires utilization of our natural resources. If the use of these resources degrades the remaining resources for future use and depletes them in a manner that these are lost to future generations, we and our governments can be accused of being shortsighted, selfish and greedy. The activities of man are thus compounded by factors of increasing technological ability, energy and material consumption and large population increases. The environmental impact of such increased activities is large and is increasingly global in character. We should be concerned not only about the disturbances to the ecological systems, but increasingly concerned seriously, in the long term, about the future of our planet. Man has now the capacity to destroy the entire living systems of the world for the first time in human history.

The close interdependencies between resources, development, population and environment seem obvious on examination. Unfortunately, quite often in the process of development, planning and the utilization of our resources wer tend to ignore the factors of environmental risks and population pressures that are involved in the developmental process. There are other social factors, neither articulated clearly nor taken into account, such as discounted resource base estimation which give a framework for calculating what is happening to our resource base with time.

Today not only do all human activities, agricultural, transport or industrial, affect human perceptions of the environment but they also have their impact on the ecological system and the resource base. The four-point focus of population, resources, development and environment gives rise to a complexity of relationships which both the development planner and the environmental manager has to take into account. However, this is not usually enought since human physiological and psychological factors such as helath, aesthetic pleasure, scientific curiosity and technological skills are important for progress of a society for caring for the environment and building a sense of purpose and satisfaction in a society. Human perceptions include aesthetic perceptions and those of human helath. Human beings and human societies have evolved a certain appreciation and communion with nature, forests and waterfalls, sunsets and seashores have fascinated man from time immemorial. The preservation of these centres of communion with nature and of nature's beauty and harmony give great satisfaction to us. Such pleasure and satisfaction cannot be measured in economic terms or in terms of technological achievement. Man likes, at least occasinally, to live in forests, hillsides and on seashores. Any society would be resentful if these are spoiled and rendered unusable, natural beauty spots are easy to destroy and they are difficult and costly, and often impossible, to regenerate. The beauty and the harmony of nature that we perceive is almost always related to the harmony and stability that exist in an ecological system. The inner harmony of an ecological system is not always visible, the outward manifestation of it is often the perceived beauty and the sense of communion which has evolved as a concrete aspect of the impact of the ecological system on man is that of health. Man's physical as well as mental health appears to be connected to his symbiotic relationship with the ecological system in which he lives. Some scientists have theorized that disease is a symptom of maladjustment with the ecological system. Man certainly protected himself from disease in a variety of ways before historical times. These have become increasingly complex and sophisticated with our increasint knowledge of nature.

The concepts of our medical system have changed from that of treatment of diseases to their prevention. Prevention of disease is often thought of as the maintenance of an equilibrium relationship of man with his environment and not as a creation of completely lifeless or sterile surroundings for human beings. In fact it is neither desirable nor possible for human beings to live in a completely abiotic system, the concept of a relationship which enables man to remain healthy without forcing him to live in a sterile atmosphere has strong support in our current knowledge of the living world.

As human population continues to increase in the next few decades, the need of adjusting ourselves to nature and of not disturbing ecological systems significantly will become even more necessary. The impact of human activities of the environment has to be taken into account and where necessary, ameliorated by suitable measures, so that human activities do not become selfdefeating in their not being able to create human well-being which was intended. Under increasing material and energy use man's relationships with his environment will become exceedingly important and complex. Such use of materials and energy will usually lead to stresses in our environment. Societies and governments will have amount of social and self-discipline that will prevent serious stesses and distoritions of the environment which might jeopardize the well being of their children and grandchildren. Quite often it might mean that a society would have to guard against depredation of ecological systems by groups or individuals, which either in the long term or even in the short term might affect humam societies to their detriment. A methodology increasingly advocated at the national level is the laying down of a comprehensive environmental policy. Such a policy will necessarily have to be updated in the light of new knowledge and changing perceptions and values. The policy will have to be carried out in terms of actions and resources allocated for activities. Its execution starts with the education of children and adults and includes monitoring, investigation, research, enactment and enforcement of laws and encouragement of social and scientific groups which concern themselves with environmental questions. The total package of policy, law, education and activities is the task of environmental organizations. Environmental management can exist at several levels. Briefly, they can be institutional, industrial, civic, co-operative and government at local, state and federal levels, depending on the nature and size of the task or the issues.

Man increasingly lives in large agglomerations in the urban settlement or small ones in rural areas. We call them, in current vocabulary, human settlements. Human settlements are usually small and population densities low in rural, agricultural, fishing or forest areas mainly because the amount of areas needed per person to sustain a population is larger. On the other hand, human beings need each other for cultural, social and economic purposes and are thus drawn together to live in proximity. Human settlements, originally drawn together for economic, social and protective purposes created civilization and culture, philosophy and education, sanitation and medicine, communication and transport. As human settlements expanded these elements developed and assumed their current character. However, now more so than in the past, human settlements which are too small find difficulty in sustaining education, health and economic activities because there are not enough people to cover the wide range of activities that a modern society needs. On the other hand, beyond a certain population size, large groups of people in rural, agricultural or forest areas are counter-productive because the land needed to support their needs of food, clothing and shelter increase beyond what the area can yield.

There tends to be an optimum population matched to a resource base and a state of technology above which poverty conditions begin increasingly to appear, similar arguments apply to the growth and limits of urban settlements which, although larger, also tend to be sustained not only by food, shelter and clothing but additionally by industry, trade, crafts, education, government activities, etc. Ancient urban centres, Thebes, Ayodhya, Athens, Rome or medieval cities of Delhi, London and Paris tended to grow but when they reached a certain population size depending on the state of technology and the resource are under the command of the city, they ceased to grow and sometimes they declined.

The growth of modern technology and industry has increasingly helped to push up the population in urban settlements. In urban areas the dynamics of population increase under the impact of modern technology in the last two centuries have been to encourage large numbers of people to live close to industries and factorries and to each other. In the last hundred years or less this trend has been served equally effectively by people tying themselves to transport systems. Industries and factories have tended to be created at convenient points where power, raw materials, transport and communications were easily available. This has led many industries to locate themselves close to each other, both for access to raw materials and transport and for the mutually supporting roles that industries tend to have with each other. The dynamics of industrialization and population increase have led to the growth of the large modern urban complexes. These have been extensively studied in recent years, the urban complexes are characterized by large per capita consumption of energy, the energy is required for the industries, for the transport of people and for providing food, domestic energy and other amenities to this large and closely knit population. The distinctive character, therefore, of urban complexes is high density of population, high concentration of energy consumption and a high quantity of wastes and effluents that both industries and the large populations produce. While energy and material requirements for rural societies are at lower densities, transport, health and other requirements may have to be provided at equivalent or higher costs.

The two situations—urban and rural—are thus completely different. The urban settlement is sometimes called urban ecosystem, in which natural linkages have been significantly or largely destroyed or perverted and a new system of social and economic linkages have developed. The urban complex as a consequence has developed new sociological, psychological and technological parameters. The urban settlements are thus not only artificial in the sense that they are man-made, but they have also a different set of interactions and dynamics and do not fall stricty within the but they have also a different set of interactions and dynamics and do not fall strictly within the definition of ecological systems. Urban ecosystem is, however, an increasingly favourite terminology to describe the urban systems. The rural complexes on the other hand, still remain closely related to their environment and the linkages with the natural ecosystem are not entirely disrupted. While rural settlements also have problems, both psychological and technological in nature, these have yet to take acute forms. Part of the reason is perhaps the safety value in rural complexes tend to be the resource base, its optimal utilization, the various options on utilization, the various options on utilization, extension of the resource base, the impact of increasing population which the resource base has to sustain and the surplus it must create for urban centres.

Energy consumption per capita is comparatively low in our rural settlements consisting of cooking, drying, irrigation, etc. for rural life. The major source in most areas is energy from the surrounding biomass. This gives strong linkages to the ecosystem. On the other hand, the technology of agriculture and animal husbandry have changed substantially in recent years due to fertilizer and persticide use to produce stresses and even changes in the ecological system. In a sense agricultural changes have further modified rural ecoystems in which new ecological relationships are being developed. Traditional agricultural, forest and tribal settlement systems have not escaped problems connected with intensive use of fertilizers and pesticides, detergents and plastics in problems connected with intensive use of fertilizers and pesticides, detergents and plastics in neighbouring areas, hence the management of agricultural wastes and resideues, as well as those due to population growth and changing pattern of rural settlements are fast becoming matters of concern, these issues have been under study to some extent and while they pose what sometimes appear to be difficult problems, they are challenging and not necessarily interactable. Energy consumption per person in rural areas is currently small perhaps around 800 watt hours per day person. However, since a part of this energy has to be transported in the form of fertilizers and pesticides as also in the form of electric power and fule, the total energy consumption and distribution in the rural areas has to be taken into account to get an understanable picture. There is local generation of energy in the rural areas from agricultural wastes, fuel woods and to small extents by local energy generative systems such as wind mills, solar energy and fuel forests, similarly nitrogen as fertilizer can be produced by culture of soil bacteria, legumes and associated rhizobia, and other methods. There are attempts in this direction in rice cultivation in Orissa. Phosphorus can be recycled through use of bone and fishmeal. These methods can, in theory, be so developed as to considerably reduce the energy needed to be imported from outside for the rural systems. In economic terms, the energy content imported is usually more than offset by the production of food and other agriculture-based products. There are however, environmental problems connected with the rural situation, porticularly since the land and water, the primary resources, are limited. Distortions are often caused by wrong agricultural practices such as intensive monoculture of hybrids, massive use of pesticides and fertilizers and bad water and soil management practices, leading to soil loss by erosion or salination.

There are other social aspects of rural areas. With the increasing local perceptions of deficiency of agricultural, educational, employment and cultural opportunities rural dissatisfaction has increased and urban migration has accelerated. Such factors underline the perception that the quality of rural life is unsatisfactory. Trends of rural to urban migration have accelerated particularly as rural population increases have brought more pressure on a fixed amount of land lending to reduce per capital income. In such migration out of rural areas, there are both rural and urban problems. It is usually the adventurous, or the ambitious or the very poor amongst the rural youth who tend to migrate to urban areas depriving a society of many of those who would be motivated and work for economic and social changes of the rural community.

Increase of urban population beyond the urban absorption capacities in terms of space, power, water, transport and handling of wastes of a city or a town on the other hand can lead to slums, social stresses and urban unemployment. The urban settlement is highly energy dependent and unlike the rural areas has very little or no local resources of enery. This high energy dependence leads to the question of the relation between energy consumption and the urban environment. The major sources of urban energy are coal, oil and hudroelectricity, although wood, dungacakes and straw are also used in the large towns. In the small towns, wood, dungcake as well as coal and oil are major sources. Monor sources of energy today are nuclear, wind and solar power of which wind and solar power are more appropriate to rural situations that in urban. The combustion of coal and oil produces electrical energy or is used directly as energy for industry or transport. In all cases, it produces large amounts of carbondioxide and particulates as well as various amounts of other gases such as carbon monoxide, sulphur dioxide, nitrogen oxides, gaseous, liquid and solid hydrocarbons, siliceous dusts and mineral oxides. Hydro-electricity is often related to dam building and change of water courses leading to silting, of dams and cause landslides. Nuclear energy has problems of disposal of radioactive wastes and hazards of accidents such as melt down of reactor core which can spill radioactivity into adjacent areas, besides being costly and technology intensive.

Whe have thus a whole range of potential and actual environmental problems ranging from small perturbations in undisturbed ecological systems into which man is now making ingress such as introduction of new plants or animals into natural lakes to entirely man-made human settlements such as cities and industrial complexes which are complex and artifical environmental systems. In man-made human settlements, man is in a sense primarily responsible for looking after the system to ensure and improve present human well-being. Man is also in a sense responsible to see that the resource base is not unduly jeopardized or badly exploited to the detriment of future people (our children and grandchildren) who will live and work in these human settlements.

In general, a good moral principle seems to be that present actions of a society should not jeopardize the future of the society or of any other neighbouring society. These considerations lead to the question of effective and good management of an environment, particularly when it is exploited for human benefit. There are many aspects-social, economic, legal, technological and biological-involved in the management of such systems. The problem is complex and difficult, rendered more so since data being incomplete. many factors are not usually fully known or understood. The problems are compounded in developing societies where the data and information base are poor, environmental awareness is lacking amongst the citizenry, pressures of rapid industrialization and urbanization are acute and there is lack of finances for environmental overheads. On the other hand, the need for taking care of the environment is often acute due to rapid deterioration of forests, agricultural lands, and waters.

Many general principles of management of any system apply equally to the management of environmental systems. However, the specific scientific, technological, social and biological aspects which relate to the environment, have also to be incorporated in the management of such systems. Some of these factors such as the effects of major engineering or technical activity may not be fully known or understood. They pose problems to which no ready answers are possible at the time that action is contemplated. There are aspects of uncertainty and risks which cannot be avoided when decisions have to be taken and action contemplated. Thereare many kinds of environmental systems and each has its own peculiarities and managerial problems. For most management purposes a broad classification of the various environmental systems is often adopted :

(a) Biosphere reserves, natural lakes or natural forests which are disired to be reserved for purposes of gene pool preservation, species protection, scientific studies such as these of evolutionary processes, aesthetic or entertainment purposes and the protection of geological or natural hydrological structures;

(b) Rivers and river valleys, forests, used as resources of timber and fuel, hill sides, beauty spots, etc;

(c) Agricultural land, agriculture waters, dams and artifical lakes; and

(d) Urban and industrial complexes and other highly disturbed zones such as mining area.

It is generally assumed that all management is based on clear perceptions of objectives. Environmental management is equally dependent on clear formulation of the objectives in as specific a manner as possible both in social and ecological terms. Unfortunately, quite often, particularly in developing scocieties, such clear articulation of objectives may not be available due to lack of information on basic characteristics of the system, ignorance of impact of developmental activities, alternatives of objectives and consequential action, etc. In a biosphere reserve or a natural forest or lake, the problem of management may be stated simply that the objective is to have minimal human interference while scientific study and utilization to the extent necessary and possible is carried out cautiously, this means that the processes, of natural evolution geological processes. energy and mass flows should be not only known and understood but also allowed to go on with as little interference, as possible. However, in most cases it is not necessarily, either possible, or desirable to fall back on a complete non-interference policy.

A legitimate questions would arise as to what are the responsibilities of management in case of a major flood or a cyclone, or an earthquake in a biosphere reserve or a natural forest a lake or any other preserved ecologicel system.

We are aware generally of the serious ecological consequences of the great Assam earthquake and occasional major floods in some depth. Similarly cyclones in the Orissa forest and coastal regions and the West Bengal and Bangladesh deltaic regions have been destructive of ecosystems. Although we are clear that we are responsible to take steps to ameliorate the situation, our ideas are still quite vaguein such cases as to what management can or should do both with regard to precautions and to actions after the event. We can and do take care of human lives in such disaster situations but elements of cost, ignorance and social organization make it difficult to enunicate policies and programmes related to animal and plant life.

Similarly, if we are dealing with the problem of a moderately disturbed ecosystem such as a river, the problems of fresh water fauna, the utilization of the river water by people who live near the banks further downstream, the problem of human settlements on the banks of the river at the location and downstream and the steps to prevent bloods have to be looked at very carefully both in terms of aesthetic value of the natural river banks and the economic utilization of the river. The conflict inherent between maintaining the natural beauty of the two banks and the building of roads and parks or human habitation (including expensive hotels) on the banks will make competitive claims into which the environmental management system has to look closely. It has to approximate to achieving the clearly defined initial objectives with wnich it was entrusted. Social and economic pressures that have developed in the meantime will, in addition, have to be taken into account to prevent serious distoration.

As far as highly disturbed or rather the man-made environmental systems such as cities and farms, dams and roads are concerned, the problems are more acute. There is usually a greater clarity regarding what is desired even though this clarity may not apply equally to state how this is to be done. If a dam is built or a water reservoir has to be preserved, one knows that within the precipitation basin of a dam or a reservoir, land erosion should be minimized so that the reservoir does not get filled up with silt in a short time. The problem often is how felling of trees or forests or other ground cover is to be regulated within the existing social parameters of energy needs and social habits of the people who previously depended on firewood or on the grazing needs of sheep, goats or cattle by nearby communities within the overall objectives of the projects.

In the case of industrial complexes, forest-based industries such as paper, mining and quarrying, the material resources which they will use and the disposal of their wastes and effluents have to be taken into account. The task is often clear if the system is well defined. However, the payoffs between the costs, sociol and economic, and benefits such as the quality of life of people who work in the industry or of the people who live in the neighbourhood are not always quantifiable. Even more vague are calculations of benefit derived from a stable ecosystem in the long and short term.

Methods in applied ecology nusing simplified models continue to be developed to provide some theoretical base to understand the complexity of the system and suggest answers to policy questions such as possible fishing quotas or mesh size of fishing nets or spraying to control agricultural pests and insect-borne diseases such as malaria or use of biological agents for control of pests. The simple models do not provide a detailed and complete theoretical structure of ecosystems or even an understanding of the complexity of internal relationships. The availability of large computers have made it possible to incorporate increasingly complex and interlinked elements in a large system of equations to simulate more complex system for purposes of study. These cannot claim in any sense to be complete studies, but can in certain regimes or circumstances elucidate some understanding of the more complex behaviour of ecological systems.

While the formulation of applied ecological models began with the classic work of Lindemann (1944) and has been pursued by Watt (1968), Jeffers (1972), Patten (1971, 1972), Van Dyne and Abransky (1973) and many others, there is a perceptible decrease in ambition in more recent work such as that of Austin and Cook (1974). Current aims are more to understand in general terms, the behaviour of ecological systems rather than to predict future behaviour of simple ecological systems such as succession or existence of multiple equalibria points or regions of instability or stability in systems, etc. A model ecological system can under constraints move from one equilibrium point to another, sometimes relatively easily. Unfortunately, such results are not quantitatively varifiable. Althoug there is evidence of ecological systems moving from one equilibrium point to another, such examples demonstrate the existence of limits of stability of ecological systems and of recovery and resilency of such systems. However, theoretical exercises have not yet been successful in making detailed predictions of behaviour of ecological systems. The studies nevertheless are very useful and continue to be made because they do give indications of precautions to be taken or actions that should be modified or avoided in the course of developmental activities.

19

Pollution : What, Where & Why?

Introduction

Today, the cry of "pollution" is heard from all the nooks and corners of the globe, and pollution has become a major threat to the very existence of mankind on this earth. It is the major challenge of our times. The pollution of various resources has gone to such an extent that we are unable to breath fresh air and drink fresh water. On one hand the advancements of Science and Technology have added to the human comforts by giving us automobiles, electrical appliances, supersonic jets, space crafts, better medicines, better chemicals to control harmful insects and other pests; etc., but on the other hand, they have given us a very serious problem to face : pollution.

In fact, it is not a new problem, except in dimensions which we see today. When man and other higher animals began their life on this earth, there was absolutely no sign of pollution. There was perfect balance in various natural processes. The air and water were pure and soil was fertile. The problem of pollution arose with the very civilization of man. As soon as man learned to use fire, the air pollution began. As human pollution increased, there was increase in our working sphere and with this also increased the pollution. The rapid unplaned industrial progress added to our pollution problem.

What is Pollution ?

Pollution can be defined as an undesirable change in the physical, chemical, or biological characteristics of the air, water, or land that can harmfully affect health, survival, or activities of humans or other living organisms. Note that according to this definition pollution does not have to cause physical harm. It may merely interfere with human activities. For example, a lake may be considered polluted if it cannot be used for boating activities.

The problem with defining pollution is specifying what constitutes "undesirable change" which requires value judgments. For example, chemicals spewed into the air or water from an industrial plant may be harmful to humans and other organisms living near the plant. However, if expensive pollution controls are required, the plant may be forced to shut down. Thus, workers who would lose their jobs may feel that the risks to them from contaminated air and water are not as serious as the benefits of having jobs.

Value judgments about short-term versus long-term risks can also cause controversy. Building a worldwide network of nuclear power plants may provide electricity for the present human generation, but it compells future generations to handle and store the radioactive wastes produced for thousands of years, even if nuclear power is abandoned or exhausted as an energy source.

Thus, we see that the determination of desirable versus undesirable effects of an environmentally altering activity is a very difficult and highly controversial process. The nature of tragedy, as the philosopher Hegel pointed out long ago, is the conflict not between right and wrong but between right and right. Neverthelss, although defining pollution is a difficult, controversial process, it must be done in order to control pollution.

Types of Pollution From a biological viewpoint, we can recognize two major types of pollutants : degradable and nondegradable. A degradable pollutant can be decomposed, removed, or consumed and thus reduced to acceptable levels either by natural processes or by human engineered systems (such as sewage trestment plants), as long as the systems are not overloaded. There are two classes of degradable pollutants : rapidly degradable (nonpersistent) and slowly degradable (persistent). Rapidly degradable pollutants, such as human sewage and animal and crop wastes, can normally be decomposed rather quickly if the system is not overloaded. For example, a rapidly flowing river can normally cleanse itself of some human sewage fairly quickly, but this natural cleansing process can be overwhelmed by too much raw sewage from a large city or a number of small cities or farms.

Slowly degradable pollutants, such as DDT and some radioactive materials, decompose slowly but eventually are either broken down completely or reduced to harmless levels. Some radioactive materials that give off harmful radiation, such as iodine-131, decay to harmless levels in a few minutes or hours and are rapidly degradable pollutants. Others, such as strontium-90 (produced by nuclear bomb blasts); can persist at harmful levels for decades; plutonium-239, produced by nuclear power plants, can remain at harmful levels for many thousands of years. Slowly degradable pollutants are often synthetic compounds, such as DDT and plastics. Because they are artificial substances, nature usually had not evolved processes for breaking them down. Potentially harmful pollutants that break down slowly must either be prevented from reaching the environment or controlled so that they do not build up to harmful levels.

Nondegradable pollutants are not broken down by natural processes. Examples of nondegradable pollutants are mercury, lead, and some of their compounds (Enrichment Study 5) and some plastics. Like slowly degradable pollutants, nondegradable pollutants must be either prevented from entering the air, water, and soil or kept below harmful levels by removal from the environment.

Air Pollution

Air is the most vital component, without which the question survival does not arise beyond a few minutes. A human being of normal health and height respires 22,000 times a day and the air utilized in course of such respiration is about 20 times more than the quantity of food consumed by him. Natural air at mean sea level contains nitrogen (78-09%), oxygen (20,94%) and argon (0.93%) as macro-constituents. The micro-components of air are carbon dioxide (0.0318%), krypton (0.0001%) and helium (0.00052%). Many other compounds *e.g.*, SO_2, NOx, O_3 etc. are also present in much lower concentration then those mentioned. Besides these, water vapour and dust particulates, in varying amounts and sizes, are present in air, depending upon the meteorological, topographical and industrial activities of the areas under examination.

Current of air is known as wind blowing with velocity ranging from 1 to 6 km/hr is known as light air. Air with velocity of 7 to 12 km/hr known as light-breeze and those from 19 to 44 km/hr as strong breeze. Air current exceeding the velocity of 90 km/hr is called storm and over 104 km/hr is known as hurricane. Such abnormal condition in air movement releases too much of dust particulates into the atmospheric air. Estimates have shown that, even now, hardly 12 to 15% of the total dust particulates released into the environment is man-made and the rest is by nature. Though man-made air pollution is statistically insignificant, it nevertheless, poses, great danger, because such pollution usually contains toxic elements as opposed to natural pollution. The second reason is that industrial air pollution is often around human habitation resulting in the living population being badly affected. The pollution caused by vehicular traffic is maximum in the heart of the urbanized areas, resulting in the constant exposure of the urbanites to lead, nitrogen oxide and unburnt particles of hydrocarbon. These released pollutants subsequently turn into the photo-chemically active species *e.g.*, ozone, PAN etc. which often produces smogy environment especially during winter.

Air prevails even upto 100 km above the surface of the earth and 11 km below, at the deepest point in the Pacific ocean around Phillipines. The pressure of air drops from one atmosphere at sea level to 3×10^{-7} atmosphere in the mesosphere region. The total mass of the atmosphere is about 5×10^{16} tons which is roughly one millionth part of the earth's total mass 80% of the total amount of the atmospheric air is confined in the troposphere, within 12 kms above the mean sea leve.

Meteorological Factors Responsible for Air Pollution

There are three major meteorological factors which govern the dispersal pollutants from the source into the surrounding atmosphere.

(i) Wind profiles are caused by the interactions of forces due to pressure, the earth's rotation and friction. On a windless day, the wind speed is zero at the earth's rotation and friction. On a windless day, the wind speed is zero at the earth's surface but gradually increases to a value representative of free atmosphere at a height which is of the order of 1 km, known as 'planetary boundry layer'.

(ii) The temperature is another profile which significantly affects the dispersion of pollutants. Super-adiabatic condition indicates unstable atmosphere which is highly desirable as it causes rapid dispersion of pollutants. Sub-adiabatic as well as inversion conditions are highly dangerous, leading to the accumulation of the pollutants and ultimately causing severe air pollution problems in the area.

(iii) Topography often influences the local wind, as for example, the mountain valley wind, where the air tends to flow down the valley at night. Similarly, the sea breeze in the coastal regions causes the stack effluents to drift over land during day and no pollution apparently appears in the night. In order to fully understand the mechanism of air pollution in the context of the roles played by meteorological factors, let us recall the different layers of the atmosphere. Of special significance are the trochosphere and troposphere regions of the atmosphere to fully appreciate the reasons of the air pollution phenomenon. Trochosphere is the region where cloud formation takes place and generally extends upto 1.5 km. The height extending upto 15 km from the surface of the earth is covered by troposphere where more than 80% of the total air of the biosphere remains confined. Within this range, the air temperature decreases with increasing altitude. This decrease in temperature with altitude is called adiabatic (no heat transfer) lapse rate and is equal to six degree centigrade for every 1000 metre height. When the temperature of air decreases faster than the adiabatic lapse rate (a condition known as super-adiabatic lapse rate), then the atmosphere is considered to be unstable. As such, the air from one altitude readily mixes with air from other altitudes which ultimately leads to the formation of looping type of plume. A fanning type of plume is caused by thermal inversion where vertical movement of pollutants is limited. The fumigation type of plume results when there is inversion above the plume and instability below. Such condition occurs when temperature instead of decreasing, increases with altitude, thus forming a very stable atmosphere. The fumigation type of plume acts as lid over the atmosphere and brings down the pollutants to ground. Thus, creating severe problem of air as well as soil and water pollution. Pollutants in air are likely to accumulate in concentration sufficiently large enough to cause such episode in the regions where several dars of low mixing height, low wind speed and no precipitation occur. The combination of these circumstances often results in air pollution episode. Studies carried out by scientists reveal that mixing height of less than 1500 metre, wind speed of less than 15 kmph (4 m/sec) and insignificant precipitation for about 4 days are very favourable conditions for creating air pollution.

Types and Sources of Air Pollution

Air pollutants are generally of two types.

(i) Primary pollutants are those which are directly produced as a result of interaction between two or more primary pollutants *e.g.* nitrogen peroxide, ozone, sulphur trioxide, aldehyde, peroxyacetyl nitrate (PAN), etc. Secondary pollutants are photochemically active and responsible for the formation for smog. Smog is a combination of smoke and fog. Smoke is an emulsion of unburnt carbon particles in air and for is the emulsion of water vapor in air. Particulates present in the atmosphere may vary in numbers from several hundred/cm^3 in very pure air to more than 10^3/cm^3. Particulate mass level ranges from 10 ugm^3 in clean air to 60-2000 ug/m^3 in the polluted air in urban areas. The particles of size between 0.1-lu govern the heat balance, fog formation and provide nuclei for ice and water droplets formation. Smog has now-a-days become a general phenomenon in big cities, especially during winter, primarily caused by automobile exhaust. The automobile exhaust contains unburnt hydrocarbon which produces photochemically active compounds known as formldehyde and causes eye irritation. Lead is another air pollutant, generated from the automobile exhaust, owing to use of tetraethyllead as anti-knock material in the petrol. Mosses are excellent indicators lead pollution from aerial sources. The lead poisioning causes elevated content of amino leveulinic acid (ALA) in the urine of tne affected person. The increased concentration of bromine in the water or soil is another indirect indication of lead contamination. Evidences have shown that consumption of dust and dirt, composed of lead outfall is a major source of unusual indigestion in urban people. Dust and smoke are commonly found in air. It is the respirable dust index (RDI) *i.e.* size of dust particulates, which affect the human health. Particulates of size 3 micron (u) reach the lungs, out of the total inhaled dust, while particulates bigger than this size are checked somewhere in the respiratory possage of nose. Particulates of much lower size 0.1 u) are removed through expelling mechanism of the human body as cough. Out of the total retained particulates, still some get dissolved in the system. Thus, being eliminated without showing any harmful effects. Even all the permanently retained particulates are not harmful; some of them are deposited indefinitely in inert state on the lungs walls without any detectable injury. Only reactive deposited particulates cause harmful effects, but they are not equally harmful. The respirable dust producti n ranges from 27 to 340 gm for one on of coal mined.

In addition to the automobile exhaust, other major sources of air pollution are mining, thermal power plants and brick kilns. The latter is a pin-point source of air pollution, especially in the rural area. The intensive coal mining regions of Jharia, Dhanbad, Katras and the adjoining areas of Bihar have been identified to be the most air polluted cities nearly 15,000 tons per year of particulate matters are released in the air in Jharia coal field areas, causing complete blanketting effect when viewed aerially. Similarly, open stack coal carbonization emitts about 35,000 tons of particulate matters annually in the atmosphere. The release of sulphur dioxide through burning of coal in the boilers in thermal power stations is causing severe damage to the plants, concrete structures and also health hazard, owing to the oxidation of SO_2 into sulphuric acid. Studies conducted by Central Mining Research Station at Dhanbad (1986) indicate the mean concentration of Fe (2.48 – ug/m^3), (0.55 ug/m^3), Ni(0.04 – 0.57 ug/m^3), Cu(0.05 – 1 ug/m^3), Cr (max. 0.55 ug/m^3) Zn (max. 2.8 ug/m^3) in ambient air in the Jharia coal field areas.

Effects of Air Pollution on Our Cultural Assets

Neon in excess of 18 ppm, helium greater than 5.2 ppm, methane higher than 1.5 ppm. krypton more than 1.0 ppm hydrogen in excess of 0.5 ppm. nitrogen dioxide more than 0.001 ppm, carbon monoxide in excess of 0.1 ppm. and sulphur dioxide greater than 0.002 ppm in air are injurious for human health.

Ozone (O_3) in concentration of over 0.3 ppm in air causes nose and eye irritation. Its presence at a level between 1 to 3 ppm causes extreme fatigue after 2 hours of exposure and at higher than 9 ppm level, causes pulmonary edema (a disease in which excessive fluid occurs) in the lungs. Ozone concentration in 0.1 ppm for 20 days exposure causes 50% reduction in pollen germination in case of tobacco plant. Exposure of plants to ozone causes yellow spot formation on leaves which subsequently affects the rate of photosynthesis considerably. Exposure to ozone at concentrations of 0 02 ppm for 65 minutes. 0.26 ppm for 5 minutes, ppm for 3 minutes, and 20 ppm for 1 second produces cracks in rubber. PAN and aldehyde found in smog are eye irritants.

Exposure to carbon-monoxide (CO) at concentration of 100 ppm even for an hour may be fatal for human kind, probably because of its affinity with haemoglobin (red corpuscles of blood) which is about 200 times more as compared to oxygen.

Sulphur dioxide (SO_2) is one of the most potential pollutants present in the air over and around industrial areas materials like papers, textiles, leathers, stones, marbles, paintings and metasl are severely attacked by sulphur dioxide in presence of atmospheric moisture. The conversion of atmospheric sulphur dioxide into sulphuric acid is known as sulphonation. The process of sulphonation is accelerated in the presence of moisture and iron. Iron is known to be present in dust, nails etc. and moisture is present in the material itself. Cellulose which is the basic building material of paper, cotton and linen is rapidly decomposed by the process of hydrolysis. As a result, the paper and textile goods become yellowish, brittle, and ultimately loose their strength and fibrous structure. Similarly, Collagen (which is the basic constituent of leather) is also disintegrated by SO_2 and thereby, the leather becomes powdery in due course of time. Stones, particularly calcarious, like marble, which are essentially calcium carbonate, are severely damaged by sulphuric acid. Calcium carbonate is converted into calcium sulphate, resulting in huge incrustation by which the beauty of marble is changed. Paintings in which lime plaster is used, become expanded in volume and look flaky and powdery, owing to their conversion into the corresponding sulphate under the action of the atmospheric sulphur dioxide. Acidity generated by atmospheric sulphur dioxide initiates the electrolytic corrosion of metal.

The over all effect of SO_2 is summarised as below :

(i) 1 ppm for 1 hr or 0.3 ppm for 8 hrs causes plant damage.

(ii) 5 ppm for 2 hrs is a high level of concentration which causes broncho-constriction in human body.

(iii) 10 ppm for 1 hr exposure is considered a distress level for human beings.

Objects made of coper and silver are much more sensitive towards hydrogen sulphide, even in much lower concentration as compared that of sulphur dioxide. Both the metals are converted into their respective sulphides and consequently loose their artistic, ornamental and antique values.

Numerous monuments and ancient temples are exposed to the sea winds, coupled with chloride impregnated fogs at the coastal sides. The chloride ions absorbed by the stones during foggy weather are carried to the surface of the stones in dry weather, causing efforesenc/ce and finally leading to the process of complete disintegration of the entity. Bronze, an alloy of copper, zinc and silver, which constitutes a large part of artifacts, is badly corroded by sea winds. The most probable mechanism of the bronze corrosion may be presented as such.

The chloride ions react with copper and form unstable cuprous chloride. The latter further reacts with environmental oxygen and moisture forming cuprous oxide, cupric chloride and hydrochloric acid. Thus, a set of chain reactions finally reflects upon the bronze objects which ultimately destroy them.

Ozone (O_3) is highly sensitive towards saturated organic materials, even at room temperature and causes their depolymerization. Resins, varnishes etc. are used in paintings and are products of saturated organic compounds. Woolen catton, paper and all other such materials that consists of cellulose are also liable to be photo-oxidised by ozone in wet condition, resulting in their rapid fading of textures and colours.

The bloom in painting is caused by environmental ammonia (NH_3) in concentration sharply detactable by human nose, due to the formation of a compound known as a ammonium sulphate. Lower concentration of ammonium is though, useful, because it takes care of free atmospheric acidity and prevents other damages. Cellulose, showing amphoteric behaviour, is hydrolysed and oxidised under both acidic and alkaline conditions. As such, the materials rich in cellulosic content are also at constant risk of being degraded in higher concentration of environmental ammonia. Furthermore, it inflames the upper respiratory passage, even in low concentration, under longer exposure.

Aldehyde (CH_3CHO) irritates nasal and respiratary tract, arsines breakdown the red cells in the blood, chlorine attacks entire respiratory tract and mucous membrane of eyes, hydrogen cyanide interfers with nerve cells, hydrogen fluoride irritates and corrodes all body passages, possibly leading to cancer.

Air Pollution Control

The safety of three M's : – Maa, material and machinery must be the ultimate object behind the measures to control the environmental pollution problems. Pollution control measures should be considered as social responsibility and obligation not only by the concerned departments, but also by all who have faith in the need for preserving the valuable natural environmental resources. The implementation of the control measures should be based on the principle of recovery or recycling of the pollutants and must be taken as an integral part of production *i.e.*, never as a liability but always as an asset. Air pollution may be controlled by physical as well as by chemical methods. Some of the most effective methos are show below ;

(*i*) *Dilution Method.* Physical methods include the suitable height and design of the stack. The minimum-height of the stack has been worked out to be ?8 metres in order to ensure proper mixing of the discharged pollutants into the adjoining atmosphere.

(*ii*) *Air Cleaning.* Air cleaning device removes or recovers suspended solids, liquid and also soluble gases and vapours from air before they are discharged into the atmosphere. The commonly used air cleaning devices are : (1) Gravity settlers, (2) Cyclone separator, (3) Wet collectors, and (4) Electrostatic precipitators.

(*iii*) *Substitution Method.* The method is based on the principle to substitute the offending materials and processed by less offending ones.

(*iv*) *Plantation.* Intensive plantation in the region considerably helps in reducing dust, smoke and other pollutants. Studies conducted by the authors indicate that the trees with simple leaf, *i.e.*, pipal pakur, banyan, teak, sal, debdaru, mango, kachnar, kadamba etc. are more effective in removing dust particles suspended in the air than those of the trees *e.g.* ashoka, palas, nim, sandal and gulmohar with compound leaf. The study further shows that the dust removal from the air is best achieved by banyan (5.35+0.12 g/m^2 of leaf), followed by debdaru (4.56 g/m^2), sal (4.50 g/m^2), pipal (4.10 g/m^2) pakur (4.09 g/m^2) and the minimum by kadomba (3.57 g/m^2) among the trees with simple leaf. Similarly, the highest proportion of dust particles settlement from the air is ensured by ashoka (3.78 g/m^2) followed by palas (3.06 g/m^2), nim (2.92 g/m^2), sandal (2.24 g/m^2) and the least is achieved by gulmohar (1.44 g/m^2) among the trees with compound leaf. Therefore, along the road side and at places where maximum dust generation is possible, trees of simple leaf should preferably by planted. Forensuring the involvement participation, of the public—plantation should be encouraged on the pattern of the social forestry. The principle of atleast one tree for each school going student, two by each teacher and three during rest of the period of life by each individual should be practised in true sense of the term. Besides, the forest conservation act of 1980 should by lup-dated and be made stringent for those who are cruel to trees and plants.

(*v*) *Chemical Methods.* This method is usually for removing sulphur dioxide from air, resulting from the burning of coal, with rich sulphur content. Several chemical methods have by now been developed to have an effective control over air pollution problems. Briefly, some of these methods are discussed below :

(*a*) *Beneficiation Process.* By this method coal is treated with lime or dolomite and processed in such a way that both the sulphur and ash content of the original coal are significantly reduced with improved yield of energy on burning. Thus, the release of sulphur dioxide is automatically reduced considerably. The other method is to fix the sulphur dioxide in she combustion chamber itself by the spraying of ammonical liquor.

(*b*) *Reduction Process.* The SO_2 emission is adsorbed by alkaline alumina, resulting in quick reduction of the former into the elemental sulphur.

(*c*) *Catalytic Oxidation Process.* Vanadium oxide is generally used as catalyst where by SO_2 is converted into H_2SO_4 or ammonium sulphate.

(*d*) *Control Measures for Smog.* Four techniques are generally used to control hydrocarbon emission from stationary sources. They are incineration, adsorption, absorption and condensation of the exhaust gases. In the adsorption method, exhaust gases are passed through a bed of a granulated adsorber where hydrocardon vapours are adsorbed into the adsorber. The absorption method is similar to adsorption, except that the exhaust is brought into intimate contact with a liquid in which the hydrocarbons dissolve or become suspended. In the condensation method, exhaust gases are passed over low temperature surfaces where they are condensed. Liquid hydrocarbons are left behind and collected.

Control of hydrocarbon emissions from automobiles is more complicated. Several engine design characteristics favour lower exhaust of hydrocarbon emission. For example, reduction of surface/volume ratio of the combustion chamber decreases hydrocarbon emission markedly.

(*e*) Ill maintained auto-vehicles with out-dated silencers should not be allowed to play on the road.

Water Pollution

Water is relatively more prone to pollution than air because it is more viscous, having lesser moment of inertia and possessing closer molecular distance than air. Air, being available in the environment in much larger proportion as compared to water, the pollutants in air are quickly diluted and dispersed away along the down wind direction. Pollution load is thus, drastically reduced in air. The unequal distribution of water on the surface of earth and the fast declining availability of fresh usuable water per capita, per year, due to increasing population, are the major causes for concern in terms of the quantity and quality of water. Water pollution problem is more delicate and severe than that of air. The fundamental difference between the pollution of air and water is as such.

Air pollution is of widespread nature whereas water pollution remains more or less confined. Therefore, the latter is localised in nature. Air pollutants are known to travel thousand of kilometres without loosing their identity and, therefore, are not matters of immediate concern. The pollutants in water undergo many reactions and subsequently sometimes turn into hazardous forms. The study of air pollution is costly but the principles involved are relatively simpie than in water pollution studies and management.

Water may be present in rivers, ponds, lakes and in ground reservoirs. The surface water bodies get easily polluted whereas ground water reservoirs become polluted only after prolonged exposure. Among the surface water bodies, the rivers are less prone to pollution than ponds and lakes, because of the current available in the former. But once the rivers become polluted, complex and costly techniques for control of such pollution are needed. The same is true in case of ground water pollution, which is difficult to treat when and where such problems exist. So far as the effects of pollution on the living biota are concerned, they are again of conflicting nature. The rivers disperse away the pollutants immediately in the downstream and thus, become free from pollution, but the effects of pollution become established and detectable more easily in river than in ponds and lakes because the biota in river and generally less tolerant to pollution whereas the biota living in the stagnant water bodies are more tolerant to pollution.

Sources of Water Pollution

There are generally two sources of water pollution :

(*i*) *Point sources* of pollution are of definite identity with almost constant volume and fixed composition of the effluents discharged. Industrial waste waters are examples of the effluents generated from point sources of pollution.

(*ii*) *Diffused sources* of pollution are points, situated far beyond the sight having wastes of unknown composition and volume, The street run-off, agricultural run-off and other such wastes are examples of the effluents produced from the diffused source of pollution.

It is much easier to design and operate the treatment plants for correcting the effluents produced from the point sources of pollution than for the effluents produced from the diffused sources of pollution. It is practically impossible to treat the effluent of the latter because of very huge investment of money, time, labour and the lack of availability of sophisticated technologies in our country.

The potential source of water pollution in all developing or underdeveloped countries are sewage and industrial wastes. For example, out of the total wastes discharged into river Ganga, volume of sewage discharged constitutes 72 per cent whereas 25 per cent discharge is contributed by industrial wastes. In term of BOD, the pollution load contributed to river Ganga by both the pollutants is in 1 : 1 ratio. The effluents from agricultural and street run-offs join the river in much larger proportion, especially during the monsoon. The excreta contributed by nomadic people living along the banks of rivers or in the respective catchment areas are the only potential diffused source for causing the bacteriological pollution of surface water bodies in India.

(i) Sewage

Sewage constitutes of 99.9 per cent solid and 0.1 per cent liquid. 70 per cent of these are organic constituents and the remaining 30 per cent are inorganic. 65 per cent of organic portion are proteins while carbo-hydrates and fats are represented by 25 per cent and 10 per cent respectively. Estimates made by Arcievalla suggest that 6 to 12g of nitrogen and 1 to 3g of phosphorous are generated by every person every day in the form of sewage. Sewage excreted by a human being contains more than 400 different species of bacteria and viruses. About 100 viral species in sewage are known to cause diseases in human beings. Out of this, 90 per cent are contained in human sewage, 100 ml of sewage contains 10^9-10^{10} bacterial cells out of which 10^6-10^9 may belong to coliform groups and 10^5-10^6 may be of faecal origin. The Salmonella typhosa variety ranges between 10^1-10^4. Sewage has been established to contain potential fish pathogens such as Acromonas, Pseudomonas etc. Myxobacteria are also present in water, polluted particularly by sewage. Well treated sewage may still contain pathogenic bacteria and viruses, unless properly chlorinated before its discharge into any water course. Sewage, on drying after treatment, is known as sludge. Sewage further diluted by kitchen and other domestic wastes are known as sullage. Sewage may be strong, average and weak. Assessment of its strength is determined on the basis of chloride content, the ratio of BOD and permanganate value and on the basis of a combination of chemical parameters, as shown below.

1. Chloride

Weak sewage contains an average of 60 ppm chloride, sewage of medium strength shows 100 of chloride and strong sewage indicates 150 ppm of chloride.

2. BOD/PV (3 minutes) ratio

Weak sewage indicates BOD/PV (permanganate value) ratio equivalent to 60, average sewage near about 100 and strong sewage shows a ratio of over 170.

(ii) Potential Pollutants in Industrial Wastes

Industrial effluents generated from sugar, distillery, tannery and paper industries are accompanied by very high organic load with BOD values ranging from 2500 to 1,22,00 ppm. The paper and pulp industrial waste contains Lignin which is made up of cellulosic material and remains unattended by microbes for long, causing depletion of fishes up to as long as 40 km distance in the downstream. The wastes from oil refinery and steel industries contain the most objectionable pollutant known as phenol, causing high odour and toxicological problem in the receiving water body. Besides, the latter also contains cyanide in appreciably high concentration to produce detrimental effects on the biota in the receiving water bodies. Fertilizer industries discharge effluents with high content of ammonia (100 ppm), urea (180 ppm), phosphate (40 ppm) and sulphate (1200 ppm), which subsequently find their ways into the adjoining water courses, causing bloom of algae. Ammonia in concentration greater than 0.1 ppm in river, stream or lake is toxic to fishes. The fertilizer waste is usually accompanied by arsenic and chromium and unfortunately both of them are equally objectionable for aquatic fauna and flora. The wastes generated from alkali manufacturing industries contain mercury, which builds up 1000 times higher concentration in fishes than found in the receiving water course. The minimata disease of Japan is a well known example of mercury hazard where thousands were poisoned and hundreds died by consuming mercurised fishes. Lead is another accumulating poison which is found in wastes generated from battery, printing, petroleum and pastes processing industries. So many of trace and toxic elements, *e.g.*, zing, boron, coppor etc. are found in large concentration in the effluents produced by non-ferrous industries. The effluents generated from mining operations are either highly alkaline or acidic and as such both are injurious for the aquatic organisms. The phosphate Pyrite mining produces wastes of very high acidic nature, accompanied by as low as pH value as 1. The mining mica and coal sometimes produces acidic wastes but these are not as strong as pyrite effluent.

All the industrial wastes generated during processing and manufacturing stages must be suitably treated before their discharge either on the land or into the water course. The cost involved in their treatment is negligible in comparison to the gain achieved.

Tolerance Limit for Pollutants Present in Water

Water is used for a variety of purposes. It may be used for drinking, bathing, cooking, irrigation and industrial purposes. Its industrial use may again be of various types. It is used in boiler, laundary, power generation, breweries, dairies and in many edible productions and processings. The standard of the quality of water vary considerably in each of the above cited industries. Detail for the quality of water standards, applicable in each of these cases is not possible to present here. However, some of the important parameters for the maximum permissible tolerance limits are below :

(*i*) *Physico-chemical Standard for drinking water.* Lead, 0.1 ppm, Cadmium and Seleninm, 0.01 ppm for each, Chromium, Arsenic and Cyanide, 0.05 ppm for each, Phenol and Mercury, 0.001 ppm for each Turbidity 10 JTU, pH 6.5 to 9.2, Total solid, 500-1500 ppm. Hardness ($CaCO_3$) : 600 ppm, Magnesium : 150 ppm, Iron : 0.3-1.0 ppm, Manganese-0.5 ppm, Copper-1.5 ppm, Chloride-1000 ppm, Sulphate-400 ppm, Flouride-1.5 ppm, Nitrate as (NO_3)-45 ppm, Nitrate as (N)-12 ppm, Zinc-15 ppm.

(*ii*) *Bacteriological Standard (WHO, 1971) for treated drinking water samples.*

1. E. Coliform should be absent.
2. 95% of the samples examined should not have any coliform count throughout the year.
3. If coliform counts were found, then in no case, it should exceed 10/100 ml.

OR

Coliform could should not be found in any two consecutive samples.

(*iii*) *Maximum permissible chemical limits for inland surface water, intended to be used as raw water for public water supply and bathing ghats.*

BOD-3 ppm, pH-6 to 9, Chloride-600 ppm, Sulphate-100 ppm, Cyanide-0.01 ppm, Arsenic-0.2 ppm, MPN count-5000/100 ml.

(*iv*) *Bacteriological standard for raw water to be used as source of Public water supply.*

Relaxation has, however been granted in case the total coliform count exceeds the maximum permissible range of 5,000 100 ml for class C water sample, as under.

Not more than 20% of the total number of water samples examined should exceed the maximum permissible range for total coliform (*i.e.*, 5000/100 ml) or almost 5 per cent of the total semples may hardly exceed 4 times the maximum prescribed limit. One more condition has been added by IS, 229-1982, which states that feacal coliform count should not exceed 40 per cent of the total coliform count.

(*v*) *Limits for industrial effluents to be discharged into water course* (*IS : 2490-1974*)

BOD-30 ppm, COD-250 ppm, Total suspended solid (TSS)-100 ppm, Sodium-0.2 per cent, pH-5.5 to 9.0, Temperature-40°C. Chloride-600 ppm, Flouride-2 ppm, Zine-5 ppm, Copper-3 ppm, Nickel-2 ppm, Cadmium-2 ppm, Chromium and Lead 0.1 ppm each, Arsenic and Cyanide-0.2 ppm each, Mercury-0.01 ppm, Selenium-0.05 ppm.

(*vi*) *Limit for industrial waste to be discharged into public sewers* (*IS : 3306-1974*)

BOD-500 ppm, pH-5.5 to 9 TSS-600 ppm, Temperature-45°C, Chloride 600 ppm, Sodium-60 per cent, Cyanide-2 ppm, Copper-3 ppm, Lead-1 ppm.

(*vii*) *Limit for the discharge of sewage into the water course* (*IS : 4754-1973*)

BOD-20 ppm, TSS 30 ppm.

(*viii*) *Maximum tolerance limit for water intended to be used for irrigation*

The suitability of water to be used for irrigation should strictly satisfy the following criteria :

(*a*) Sodium Absorption Ratio (SAR)

$$SAR = \frac{Na}{(Ca + Mg/2)^{1/2}}$$

Note : The values of SAR should not exceed 26.

(*b*) Soluble sodium percentage (SSP)

$$\frac{Na \times 100}{Ca + Mg + Na}$$

Note : SSP must not exceed 60, otherwise the water would become unfit.

(*c*) Potential soil salinity (PSS) :

Cl (epm) + 1/2 SO_4 (epm) where, epm is equivalent part per million which is equal to ppm/ev and ev is equivalent weight.

The water having 6 epm or more chloride is injurious for acidic plants.

(*d*) Residual sodium carbonate (RSC) : (CO_3 + HCO_3) − (Ca + Mg) in epm. Water with RSC greater than 2.5 is unfit, and those with RSC between 1.25 and 2.5 is marginal and water with less than 1.25 epm of RSC is good.

(*ix*) *Chemical quality criterials of grounwater.* The water exceeding the maximum permissible concentration of any single or a number of constitutents, not only unfit for drinking, rather creates many problems. For example, waters having more than 400 ppm of bicarbonate, 100 ppm sulphate or 40 ppm silicate may contribute incrustation to the carrying pipes. Waters showing 2 ppm or more of iron or 1 ppm or more of manganses tend to accentuate the process of incrustration, especially when sulphate, silicate and bicarbonate are at optimum concentration in it. Similarly, waters with low total dissolved solid (TDS) and low PH value contribute to chemical corrosion whereas, the waters with high TDS, usually greater than 1,000 ppm of high pH value lead to the electro-chemical corrosion.

(*x*) *Biological quality criteria for limnological study*

1. Biotic Index (BI)

A stream, pond of lake with a biotic index of more than 15 is clean and when it is lower than 10, is considered to be polluted.

Biotic Index = 2 (number of class 1 organisms) + (number of class II organisms).

Class 1 organisms include caddis fly, may fly etc. and are pollution intolerent. The organisms belonging to class II category *viz.* dragon fly, shrimps, plenoria etc. are indicators of mid pollution Class III organisms *e.g.* mosquito larva, chironomoids, tubifex etc. are indicators of heavy pollution and their presence marks zero biotic index.

2. Sequential Comparison Index (SCI)

The determination of the values for SCI does not require much of a background of biological knowledge, rather it is done through simple technique, even by persons of ordinary skill. The mathematical expression for SCI values are as such.

$$SCI = \frac{\text{Number of runs}}{\text{Number of organisms}} \times \text{No. of equivalent species.}$$

The SCI values decrease with increasing intensity of pollution because the latter leads to a significant reduction in the number of community species at the cost of the increase in the number of individuals of a particular tolerant species. Generally, SCI value less than 5 indicates severe pollution, between 5 to 10 mild pollution whereas a value above 15 is indicative of clean water zone. For confidence limit of 90 per cent with mean difference of 100 per cent, requires a minimum of 7 to 8 times determination for each of all the samples drawn from the bottom.

(*xi*) *Algal Genus Index* (*AGI*). Score of 20 or more secured by the algae belonging to different general in water sample is indicative of organic pollution. Score lying between 15 and 19 is taken a probable indication of high organic pollution. AGI value lower than 5 is indicative of clean water zone. The presence of algae means the presence of 50 or more individuals of a single genus in one ml of a water sample.

Soil Pollution

Soil sustains innumerable, microbes and a large number of plants and animals along with plenty of mineral reserves for purposeful exploitation. Soil has different meaning to different people. Chemists consider the soil like an object endowed significantly with high buffering capacity while to a geologist, the soil is a means for ascertaining the age of the earth from a large number of isotopes present in it. The soil to a limnologist is a dependable source of nutrition for the biota, and to a farmer, the soil in the form of land is a source of living whereas to a sanitary engineer, it is nothing more than dumping ground for all kinds of wastes. Above all, the soil is a living record from where originated all major ecological events of the past.

Land is a valuable but fast depleting resource with increasing population. It has been estimated that the total surface area of the earth is 3, 15, 14,640 km^2. About 30 per cent or 91.56,800 km^2 is land surface. Of this nearly 80 lakh km^2 forms the Antarctica, which is too cold to be inhabited by human population. Another 1 crore 44 lakh km^2 area is covered by desert and 1 crore 75 lakh km^2 is mountaneous regions (*e.g.* 80 lakh km^2) are not habitable. Thus, human population is left with hardly 75.2 m km^2 (7 crore 52 lakh km^2) of land surface to live in. Presently, there are more than 75 people living per km^2 of land. The per capita availability of habitable land is likely to be further reduced to less than 50 per cent of its present proportion by 2000 A.D. (Many of the highly urbanized centres of the world are faced with acute space problem. For example, there is hardly 20 sq. ft. of living space per capita available in a crowded city like Calcutta, thereby, causing a serious ecological concern. The decreasing availability of living space is not the only problem faced by the world; rather the fast degradation of land is a more serious aspect of pollution.

India is loosing nearly 6,000 million tons of top soil per year in Ganga basin alone, caused mainly by ruthless deforestation, thus leading to severe erosian problem. Statistics prepared by the Ministry of Agriculture, Government of India (Referred from the World Environment, 1985), indicate that out of 304 million hectare (mh) of land, nearly 175 mh is becoming less and less productive year after year. Such increasing trend of desertification is really devastating and needs our urgent attention towards waste land management. Besides, increased intake of pesticides fertilizer along with radioactive out-fall, is putting additional poisonous ingredients into the land surface and causing serious pollution problem.

Sources of Soil Pollution

There are numerous sources for soil pollution, but the important, among them, are the following three :

(i) Indiscriminate discharge of industrial effluents on land and in water course.

(ii) Open defecation by animals and human beings especially in rural areas of the country.

(iii) Unscientific disposal of solid waters on land.

In developing countries like India, the major source of soil pollution is of the third type, and is prevalent in both urban as well as in rural regions. The garbage generated by the community in India contains 30-40 per cent of compostable matter from which good qualities of manures can be obtained through anaerobic digestion. Indian refuse, usually contains less than 1 per cent of plastic and glass materials. The paper content of the refuse varies between 3-7 per cent. Further, among the physical characteristics shown by them, the density of the refuse varies from 330-560 kg/m^3 and the per capita refuse production in Indian cities varies from 0.15 to 0.35 kg/day, depending upon the social, economic and educational status of the people living in the community. Generally, the people with modern scientific vision produce less refuse than those with poor and dirty living habits. Based on the comprehensive survey conducted by it on the overall physico-chemical composition of the refuse, NEERI has recommended scientific conversion of refuse into compost. Accordingly, a compositing plant with a capacity of 200 tons per day of manure production has been installed at Calcutta. All other big and medium size Indian cities must follow similar practice to keep their cities clean.

Radioactive substances produced during fission and fusion enter into the life supporting environmental cycles. Nuclear fission is a process in which heavy nuclides are broken into lighter ones whereby energy of the order 10^9K. Cal is released. On the other hand, a process in which lighter nuclides combine together to form a heavy nuclide is known as nuclear fusion. The high temperature of sun is maintained through nulear fusion whereby, hydrogen atom is converted into helium. All elements released into environment do not enter into the food chain, rather many of them go into marine ecosystem and many of them are so diluted by nutrient rich terrestrial ecosystem that they hardly remain harmful. Examples of such harmless radioactive elements are Manganess-54, Iron-59, Cobait-60 and Zinc-65, those that are present in particulate or collodial from *e.g.* Parasedodymium-144, Zirconium-95 and Rhodinm 106, which get transferred into the marine ecosystem in the highest amount. In contrast, the soluble fission products such as Strontium-90, Cesium-137 are found in highest possible amount in land and plant. Further, Cesium-137 gets accumulated only in certain plants like mushrooms and lichens and hence does not pose as great a hazard as Strontium-90. The greatest hazard of Sr-90 lies in its accumulation in the human system through various plants and foods like milk, grass, vegetables etc., and creates a negative balance of calcium present in the human bone.

Food Pollution

Food ranks third in position, after air and water, for the maintenance of the process of life on this earth. Strength to fight against diseases depends on the intake and purity of the vital constituents of food, such as proteins, fats, carbohydrates, vitamins and minerals. Removal or denudation, in part or in whole of any of the nutrients from food articles, or addition of adulterants adversely affect our health. The strength of a nation is by and large dependent on the health of its population. The proverb "Health is Wealth" and "sound mind in sound body" are well known to us. Therefore, maintenance of the purity of food and preventing adulteration dy harmful ingredients are direct contributions to the national health development programme. Food adulteration act has long been enacted

(1954) by the Government of India, under which food adulteration is an offensive criminal act and is liable for deterrent punishment.

Adulteration, Pollution and Contamination of Food

International mixing of foreign substances with food materials, for undue economic gain is known as adulteration. The entry into the food chain, of undesirable substances through environmental agencies *i.e.*, air, water and soil is termed as food pollution. The production of toxins in food through microbial agencies is known as food poisoning. The degradation in quality of food materials under the microbial influence to such an extent so as to produce diseases on consumption is termed as contamination of food.

The pollutants may be of chemical and biological types. Both types of pollutants enter into the food materials either through environmental routes or by adulteration. Some of the pollutants have been reported to be naturally present in harmful concentration in some of the food materials. The potential chemical ingredients present in food, as pollutants are flourides, nitrate, pesticides, radioactive isotopes and many trace element, *e.g.*, selenium, lead, mercury, chlomium etc. Naturally present organic pollutants in some of the food are-solanine, gossypol, oxalic acid and erucic acid. Commonly used food adulterants are argemone oil, tricresyl phosphate and white oil for mustard oil, vanaspat (hydrogenated oils) for butter, and chromium or lead for turmeric (haldi). Toxins are produced in food by clostridium batulinum, cl. perfringens staphylococcus aureus, salmonella and bucillus. The detail for the above mentioned, pollutants, adulterants, food addatives and bacteriological toxins are briefly presented below.

CHEMICAL POLLUTANTS

(i) Flouride

Tea, spice, betel, gram, french beans and fishes are rich sources of fluoride. Fruits and vegetables are generally low in fluoride content. Among cereals, fluoride content is minimum in rich preparations while it is maximum in chapati made of wheat. Which contains 0.8-1.8 ppm of fluoride. A high intake of calcium, magnesiun and vitamin C is known to play an important role in preventing chronic fluoride intoxication. People, living in the regions where water contains more than 1.5 ppm of fluoride, are advised to take citrus fruits which are rich in vitamin C Besides, they are advised to avoid spicy food because it increases the flouride.

(ii) Lead

Presence of lead in human blood in more than 40 microgram per 100 grams blood can inhibit enzymes such as ALA which are essential for biosynthesis of haem (a basic unit of blood). Soft water, paints etc., also contribute lead to our food materials.

(iii) Cadmium

It is not required by the human body even in the trace amount and therefore is completely absent in an infant. The cadmium intoxication can considerably be prevented by increasing the intake of vitamin C Cadmium burden in excess of 20 to 30 mg in the human body may create a negative calcium balance, leading to osteomalcia (softening of the bones. Multiple factures resulting from osteomalcia are responsible for the painful disease known as Itai-itai.

(iv) Tin

Oils, fats, ghee, beverages, jams, fruits, vegetables, biscuits confectionery, coffee, tea etc. are marketed and stored in consumer's houses in containers generally made of tin. Acidic food stuff are more prone to be contaminated by tin. Canned moist food *i.e.*, butter, grape fruit juice etc. on opening starts dissolving tin. For example, canned tins apple juice when first opened may contain 20.3 mg of tin per kg of juice while after 2 days, the quantity may go upto 87 mg/kg and after 4 days, upto 364 mg/kg. The food stuff and beverages normally contain tin in amounts which are usually below 100 mg/kg and are generally well tolerated. The toxic dose in food which are causes acute toxication is believed to be around 2 to 3 grams/day for weeks together.

ADULTERANTS

(i) Argemone Oil

The argemone sects are black in colour and similar in size as mustard seeds any latter is therefore very been adulterated with the former, it has been reported that regular consumption for 1-3 months of and edible oil, contaminated with as low a concentration as 0.1% of argemone oil can give rise to epidemic dropsy In human beings. Dropsy is a disease, causing swelling of feet, thighs and face. If the consumption of adulterated oils is not discontinued, it may cause glaucoma (disease of the eye) leading to blindness or swelling of liver which may be fata.

(ii) Tricrecyl Phosphate (TCP)

TCP is practically colourless, odourless, non-volatile and stable liquid, insoluble in water but is freely miscible in vegetables oils. It is used as adulterant with the mustard oil. Consumption of oils containing TCP for at most 20 days may cause flabbly paralysis. Like parathion (a phosphorous pesticide), it may also impair the nuero-muscular control of the body.

(iii) Vanaspati (Vegetable Oil)

The fundamental difference between ghee (butter, oil, vanaspati is that the former contains saturated fatty acids whereas the latter contains such in free state. Vanaspati unlike ghee, contains trans-acid which may cause adverse effects on the consumer's health on prolonged use.

Vanaspati is not only cheaper than ghee, but also closely resembles the latter in appearance. Vanaspati is, therefore, commonly used for adultering ghee. Such adulteration can very easily be detected by the following method even without going through any complicated chemical tests as such :

Gheè is taken in a test tube and is method in a burner or spirit lamp. One spoon full of sugar and 10 cc of concentrated hydrochloric acid are added into it. The mixture is then warmed for two minutes. Development of red colour confirms the adulteration of the ghee with vanaspati.

MICROBIAL TOXINS

(i) Botulinus Poisoning

Such poisoning is caused by colostridium botulinum (an anaerobic spore forming organism) in improperly sterilized preserved foods which have a near neutral or slightly alkaline condition. The poison caused by cl. botulinum is most potent because ingestion of even minute quantities would result in a severe syndrome that may be fatal within 2 to 10 days.

(ii) Staphylo-coccus Poisioning

Stapylo-cocci bacteria grow in cream, puddings, sauce and other starchy preparations. Besides these, milk, egg and meat are also prone to their attack, especially during summer, warm foods, kept for long, at room temperature, in summer months are open invitations for such microbes to make the food poisonous. In contrast to salmonellosis or butulism the attack does not last more than one day and usually wanes within 6 hours or less.

NATURAL ORGANIC TOXICANTS PRESENT IN SOME OF THE FOODS

(i) Solanine

It is a common name used collectively for a group of tozic glycoalkaloids found in a number of plants. Potatoes damaged or black-spotted potatoes may contain upto 400-500 ppm of this toxicant. Consumption of such potatoes may cause abdominal pain and diarrhoea.

(ii) Gosypoll

It is a toxicant made up of phenolic compounds and may be found in free state as well in combination with protein, in cotton seed and its flour, but never in cotton seed oil. It is toxic to man, pig sheep, chicken and fish, but not to the ruminants (cows and buffaloes). The maximum allowable concentration of this toxicant is 1.1 per cent in cotton seed flour used for making biscuits.

(iii) Oxalic Acid

It is contained in chana dal (180 mg) piyas (407 mg), amla (296 mg) and khesari oil (lathyrus 122 mg) per 100g. Consumption of heavy dose of this food produces acid cramps, haemohiliae and even death. But such poisoning is generally rare. However, it is advisable to restrict the intake of foods containing more than 0.5 per cent of the acid. Incidence of formation of stones in kidney is nigh among the poor population of Rajasthan, presumably because of their habitual heavy consumption of the above mentioned food items.

(iv) Erucic Acid

It is a long chain unsaturated fatty-acid normally present in mustard oil mixed with rape-seed oil in concentration exceeding to about 45 per cent. Habitual use of such mixed oil has been found to cause myocardial fibrosis (reduction in eye sight) in human being.

Noise Pollution

Noise pollution is often misinterpreted by most of the people as sound pollution. As a matter of fact, sound is a pure tone, harmonically related, having fixed frequencies and amplitudes, occurring at regular intervals and produces meaningful communication and pleasure in hearing. Musical lectures given by teachers, leaders and the preaching by saints are the examples of sound. On the other hand, 'unwanted sound' is termed as noise, having a complex mixture of a number of pure tones of various frequencies and amplitudes. These sound waves fluctuate and repeat themselves in highly haphazard manner. Noise pollution has become a part of industrial life in modern times Urban people are at constant exposure to such pollution contributed by disorderedly maintained vehicles Rural population are increasingly being exposed to noise pollution by 'loudspeakers', an amplifier of music especially during social functions. The people working in non industrial establishments are also vulnerable to noise pollution, significantly contributed by ill-maintained type-writers and electronic instruments.

Units of Noise Pollution Measurement

The unit for the expression of the Noise pollution intensity is decibel (db) *i.e.* the logarithmic ratio of noise intensity (I_0) to reference level (I). The minimum intensity of sound audible by man is termed as reference level (I), approximately equal to the sound pressure of 0.0002 microbars (dynes per cm^2 of an energy of about 10^{-16} watts). Mathematically the db is expressed as ;

$$db = 10 \log_{10} \frac{I_0}{I}$$

Accordingly, 10, 20 and 100 db represent 10, 10^2 and 10^{10} times the threshold intensity respectively. The threshold intensity of sound required for human hearing is zero db, the rattling of leaves produces 10 db, and sounds generated during broadcasting is 20 db. Sound level of 30 db is generated is bed room at night pillow talk. 40 db in library and 60 db in normal conventional speeches. The play of the radio at average volume induces a sound level of 70 db, light traffic produces noise at 74 db and 80 db of noise is produced by a typical factory. Running train generates a noise of 90 db, orchestra of 100 db, rock band of 110 db while an aircraft take off produces noise of 120 db. Threshold of pain is of 140 db.

Measurement of Noise Pollution

The instruments needed for the measurement of noise pollution are many and varied, depending upon the types of experiments to be done. For examples, sound level meter, noise dosimeter and frequency analyser are the instruments, commonly used for measuring the total intensity of noise, the actual noise that the worker receives during working hours and the frequency bands respectively. Brief description of each of the instruments is given below :

(i) Sound Level Meter

A sound level meter consists of a microphone, amplifier, rectifier and a display meter. The sound pressure level in db is directly read out from this instrument.

(ii) Noise Dosimeter

The different noise levels, existing close to or away from the machines are measured directly by this instrument. The total levels of noise that wearers receive during their working periods are indicated whereas the total intensity of sound produced in the campus are shown by the sound level meter.

(iii) Frequency Analyser

The frequency-cum-noise level distribution is measured by such type of instrument which helps to devise any control measures to reduce the potentially effective part of the noise level.

Acceptable Noise Standard

Sound is essential for life. Organisms are at constant exposure to some other forms of energy, without which the brains may not function. At the same time the adage, "Too much of every thing is bad", is also applicable in this case. The 'unwanted sound', what is called as noise, is not at all ecologically desirable at any level. However, the permissible noise levels in db is dependent upon the time of exposure, 80 db of noise level for 8 hours days, 90 db of noise for 4 hours per day, 95 db of noise for 2 hours per day, 100 db of noise for 1 hour/day 105 db of noise for 1/2 hour per day, 110 db of noise for 1/4 hour (15 minute)/day and 115 db of noise for 1/8 hour per day have been permitted by almost all the countries of the world.

Exposures to continuous or intermittent noise level higher than 115 db should not be permitted. However, 90 db of noise level for 8 hours/day of exposure has been granted in case of mining occupations for workers in India. The maximum desirable limit for noise levels in residential area and recreational activities centres around the habitations has been fixed as 40 db in night, 45 db in evening and 50 db during day time.

Effect of Noise Pollution on Human Health

Noise level above 80 db for more than 8 hours a day has been found to increase tension in a human being of normal mental make-up. Noise level above 80 db during intermittent exposure produces a change in breathing pattern. Continued exposure to high levels of noise results in annoyance, fa igue, shifting of threshold limit of hearing and the worst may be the total loss of hearing. The bad effects of noise pollution have been seen to interrupt the process of blood circulation, to increase breathing amplitude to bring changes in size of the pupils of the eye and in the rate of saliva and gastric secretions. Noise pollution above 120 db produced by air craft and turbine have been shown to induce many adverse biochemical changes in a human being for 3.88 hours of exposure per day for a period of one to 6 months. The concentration of cholesterol in blood becomes elevated alongwith increased white cell counts in peoples living around close vicinities of busy aerodromes. Industrial workers, exposed to noise level of more than 100 db, reveal the tendency of reduced intra-ventricular conductivity besides hypertension.

Remedies of Noise Pollution

(i) Decibel meters should be installed along highways and in places of public gatherings such as picture halls, libraries, offices and hostels. Each industrial establishment must have such facilities in order to have a check on the intensity of noise pollution, being produced throughout the working period.

(ii) Greenbelt vegetation and open space, at regular intervals, within the densely populated areas of big towns, must be developed. A dense evergreen hedge can reduce the noise by 10 db during garbage collection in the town. Border planting along highways can very effectively control the noise pollution if plantings be done in lower side with respect to the source of noise production and be on high side towards the listener.

(iii) A strip of 20 feet wide plantation inside the compound wal l, effectiveiy protects the houses from sound pollution, contributed by vehicular traffic.

(iv) Technological modification is suggested to be achieved so that a sound associated with minimum possible frequency be produced from the machines,